I0055627

Introduction to Time-Dependent Quantum Mechanics with Python

IISc Lecture Notes Series **ISSN: 2010-2402**

Editor-in-Chief: Diptiman Sen

World Scientific Publishing Company Singapore and Indian Institute of Science (IISc), Bangalore, will co-publish a series of prestigious lectures delivered during IISc's centenary year (2008–09), and a series of textbooks and monographs, by prominent scientists and engineers from IISc and other institutions.

This pioneering collaboration will contribute significantly in disseminating current Indian scientific advancement worldwide. In addition, the collaboration also proposes to bring the best scientific ideas and thoughts across the world in areas of priority to India through specially designed India editions.

The "IISc Lecture Notes Series" is based on postgraduate courses developed and taught in IISc, and in other major postgraduate programs in India. These are class tested, compact, can be used directly as lectures.

Published:

Vol. 7: *Introduction to Time-Dependent Quantum Mechanics with Python*
by Atanu Bhattacharya & Elliot R Bernstein

Vol. 6: *Ultrafast Optics and Spectroscopy in Physical Chemistry*
by Atanu Bhattacharya

Vol. 5: *Introduction to Pattern Recognition and Machine Learning*
by M Narasimha Murty & V Susheela Devi

Vol. 4: *Game Theory and Mechanism Design*
by Y Narahari

Vol. 3: *Noise and Vibration Control*
by M L Munjal

Vol. 2: *Schwarz's Lemma from a Differential Geometric Viewpoint*
by Kang-Tae Kim & Hanjin Lee

Vol. 1: *Introduction to Algebraic Geometry and Commutative Algebra*
by Dilip P Patil & Uwe Storch

IISc Lecture Notes Series

Introduction to Time-Dependent Quantum Mechanics with Python

Atanu Bhattacharya
Gandhi Institute of Technology and Management, India

Elliot R Bernstein
Colorado State University, USA

NEW JERSEY • LONDON • SINGAPORE • BEIJING • SHANGHAI • HONG KONG • TAIPEI • CHENNAI • TOKYO

Published by

World Scientific Publishing Co. Pte. Ltd.
5 Toh Tuck Link, Singapore 596224
USA office: 27 Warren Street, Suite 401-402, Hackensack, NJ 07601
UK office: 57 Shelton Street, Covent Garden, London WC2H 9HE

Library of Congress Control Number: 2023039670

British Library Cataloguing-in-Publication Data
A catalogue record for this book is available from the British Library.

IISc Lecture Notes Series — Vol. 7
INTRODUCTION TO TIME-DEPENDENT QUANTUM MECHANICS WITH PYTHON

ISBN 978-981-127-716-0 (hardcover)
ISBN 978-981-127-717-7 (ebook for institutions)
ISBN 978-981-127-718-4 (ebook for individuals)

For any available supplementary material, please visit
https://www.worldscientific.com/worldscibooks/10.1142/13434#t=suppl

Desk Editors: Aanand Jayaraman/Sandhya Devi

Typeset by Stallion Press
Email: enquiries@stallionpress.com

Series Preface

World Scientific Publishing Company - Indian Institute of Science Collaboration

IISc Press and WSPC are co-publishing books authored by world renowned scientists and engineers. This collaboration, started in 2008 during IISc's centenary year under a Memorandum of Understanding between IISc and WSPC, has resulted in the establishment of three series: IISc Centenary Lecture Series (ICLS), IISc Research Monographs Series (IRMS), and IISc Lecture Notes Series (ILNS).

This pioneering collaboration will contribute significantly in disseminating current Indian scientific advancement worldwide.

The **"IISc Centenary Lecture Series"** will comprise lectures by designated Centenary Lecturers - eminent teachers and researchers from all over the world.

The **"IISc Research Monographs Series"** will comprise state-of-the-art monographs written by experts in specific areas. They will include, but not limited to, the authors' own research work.

The **"IISc Lecture Notes Series"** will consist of books that are reasonably self-contained and can be used either as textbooks or for self-study at the postgraduate level in science and engineering. The books will be based on material that has been class-tested for most part.

Diptiman Sen, Editor-in-Chief (diptiman@iisc.ac.in)

To Our Families, Friends, Mentors, and Loving Students

Preface

Professor David J. Tannor (Weizmann Institute of Science, Israel), in his classic book, *Introduction to Quantum Mechanics: A Time-Dependent Perspective*, has very rightly commented that time-dependent quantum mechanics is "the orphan child in the standard curriculum of chemistry, both on the undergraduate and graduate levels ... Traditional quantum chemistry courses start with the Time-Dependent Schrödinger Equation (TDSE); however, generally within half a lecture a separation of space and time variables is invoked and the rest of the course deals with the Time-Independent Schrödinger Equation (TISE)." This creates a big knowledge gap, which causes the development of many absurd questions or concepts in chemistry students' minds whenever they begin thinking about time-dependent quantum phenomena. This is why it is quite instructive that physical chemistry students are introduced to the time-dependent concepts of quantum mechanics as soon as they learn that matter is a wave and a particle is nothing but a wavepacket (a localized matter wave).

In the past, perhaps, many teachers and textbook writers of quantum mechanics realized the need for a balanced teaching approach involving both time-dependent and time-independent quantum mechanics. The most notable example includes Cohen-Tannoudji and co-workers' textbook *Quantum Mechanics*, which rigorously presents several important concepts of time-dependent quantum mechanics, including the wavepacket and the time-dependent view of quantum scattering, along with the time-independent concepts. However, to the best of our belief, for the first time, almost all pillars of time-dependent quantum mechanics, which include not only the wavepacket, correlation function, semiclassical methods, and numerical methods but also femtosecond or attosecond excitation, the treatment of strong fields, coherent control, and reactive scattering, have been discussed in sufficient detail by Tannor's book, *Quantum Mechanics: A Time-Dependent Perspective*. Both Cohen-Tannoudji's and Tannor's books are undoubtedly needed vehicles for training the current generation of physical chemistry students who are interested in modern chemical dynamics, spectroscopy, and ultrafast spectroscopy.

Despite the availability of the needed books, perhaps everybody familiar with the Indian course curriculum (both at the undergraduate and graduate levels) will agree that, in general, a chemistry student who has obtained quantum education at the bachelor's or master's level in India finds it difficult to go through and comfortably discuss the topics covered by both Cohen-Tannoudji's and Tannor's books. The reason is multifaceted. Even to date, quantum chemistry is taught at both the undergraduate and graduate levels in India following the traditional approach, with almost the entire focus on time-independent systems with known analytical derivations. Particle in a one-dimensional box, simple harmonic oscillator, hydrogen atom, and Hartree–Fock theory (and, on rare occasions, the density functional theory) are very popular examples. This traditional approach develops a profound "thinking momentum" along the time-independent viewpoint of quantum mechanics in chemistry students' minds. It takes quite an effort to nudge this "thinking momentum" toward the time-dependent viewpoint. The level of difficulty is further magnified by the fact that, even to date, most of the class-room discussions on quantum mechanics in India do not involve hands-on exercises for numerical implementation of quantum chemical problems using a suitable programming language. Most of the practical problems in quantum mechanics, even at the very preliminary research level, do not have analytical solutions. They must be solved numerically using a suitable programming language.

Realizing the above-mentioned problems, recently, a course on time-dependent quantum mechanics was formulated and offered to the integrated and regular PhD students at the Indian Institute of Science (offline) and at the NPTEL platform (online) to bridge the gap between the current quantum chemistry curricula prevailing in India (which are primarily at the level of *Quantum Chemistry* by Donald A. McQuarrie) and Tannor's classic book using a modern programming language. A very favorable reception to the course by the students from all over the country confirmed the existence of the need. One component of the course which all the students have enjoyed the most was the programming part, which demonstrated Python-based numerical approaches to directly solve simple time-dependent quantum mechanical problems. Currently, to the best of our knowledge, no textbook provides a hands-on compilation of time-dependent quantum mechanical methods and Python-based numerical routines.

The purpose of this textbook is to publish the lecture notes of the above-mentioned course after recasting the theories pertaining to time-dependent quantum mechanics in a more consolidated but easily perceptible form so that the eventual process of going through Tannor's or Cohen-Tannoudji's classic books becomes easier for chemistry students. Furthermore, each discussion of analytical theory is augmented by hands-on Python-based numerical recipes so that students can comfortably start writing programs to find solutions to simple time-dependent quantum mechanical problems. For this textbook, the Python programming language is selected due to many favorable reasons: It is very easy to learn, it is becoming increasingly popular within the modern scientific community for its rich scientific libraries, and it is freely available.

We strongly hope that chemistry students interested in time-dependent quantum mechanics will find this textbook useful. Finally, immeasurable thanks will be always due to our past and present collaborators, Prof. Todd Martinez (Stanford University), Dr. Ganga Periyasamy (Bangalore University), Dr. Gopal Dixit (IIT-Bombay), Dr. K. R. Shamasundar (IISER Mohali), and Dr. Sai G. Ramesh (IISc), who have clarified many doubts on different aspects of time-dependent quantum mechanics at different stages of our joint research work on the decomposition dynamics of energetic molecules and the attosecond dynamics of chemical bonding. In the end, we gratefully acknowledge the financial support received from USIEF (United States-India Educational Foundation) for the Fulbright Fellowship which has ignited our joint efforts breaking the boundaries of two countries.

About the Authors

Atanu Bhattacharya was born at Matiari, Nadia, in West Bengal, India, in 1983. He received the BSc degree in Chemistry from R. K. Mission Vidyamandira, Calcutta University, India, and the MSc degree in Physical Chemistry from the Indian Institute of Technology, Bombay, India, in 2003 and 2005, respectively. He received the PhD degree in Physical Chemistry from Colorado State University, USA, in 2010. His doctoral research involved the time- and frequency-resolved spectroscopy of energetic molecules in a molecular beam under the supervision of Prof. Elliot R. Bernstein. In 2010, he joined the Department of Chemistry at Brookhaven National Laboratory, USA, as a postdoctoral fellow to study catalytic reaction dynamics in the femtosecond to picosecond time domain under the supervision of Dr. Nicholas Camillone III. Then, in 2012, he joined Kyoto University, Japan, as a program-specific researcher to work on liquid beam time-resolved photoelectron spectroscopy and liquid beam X-ray absorption spectroscopy under the supervision of Prof. Toshinori Suzuki. In 2013, he joined the Indian Institute of Science in Benaluru, India, as an assistant professor at the Department of Inorganic and Physical Chemistry. Currently, he is an associate professor at the Department of Chemistry, Gandhi Institute of Technology and Management (GITAM, Deemed to be University), in Visakhapatnam, India. Dr. Atanu Bhattacharya, in his academic career, has specialized in attosecond chemistry, the femtosecond chemistry of catalysis and explosives, femtosecond X-ray spectroscopy, quantum dynamics, and chemical applications in quantum computers.

Dr. Atanu Bhattacharya is also an acting advisor for the IBM-India quantum computing initiative for chemical applications. In the past, he was a young associate of the Indian Academy of Science (2016–2018) and a Liverpool-India fellow (2014). He received the Teresa Fonseca Memorial Prize (2009) from Colorado State University as an outstanding researcher in physical chemistry.

Elliot R. Bernstein was born in New York City (NYC) in 1941. He attended public schools in NYC, graduating in 1959 from Stuyvesant High School. He received his A. B. degree in chemistry from Princeton University in 1963. As an undergraduate, he wrote a junior thesis on NMR and its applications and a senior research thesis on optical rotation in weakly bound, non-covalent crystals. Elliot's mentor at Princeton was Prof. Walter Kauzmann. These experimental and theoretical studies were developed in his research group. Elliot received his PhD from the California Institute of Technology (Caltech) in 1967. His research involved theoretical and experimental studies of vibrational and electronic excitons in weakly bound crystals, with special emphasis on the benzene system. Energy and excitation transfer, exciton annihilation, and isotopically mixed crystal energy levels were some of the topics covered in this work. His postdoctoral studies at the University of Chicago (Fermi Institute) with Prof. Clyde A. Hutchison Jr. from 1967 to 1969 dealt with electron–electron double resonance of rare earth ions in crystals. Prof. Bernstein was an assistant professor at Princeton University from 1969 to 1975 and an associate (1975–1980) and full (1980–2019) professor at Colorado State University (CSU). He became Professor Emeritus in

2019. Prof. Bernstein's research interests have been wide ranging, from zero-field magnetic resonance, molecular crystal phase transition, cryogenic liquids, gas phase clusters (pure and solvated), organic energetic molecules and crystals, various sugars, ultrafast (attosecond and femtosecond) electron dynamics, and state evolution: These research areas have been explored through many spectroscopic techniques, including IR/UV double resonance, Brillouin scattering, Raman spectroscopy, photoelectron spectroscopy, mass spectrometry, multiphoton spectroscopy, VUV single photon ionization spectroscopy, and ultrafast dynamic pump–probe spectroscopy.

Prof. Bernstein is a fellow of the American Association for the Advancement of Science (AAAS), the American Physical Society (APS), and the Japanese Society for the Preservation of Science (JSPS). He is additionally a JSPS Bridge Fellow and a Fulbright Specialist Program Fellow (IISc Bengaluru, IISER Pune, TIFR, and IIT Bombay). He is a recipient of the American Chemical Society (ACS) Lifetime Career Research Chemist and the VA Commonwealth University Mary E. Kapp awards. He was also an NSF Summer Fellow, a Woodrow Wilson Fellow, an E. Fermi Fellow, a Third Cycle in Chemistry Lecturer in Switzerland (Bern, Basel, and Lausanne), and a Chinese Academy of Sciences (CAS) invited scholar.

Contents

Introduction

Highlights: *Targets and Structure of the Present Textbook.*

Currently, the importance of time-dependent quantum mechanics in physical chemistry is evident to both theoreticians and experimentalists alike due to the impetuous growth of ultrafast spectroscopic techniques over the past several decades. The usage of ultrafast optical, X-ray, or electron pulses (with picosecond, femtosecond, and attosecond durations) to study photoinduced dynamical processes in atoms, molecules, nanostructures, solvated species, and solids is very common in physical chemistry research.[1] Several hitherto-developed sophisticated pump–probe techniques, such as transient absorption spectroscopy, stimulated Raman spectroscopy, high harmonic generation spectroscopy, ultrafast two-dimensional spectroscopy, and ultrafast (X-ray and electron) diffraction, are capable of providing unprecedented time-dependent structural and dynamical information of immense chemical importance.[2] However, these sophisticated time-dependent spectroscopic techniques alone cannot provide the necessary chemical and dynamical information. Almost without exception, these experimental investigations must be augmented by computational spectroscopic and computational quantum chemical dynamic studies for the interpretation of the complex signatures delivered by the time-resolved techniques.[3,4]

Computational spectroscopy and computational quantum chemical dynamics together constitute a vast field in physical chemistry. A significant part of this field is developed based on the concepts of time-dependent quantum mechanics and their numerical implementations. This textbook attempts to introduce the subject of **time-dependent quantum mechanics and its Python-based implementations** with the assumption that students have completed their first quantum mechanics course at the level of *Quantum Chemistry* by Donald A. McQuarrie (covering basic topics, such as the postulates of quantum mechanics, the one-dimensional box problem, the simple harmonic oscillator problem, the hydrogen atom problem, and the Hartree–Fock theory) and, subsequently, would like to reach the level of *Quantum Mechanics: A Time-Dependent Perspective* by David J. Tannor (where advanced time-domain numerical methods, femtosecond excitation, nonadiabatic dynamics, the treatment of strong fields, coherent control, reactive scattering, etc., are discussed in adequate detail).

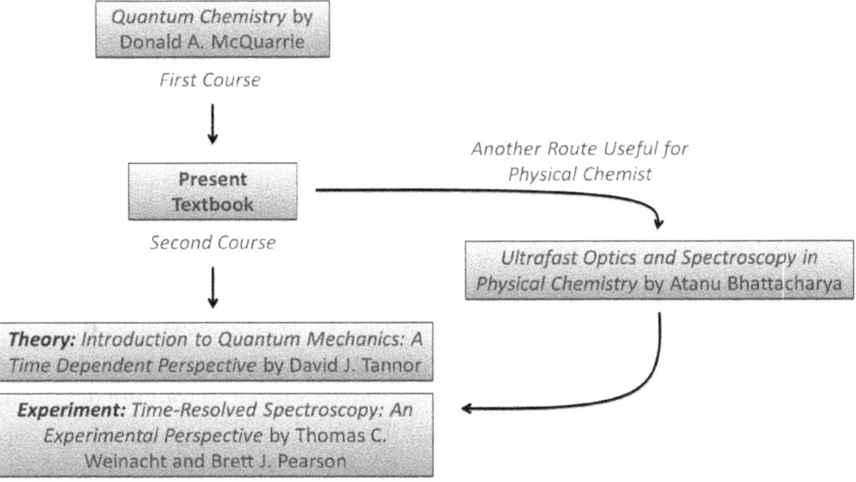

Suggested Learning Chart for Time-Dependent Quantum Mechanics

This textbook includes the contents of a one-semester course on time-dependent quantum mechanics, in which **the dynamics of a quantum particle is discussed under the influence of only a time-independent Hamiltonian** (dynamics under the influence of a time-dependent Hamiltonian, which is not discussed here, can be another semester course). In this textbook, different concepts of time-dependent quantum mechanics are systematically presented by first giving emphasis on the contrasting viewpoints of classical and quantum mechanical motions of a particle (in Chapters 1 and 2), then by demonstrating the ways to finding a classical flavor in quantum dynamics (through the Ehrenfest theorem, the equation of continuity, and Bohmian mechanics in Chapter 2), thereafter by formally defining the wavepacket, which represents a quantum particle (Chapter 3), and finally by demonstrating numerical methods to explore the wavepacket dynamics (the quantum dynamics of a particle) in one dimension. The last topic requires knowledge of linear algebra (which is contextually presented in Chapter 4) and of the time propagator (which is contextually introduced in Chapter 5). In the end, Chapter 6 features some research-front aspects of wavepacket dynamics to raise readers' curiosity about chemical applications.

Along with the above chapters, where students realize the theory of wavepacket dynamics in one dimension thoroughly, accompanying five PythonChapters (from A to E) take them on a hands-on tour with Python programming so that they can finally solve one-dimensional wavepacket dynamics numerically. Giving a thorough hands-on experience, even with the simplest quantum dynamics problem, to students with no prior experience with programming or numerical methods is a very tricky task. This, however, has been accomplished here by first giving them a quick introduction to Python programming (in PythonChapter A), then by introducing the position-space grid representation of the wavefunction (in PythonChapter B), thereafter by making them familiarized with Fourier transforms to represent discretized wavefunctions in momentum space (numerical method of implementing the time propagator requires a wavefunction to be represented in momentum space: a concept which is developed in Chapter 5), subsequently by showing the Python-based methodologies to express the Hamiltonian operator in matrix form (in PythonChapter D), and finally by demonstrating an entire Python program which solves the wavepacket dynamics in one dimension under the influence of a time-independent Hamiltonian following the split-operator approach (in PythonChapter E).

To reach the final goal of exploring wavepacket dynamics in one dimension, as discussed in PythonChapter E, one has to systematically go through the material given in PythonChapters A–D.

Rigorous class testing of the presented lecture notes at the Indian Institute of Science, Gandhi Institute of Technology and Management (GITAM, Deemed to be University), and the NPTEL platform reveals that physical chemistry students, after thoroughly going through all chapters, not only develop an in-depth understanding of the wavepacket dynamics and its numerical implementations but also successfully start writing their own Python code for solving any one-dimensional wavepacket dynamics problem.

Notes, References, and Further Reading

1. (a) A. Bhattacharya, *Ultrafast Optics and Spectroscopy in Physical Chemistry*. World Scientific, Singapore (2018); (b) T. C. Weinacht and B. J. Pearson, *Time-Resolved Spectroscopy — An Experimental Perspective*. CRC Press: Florida (2019).
2. M. Maiuri, M. Garavelli, and G. Cerullo, Ultrafast spectroscopy: State of the art and open challenges. *J. Am. Chem. Soc.* 142, 3 (2020).
3. I. Conti, G. Cerullo, A. Nenov, and M. Garavelli, Ultrafast spectroscopy of photoactive molecular systems from first principles: Where we stand today and where we are going. *J. Am. Chem. Soc.* 142, 16117 (2020).
4. X. Li, N. Govind, C. Isborn, A. E. DePrince III, and K. Lopata, Real-time time-dependent electronic structure theory. *Chem. Rev.* 120, 9951 (2020).

Chapter 1

Introduction to Quantum Dynamics with Quantum Superposition

Highlights: *Time-Dependent Schrödinger Equation, Variable Separation, Stationary and Superposition States, Electronic and Vibrational Superposition States, Optical Analogy to Quantum Superposition.*

1.1 Introduction

Quantum mechanics governs the structure, dynamics, and physical and chemical properties of matter at the atomic and molecular levels. In quantum mechanics, the wavefunction, denoted as $\psi(x,t)$ (let us choose a single dimension to obtain a simple picture), provides a complete description of all the properties of matter at any given time and gives information about a matter's dynamical observables (the physical quantities which can be experimentally measured or observed, such as position, momentum, and energy). The wavefunction of a quantum system evolves in time according to the time-dependent Schrödinger equation (abbreviated as TDSE). The TDSE is a differential equation of first order in time postulated by Schrödinger.[1]

1.2 Time-Dependent Schrödinger Equation

The TDSE for a (nonrelativistic) single particle (whose speed is assumed to be much less than the speed of light) in one dimension is written as

$$i\hbar\frac{\partial}{\partial t}\psi(x,t) = \hat{H}\psi(x,t) \tag{1.1}$$

where $i = \sqrt{-1}, \hbar$ is the Planck's constant divided by 2π (called reduced Planck's constant, which is equal to 1.054572×10^{-34}J s $= 0.65821 \times 10^{-15}$eV s), and $\hat{H}$ is the Hamiltonian operator: $\hat{H} = [-\frac{\hbar^2}{2m}(\frac{\partial^2}{\partial x^2}) + V(x)]$, in which the first and second terms are, respectively, kinetic energy and potential energy operators (more details of the quantum mechanical operators can be found in Chapter 2). As considered here, $\hat{H}$ is a function of x only (**throughout this book we have assumed that it has no explicit dependence on time**). The function $\psi(x,t)$, which satisfies the above equation, is called the wavefunction (more specifically, the time-dependent wavefunction).[2] Nature and time evolution of $\psi(x,t)$ for a quantum particle is solely determined by the Hamiltonian associated with it. $|\psi(x,t)|^2dx$ represents the probability of finding the particle between x and $(x+dx)$ at time t. If $\psi(x,t)$ is normalized, one can write $\int_{-\infty}^{+\infty} |\psi(x,t)|^2dx = 1$. Furthermore, $|\psi(x,t)|^2$ is called the probability density of finding the particle at position x and at time t. Before proceeding further to solve the TDSE using the variable separation method, contrasting meaning of a moving particle in the classical and quantum mechanics is presented below.

Important Note

In mathematical language, TDSE is an initial value problem: In order to express the state of a quantum particle at a later time, t(using $\psi(x,t)$), we must have prior knowledge of the state of the quantum particle at $t = 0$ (i.e., $\psi(x,0)$ which represents the moment just before the onset of the time-evolution process). Therefore, an experimental test of a quantum dynamical prediction can only be successfully performed when the experimental conditions (both the initial conditions and the conditions under which the system is undergoing the time-evolution process) are well understood and are correctly represented by $\psi(x,0)$ and $\hat{H}$.

1.3 Classical versus Quantum Mechanical Picture of Motion of a Particle

If a particle of mass m under the influence of a conservative force[3] F is moving along the x-axis (our favorite one dimension is chosen to obtain a simple picture), classical mechanics gives us the position of the particle, $x(t)$, at any given time using Newton's equation of motion: $m\frac{d^2x}{dt^2} = F = -\frac{dV}{dx}$ if the initial condition is already given (i.e., if the position and velocity at $t = 0$ are known). When $x(t)$ is known, one can easily figure out the velocity $\left(v = \frac{dx}{dt}\right)$, the linear momentum $(p = mv)$, the kinetic energy $\left(= \frac{1}{2}mv^2\right)$, or any other classical dynamical variable associated with the motion of the particle at any later time t. Our classical or **local notion** of the motion of a particle is depicted in Figure 1.1(a).

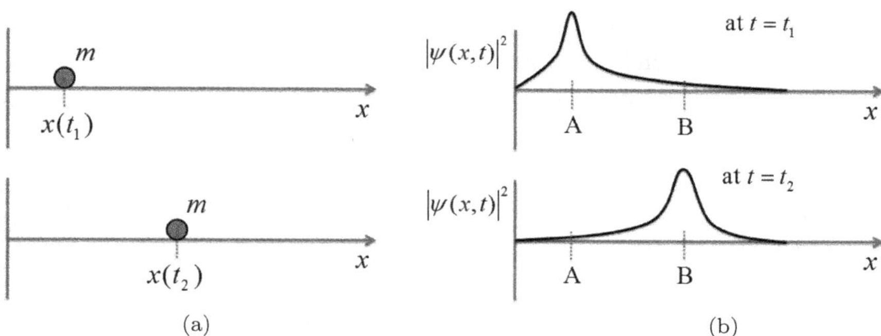

Figure 1.1. (a) Classical picture of the motion of a particle. Two different positions are shown for two different times. We get point-by-point information of the motion of the particle in classical mechanics. (b) Quantum mechanical picture of the motion of a particle does not directly provide point-by-point information about the motion of the particle. The probability density, which represents the probability of finding the particle at a point x and a time t, evolves over time. This temporal evolution of the probability density represents the motion of a quantum particle. In the present example, the most probable positions of the particle at two different times, t_1 and t_2, are represented by positions A and B, respectively. Thus, the deterministic nature of a classical particle is clearly evident in (a), and the probabilistic nature of a quantum particle is clearly evident in (b).

In quantum mechanics, all the physical and chemical descriptions of a particle are given by its wavefunction $\psi(x,t)$. Note that a time-dependent wavefunction may refer to either a stationary or a non-stationary state. At this point of discussion, these two states are not distinguished, which will be done later in this chapter. For a given initial condition, $\psi(x,0)$, the TDSE can determine $\psi(x,t)$ for any later time. The wavefunction, by its nature, is spread out in space (**delocalized over the space**) for any given time t. The global (delocalized) nature of the wavefunction of the particle at any given time contradicts our classical (or local) notion that, as the wavefunction represents a tiny little particle, it must be localized at a point on the x-axis. **How can a delocalized mathematical function (wavefunction) represent a particle which is supposed to be localized?**

Quantum mechanics does not give any direct answer to the above question. Only a statistical or probabilistic answer to the above question is provided by quantum mechanics. The probability of finding the particle at point x and at time t (or, more specifically, the probability density or probability amplitude for the particle) is given by $\rho(x,t) = |\psi(x,t)|^2 = \psi^*(x,t) \cdot \psi(x,t)$. Figure 1.1(b) illustrates schematically this probability amplitude or probability density of a moving quantum particle at two different times. According to the definition, $|\psi(x,t)|^2 dx$ is the probability of finding the particle between x and $(x + dx)$. As $|\psi(x,t)|^2 dx$ represents the area under the curve $|\psi(x,t)|^2$ for an interval of dx, the statistical interpretation of the wavefunction illuminates an important fact that at $t = t_1$, the particle is more likely (i.e., has a higher amplitude) to be found near the point A than near the point B; and at $t = t_2$, the particle is more likely (i.e., has a higher amplitude) to be found near the point B than near the point A. In addition, the total probability of finding the particle over the entire space at any given time must always be 1 (because it is a single particle). This is represented by the integral

$$\int_{-\infty}^{+\infty} |\psi(x,t)|^2 dx = 1 \tag{1.2}$$

Important Note

Every wavefunction in quantum mechanics is spread out in space (**delocalized over the space**). Readers who are new to quantum mechanics may have learned quantum mechanics without even noticing it. This is why this point is further highlighted here by taking a very popular example of a particle in a one-dimensional box problem. The ground state wavefunction of the particle in a one dimensional box is given by

$$\psi(x) = \sqrt{\frac{2}{L}} \sin\left(\frac{\pi x}{L}\right)$$

Here, L represents the length of the box. This wavefunction is depicted schematically in the following figure.

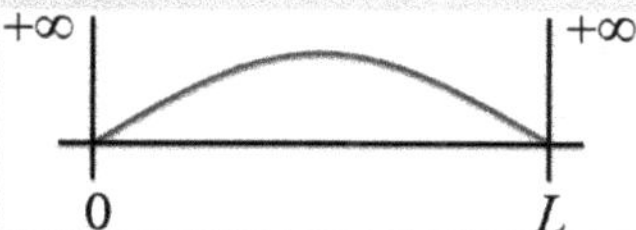

It is clearly evident from the above figure that the ground state wavefunction of a particle in a one-dimensional box is spread out in the given problem domain (in the one-dimensional position space between 0 and L). A delocalized wavefunction has a statistical interpretation for the experimental observation. This interpretation is discussed, in detail, in Chapter 2. Note that the above ground state wavefunction is also called a stationary state wavefunction. Stationary and non-stationary states are defined in this chapter after introducing the variable separation method.

When a wavefunction satisfies the condition given in equation (1.2), it is called a **normalized wavefunction**. Therefore, a physically realizable wavefunction, which can represent the particle and is a solution to the TDSE, must be a normalizable wavefunction. If a wavefunction cannot be normalized, it is not acceptable as a solution to the TDSE because the statistical interpretation (physically realizable interpretation) of such a wavefunction fails. A more detailed discussion on normalizable wavefunctions is given in Chapter 4.

We are being careful about the language used here to describe the particle in question because one should not think of an electron as a small, hard particle that moves within the Coulomb potential of the proton. Under such circumstances, the electron assumes a $1s, 2s, 2p, \ldots, 3d, \ldots$ *probability density distribution: That is all we know about it, based on the quantum mechanics of a hydrogen atom (see L. Pauling and E. B. Wilson, Introduction to Quantum Mechanics: With Application to Chemistry, Dover Publications Inc. 1985).*

Two Important Notes

1. Unfortunately, the wavefunction of a quantum particle cannot be observed experimentally.

   $\not\rightarrow \psi(x,t)$

   Repeated experiments can only measure the probability density distribution of a particle.

   $\longrightarrow |\psi(x,t)|^2$

   This is an unavoidable limitation of the theory and the experiment.
2. **Size and shape of a quantum particle**: It is not possible to determine the size and shape of a quantum particle (it is so small that quantum mechanics governs the physical and chemical properties of the particle). However, probability density distribution function, $|\psi(x,t)|^2$, associated with the particle exhibits a certain size and shape. For example, an electron in a p-atomic orbital state of a hydrogen atom shows an angular probability density distribution resembling two hollow spheres touching each other, as shown in the following figure.

   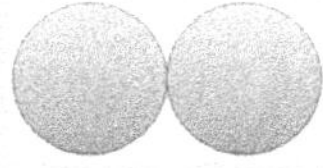

   See L. Pauling and E. B. Wilson, *Introduction to Quantum Mechanics: With Applications to Chemistry* (Dover Publications Inc., 1985) for more details. The above probability density distribution represents possible outcome of many repeated experimental measurements of the position of the electron, as shown below (for more discussion, see Chapter 2).

   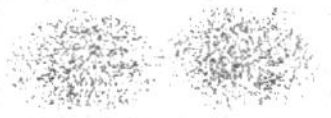

(*Continued*)

(Continued)

Here, each dot in the above figure represents the result of a single experimental measurement of the position of the electron. The size and shape of the probability density distribution of a quantum particle depend primarily on the potential $V(x)$ experienced by the particle.

1.4 Variable Separation

The TDSE is solved by a separation of variables in position and time.[4] This is done under the **assumption** that $\psi(x,t)$ can be written as

$$\psi(x,t) = \psi(x) \cdot \psi(t) \tag{1.3}$$

and $\hat{H}$ does not depend on time. Inserting this trial solution into equation (1.1), we get

$$i\hbar\psi(x)\frac{d}{dt}\psi(t) = \psi(t)\hat{H}\psi(x)$$

$$\text{or} \quad i\hbar\frac{1}{\psi(t)}\frac{d}{dt}\psi(t) = \frac{1}{\psi(x)}\hat{H}\psi(x) \tag{1.4}$$

We immediately notice that the left-hand side of equation (1.4) is a function of only time, while the right-hand side is a function of only position (as mentioned earlier, $\hat{H}$ does **not depend on time**). Therefore, both sides must be equal to a constant (say E). Consequently, two equations are obtained by separating the variables:

$$\hat{H}\psi(x) = E\psi(x) \tag{1.5}$$

$$\text{and} \quad i\hbar\frac{d}{dt}\psi(t) = E\psi(t) \tag{1.6}$$

Equation (1.5) is called the time-independent Schrödinger equation (TISE). This equation is an "eigenvalue" equation: The constant E is called the eigenvalue, and $\psi(x)$ is called the eigenfunction (eigenstate). Eigenvalue equations will be discussed in Chapter 4 in detail.

A solution to equation (1.6) can be given as

$$\psi(t) = \psi_0 e^{-\frac{iEt}{\hbar}} \tag{1.7}$$

As a result, using equations (1.3) and (1.7), a solution to the TDSE can be obtained as

$$\psi(x,t) = \psi(x)\psi_0 e^{-\frac{iEt}{\hbar}}$$

Here, ψ_0 is just a multiplicative factor; therefore, considering the normalization condition, $\int_{-\infty}^{+\infty} |\psi(x,t)|^2 dx = 1$, without loss of any information, one can write a solution to the TDSE as

$$\psi(x,t) = \psi(x) e^{-\frac{iEt}{\hbar}} \tag{1.8}$$

In general, the TISE (equation (1.5)) gives a set of solutions. The set of energies supporting physically meaningful solutions to this equation represents the spectrum of the quantum system (or of the respective Hamiltonian). Each solution is represented by $\psi_n(x)$, which is called a stationary state wavefunction, wherein n denotes the nth stationary state. Often, in chemistry, different spectroscopic properties of an atom or a molecule are expressed using these stationary states. Therefore, a more appropriate way of writing a solution to the TDSE is

$$\psi_n(x,t) = \psi_n(x) e^{-\frac{iE_n t}{\hbar}} \tag{1.9}$$

Here, $\psi_n(x)$ represents the nth stationary state (e.g., $1s$, $2s$, $3d$ state of a hydrogen atom) with energy E_n, and $e^{-\frac{iE_n t}{\hbar}}$ is the associated **phase factor**. The name "phase factor" originates from the fact that for any complex number written in polar form ($re^{i\theta}$), the phase factor is the complex exponential factor ($e^{i\theta}$).

To obtain a clear meaning of equation (1.9), recall the classic problem of a particle in a one-dimensional box again. When we solve the TISE with the Hamiltonian $\hat{H}(x) = \left[-\frac{\hbar^2}{2m}\frac{d^2}{dx^2} + V(x)\right]$ with the condition $V(x) = 0$ when $0 \leq x \leq L$ and $V(x) = +\infty$ otherwise, we obtain a set of solutions which represent possible energy states of the particle. The nth state of the particle is described by the wavefunction $\psi_n(x) = \sqrt{\frac{2}{L}}\sin(\frac{\pi x n}{L})$, where the $n = 1$ state represents the ground state wavefunction of the particle. The energy of the corresponding nth state of the particle is given by $E_n = \frac{n^2\pi^2\hbar^2}{2mL^2}$. The wavefunction $\psi_n(x) = \sqrt{\frac{2}{L}}\sin(\frac{\pi xn}{L})$ is also called the nth stationary state wavefunction of the particle because this wavefunction is obtained by solving the TISE (it is stationary with respect to time because no time-dependent term is present in the respective mathematical expression).

According to equation (1.9), the time-dependent wavefunction of the nth state of the particle in a one-dimensional box is given by $\psi_n(x,t) = \sqrt{\frac{2}{L}}\sin(\frac{\pi xn}{L})\ e^{-\frac{iE_nt}{\hbar}}$, where $E_n = \frac{n^2\pi^2\hbar^2}{2mL^2}$. Carefully note that the wavefunction $\psi_n(x,t) = \sqrt{\frac{2}{L}}\sin(\frac{\pi xn}{L})\ e^{-\frac{iE_nt}{\hbar}}$ is time-dependent; nonetheless, the state represented by the wavefunction $\psi_n(x,t) = \sqrt{\frac{2}{L}}\sin(\frac{\pi xn}{L})e^{-\frac{iE_nt}{\hbar}}$ features a stationary state of the particle. This statement, which may appear puzzling at face value, will be further clarified in the following.

Guiding Questions

1.1. One key method of solving the TDSE discussed above is "variable separation." We have seen that the TDSE is separable when $\hat{H}$ is independent of time. $\hat{H}$ is the Hamiltonian operator, which includes the kinetic $\left(-\frac{\hbar^2}{2m}\frac{\partial^2}{\partial x^2}\right)$ and potential (V) energy terms. By construction, the kinetic energy term does not depend on time; however, the potential energy term V can be a function of both time and space. We have learned that the TDSE is separable when V is independent of time (only depends on space). This argument is not entirely correct. The variable separation method can also be employed when V is a function **only of time** everywhere in space. If $V(t)$ is expressed as $V_0\cos(\omega_0 t)$, find an expression for $\psi(x,t)$ by solving the TDSE.

1.2. For a certain one-dimensional particle of mass m, the wavefunction is given by $\psi(x,t) = Ae^{-\frac{amx^2}{\hbar}}e^{-iat}$, where A and a are real positive constants. Find the potential energy for which the above wavefunction satisfies the TDSE.

1.5 Stationary and Superposition States

Given a solution to the TDSE (equation (1.9)), as illustrated above, one can explore the probability density of finding the particle in the

nth stationary state:

$$|\psi_n(x,t)|^2 = \psi_n^*(x,t) \cdot \psi_n(x,t) = \psi_n^*(x) e^{\frac{iE_n t}{\hbar}} \cdot \psi_n(x) e^{-\frac{iE_n t}{\hbar}}$$
$$= \psi_n^*(x) \cdot \psi_n(x) = |\psi_n(x)|^2$$

We immediately realize from the above expression that the probability density of finding the particle represented by the wavefunction $\psi_n(x)e^{-\frac{iE_n t}{\hbar}}$ is independent of time! If the probability density does not change with time, there is no time-dependence; in other words, there is no effective motion of the particle (wavefunction is not an experimentally measurable quantity but the probability density is). As a result, this state is called a stationary state. This is why, as mentioned earlier, the time-dependent nth state wavefunction of the particle in a one-dimensional box, $\psi_n(x,t) = \sqrt{\frac{2}{L}} \sin(\frac{\pi x n}{L}) e^{-\frac{iE_n t}{\hbar}}$, represents a stationary state of the particle because the probability density or amplitude $|\psi_n(x,t)|^2 = \frac{2}{L}\sin^2(\frac{\pi x n}{L})$ of the particle at the nth state is time-independent.

Stationary State of the Particle in One-Dimensional Box

The time-dependent ground state wavefunction of the particle in a one-dimensional box, $\psi_1(x,t) = \sqrt{\frac{2}{L}} \sin(\frac{\pi x}{L})\ e^{-\frac{i\frac{\pi^2 \hbar^2}{2mL^2}t}{\hbar}}$, changes its phase as a function of time, as schematically depicted in the following figure ($e^{i\theta} = \cos\theta + i\sin\theta$ represents an oscillatory function).

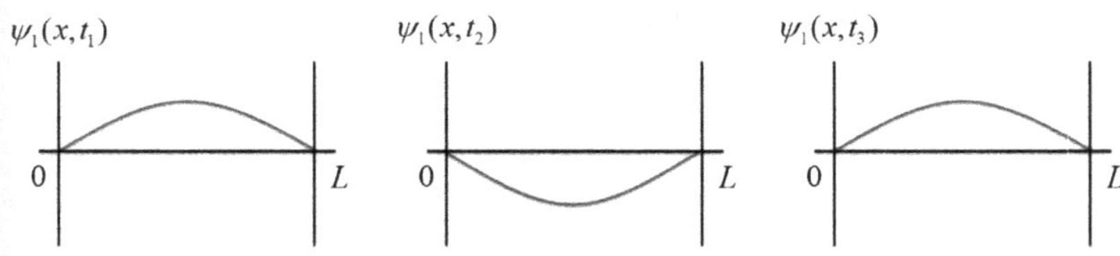

However, the corresponding probability density $|\psi_1|^2 = \frac{2}{L}\sin^2(\frac{\pi x}{L})$ does not change as a function of time, as schematically depicted in the following figure.

(*Continued*)

(*Continued*)

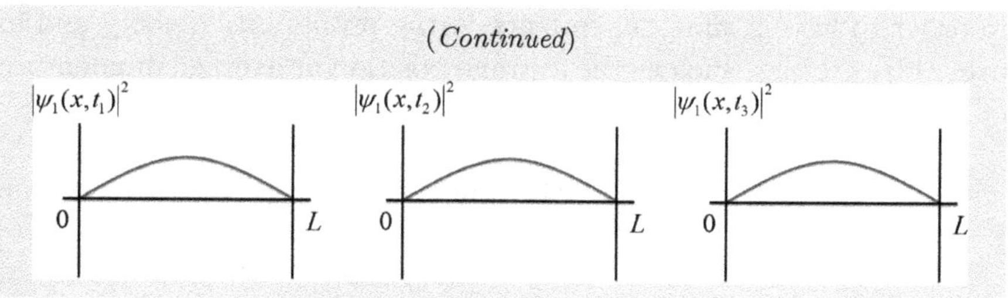

Therefore, if we want to describe the motion of a quantum particle, we have to describe the quantum system not by a particular solution (stationary state) given in equation (1.9), but rather by a linear combination of particular solutions (stationary states). Note that any linear combination of particular solutions is also a solution to the TDSE (prove it yourself). When a linear combination of particular solutions (stationary states), each with its own characteristic time-dependent phase factor, is used to describe a quantum system, we obtain time-dependence in the probability density. Such a state is called a superposition state.[5]

For example, let us assume that a particle can be represented by the following wavefunction:

$$\psi(x,t) = a_1\psi_1(x)e^{-\frac{iE_1t}{\hbar}} + a_2\psi_2(x)e^{-\frac{iE_2t}{\hbar}}$$

As a result, the probability density of finding the particle is given by

$$\begin{aligned}|\psi(x,t)|^2 &= |a_1|^2|\psi_1(x)|^2 + |a_2|^2|\psi_2(x)|^2 \\ &\quad + a_1a_2^*\psi_1(x)\psi_2^*(x)e^{\frac{-i(E_1-E_2)t}{\hbar}} + c.c.\end{aligned}$$

(where *c.c.* refers to complex conjugate.)

The above equation has three important terms. The first one originates from the pure $\psi_1(x)$ state, the second one comes from the pure $\psi_2(x)$ state, and the third one develops from interference between two stationary states. The interference term is a result of having a superposition of eigenstates with different energies — called a wavepacket. All the time-dependence of the probability density is contained in this interference term. Therefore, a wavepacket, which originates from a superposition of stationary states with different energies, is required in order to have a time-dependence in the probability density and in other observables, such as the average position or average momentum of the particle.

Metaphor

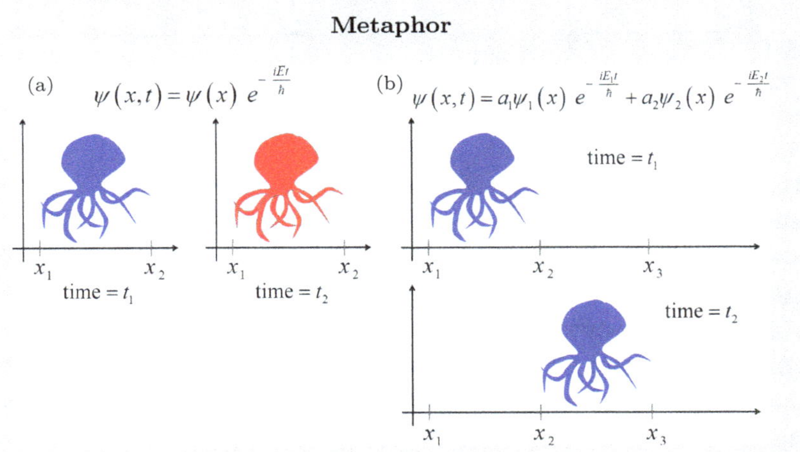

An octopus can change its skin color. Let us assume that an octopus, localized (not moving) in the $(x_1 - x_2)$ space, changes its skin color periodically (see Figure (a)). In this metaphorical example, the octopus represents a stationary state wavefunction of a particle. A wavefunction of a stationary state contains a time-dependent phase factor which changes the phase of the wavefunction periodically (equivalent to a periodic change in color); however, the particle does not move in space merely due to the presence of its own time-dependent phase factor because its probability density is independent of time. Only a superposition state of the particle allows the particle to move in space (see Figure (b)).

Obtaining a visually appealing meaning of the phase (more specifically, the complex time-dependent phase) of a wavefunction is a difficult task. As metaphorically defined above, the phase of a wavefunction represents a color which keeps changing periodically. Thus, one can view the angular part of the time-dependent wavefunction of an electron in the *p*-atomic orbital of a hydrogen atom as given in the following figure.

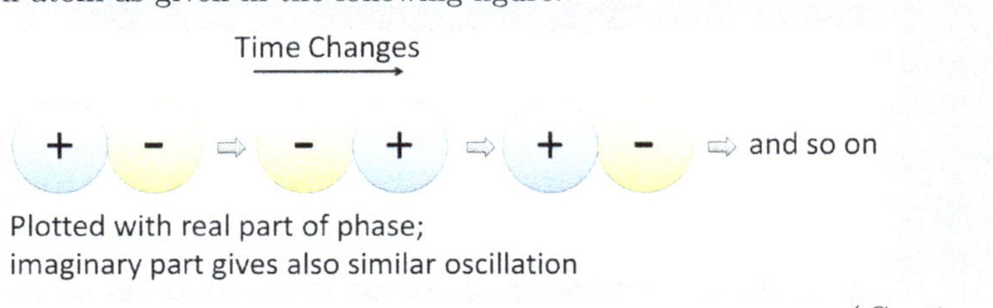

(*Continued*)

(Continued)

This is because the wavefunction undergoes a periodic change in color (or phase) at different times. Relative phase becomes an important factor when two wavefunctions interact. For example, relative color should match to render a favorable (bonding) interaction, as shown in the following figure.

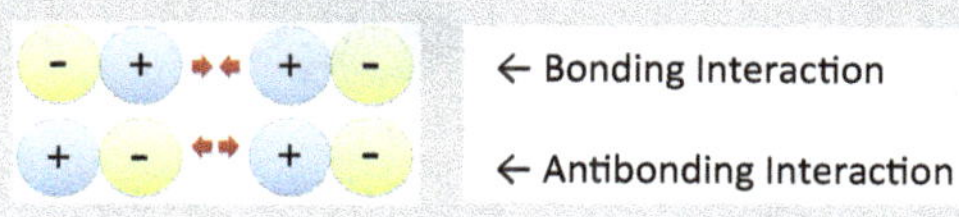

Thus, the general solution to the TDSE should be given more appropriately by

$$\psi(x,t) = \sum_{n=0}^{\infty} a_n \psi_n(x) e^{-i\frac{E_n t}{\hbar}}$$

(if stationary states have discrete spectrum/ energy states)

$$\text{or } \psi(x,t) = \int_0^{\infty} a_E \psi_E(x) e^{-\frac{iEt}{\hbar}} dE$$

(if stationary states have a continuous spectrum)

Here, in the integral form, a_E is the amplitude (which refers to the extent of the contribution of the E-th stationary state to the construction of the wavepacket) and $e^{-\frac{iEt}{\hbar}}$ is the corresponding phase factor. Therefore, to observe quantum dynamics, a wavepacket must be constructed with characteristic amplitudes and phase factors.

1.6 Simple Examples of Superposition States

1.6.1 *Electronic superposition state*

One of the simplest examples of an electronic superposition state, which can be easily realized, is perhaps a superposition state created by the two lowest lying ${}^2\Sigma_g^+$ and ${}^2\Sigma_u^+$ electronic states of H_2^+ (see Figure 1.2(a)). The electronic wavefunctions of these two states

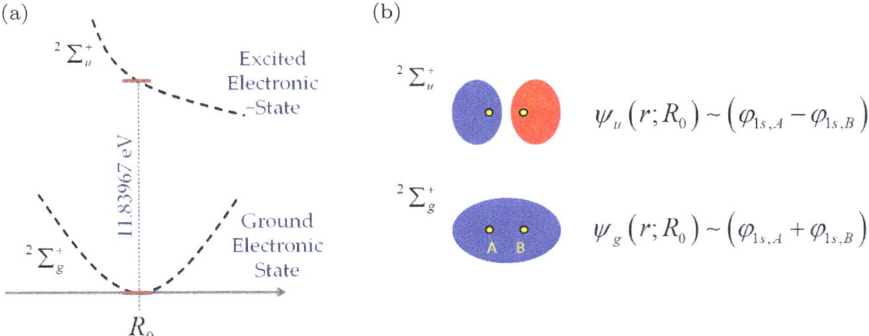

Figure 1.2. (a) Ground electronic state ${}^2\sum_g^+$ and first excited electronic state ${}^2\sum_u^+$ of H_2^+ are depicted. Here, R_0 represents the equilibrium internuclear distance of H_2^+ in its ground electronic state. (b) Electronic wavefunctions of the ground electronic state ${}^2\sum_g^+$ and the first electronically excited state ${}^2\sum_u^+$ of H_2^+ are depicted. As H_2^+ contains only one electron, its electronic wavefunction is represented approximately by the respective molecular orbital.[6] Note that each molecular orbital (${}^2\sum_g^+$ or ${}^2\sum_u^+$) is expressed as a linear combination of atomic orbitals (LCAO). LCAO, by its nature of construction, carries the sense of "superposition;" however, the (time-dependent) phase factor associated with each atomic orbital is not taken into account in its construction.

(which are depicted in Figure 1.2(b)) can be written as

$$\psi_g(r,t;R_0) = \psi_g(r;R_0)e^{-\frac{iE_g t}{\hbar}}$$
$$\text{and} \quad \psi_u(r,t;R_0) = \psi_u(r;R_0)e^{-\frac{iE_u t}{\hbar}}$$

Here, we have assumed that the nuclear positions are fixed at the equilibrium geometry (R_0) of the ground electronic state of H_2^+; however, the electron has the appropriate probability density distribution for each of its respective states. Thus, the total electronic wavefunction of the superposition state can be expressed as

$$\begin{aligned}\psi(r,t;R_0) &= \psi_g(r,t;R_0) + \psi_u(r,t;R_0)\\ &= \psi_g(r;R_0)e^{-\frac{iE_g t}{\hbar}} + \psi_u(r;R_0)e^{-\frac{iE_u t}{\hbar}}\end{aligned}$$

Finally, the time-dependent electron density is given by

$$
\begin{aligned}
|\psi(r,t;R_0)|^2 &= [\psi^*(r,t;R_0)\cdot\psi(r,t;R_0)] \\
&= \left[\psi_g^*(r;R_0)e^{\frac{iE_gt}{\hbar}} + \psi_u^*(r;R_0)e^{\frac{iE_ut}{\hbar}}\right] \\
&\quad \cdot \left[\psi_g(r;R_0)e^{-\frac{iE_gt}{\hbar}} + \psi_u(r;R_0)e^{-\frac{iE_ut}{\hbar}}\right] \\
&= |\psi_g(r;R_0)|^2 + |\psi_u(r;R_0)|^2 + \psi_g(r;R_0) \\
&\quad \times \psi_u^*(r;R_0)e^{\frac{i(E_u-E_g)t}{\hbar}} + c.c. \\
&= |\psi_g(r;R_0)|^2 + |\psi_u(r;R_0)|^2 + 2\psi_g(r;R_0) \\
&\quad \times \psi_u(r;R_0)\cos\left[\frac{(E_u-E_g)t}{\hbar}\right]
\end{aligned}
$$

because $\psi_g(r;R_0)$ and $\psi_u(r;R_0)$ are real and are given, respectively, by

$$
\begin{aligned}
&\psi_g(r;R_0) \sim (\varphi_{1s,A} + \varphi_{1s,B}) \\
\text{and}\quad &\psi_u(r;R_0) \sim (\varphi_{1s,A} - \varphi_{1s,B}).
\end{aligned}
$$

Here, $\varphi_{1s,A}$ and $\varphi_{1s,B}$ refer to the $1s$ orbitals of two hydrogen atoms denoted as A and B, respectively.

As a result,
$$
\begin{aligned}
|\psi(r,t;R_0)|^2 &= |\psi_g(r;R_0)|^2 + |\psi_u(r;R_0)|^2 + 2\psi_g(r;R_0) \\
&\quad \times \psi_u(r;R_0)\cos\left[\frac{\Delta E_{ug}t}{\hbar}\right]
\end{aligned} \tag{1.10}
$$

where ΔE_{ug} represents the energy separation between the ${}^2\sum_g^+$ and ${}^2\sum_u^+$ electronic states of H_2^+. The above equation shows that time-dependent electron density oscillates with a period of $T = \frac{1}{\nu} = \frac{2\pi}{\frac{\Delta E_{ug}}{\hbar}} = \frac{h}{\Delta E_{ug}}$.

For H_2^+, ΔE_{ug} is 11.83967 ev[7] at the equilibrium geometry of the ground electronic state; therefore, the oscillation time period of the time-dependent electron density is calculated to be 349 attoseconds (where 1 attosecond $= 1 \times 10^{-18}$ s).[8] But how does this electron density change after the creation of the superposition state?

We know that a cosine function can take any value between +1 and −1. To visualize the temporal evolution of the electron density for the electronic superposition state, we may consider $+1, 0$ and -1 values for equation (1.10):

(a) $\cos\left[\frac{\Delta E_{ug} t}{\hbar}\right] = +1$; or, at $t = 0$,

$$\begin{aligned}|\psi(r,t;R_0)|^2 &= |(\varphi_{1s,A} + \varphi_{1s,B})|^2 + |(\varphi_{1s,A} - \varphi_{1s,B})|^2 \\ &\quad + 2(\varphi_{1s,A} + \varphi_{1s,B})(\varphi_{1s,A} - \varphi_{1s,B}) \\ &= 4\varphi_{1s,A}^2\end{aligned}$$

Thus, at $t = 0$, when the superposition state is just created, the total electron density remains localized at one hydrogen atom (at atom A, as depicted in Figure 1.3).

(b) $\cos\left[\frac{\Delta E_{ug} t}{\hbar}\right] = 0$; or, at $t = \frac{\pi}{2}\frac{\hbar}{\Delta E_{ug}}$,

$$\begin{aligned}|\psi(r,t;R_0)|^2 &= |(\varphi_{1s,A} + \varphi_{1s,B})|^2 + |(\varphi_{1s,A} - \varphi_{1s,B})|^2 \\ &= 2\varphi_{1s,A}^2 + 2\varphi_{1s,B}^2\end{aligned}$$

Thus, at $t = \frac{\pi}{2}\frac{\hbar}{\Delta E_{ug}}$ after the superposition state is created, the total electron density remains delocalized over both the hydrogen atoms (see Figure 1.3).

(c) $\cos\left[\frac{\Delta E_{ug} t}{\hbar}\right] = -1$; or, at $t = \pi\frac{\hbar}{\Delta E_{ug}}$,

$$\begin{aligned}|\psi(r,t;R_0)|^2 &= |(\varphi_{1s,A} + \varphi_{1s,B})|^2 + |(\varphi_{1s,A} - \varphi_{1s,B})|^2 \\ &\quad - 2(\varphi_{1s,A} + \varphi_{1s,B})(\varphi_{1s,A} - \varphi_{1s,B}) \\ &= 4\varphi_{1s,B}^2\end{aligned}$$

Thus, at $t = \pi\frac{\hbar}{\Delta E_{ug}}$ after the superposition state is created, the total electron density remains localized at the other hydrogen atom (at atom B, as depicted in Figure 1.3).

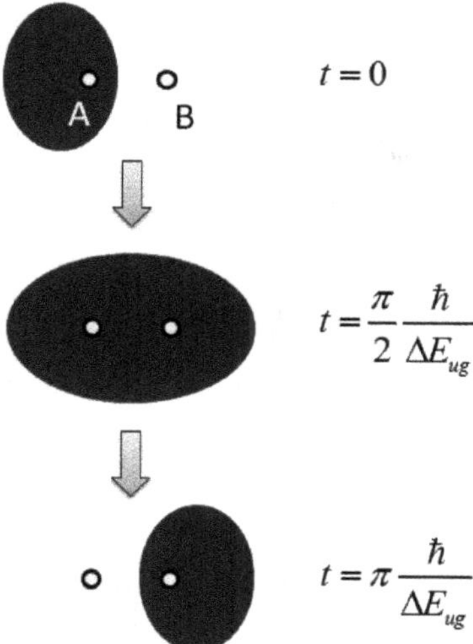

Figure 1.3. Time-dependent electron densities are schematically depicted at three different times following the preparation of the electronic superposition state, created by the two lowest lying electronic states of H_2^+.

Guiding Question

1.3. Effect of vibration on an electronic superposition state: In the above analysis of electronic superposition state, we have assumed that the H_2^+ ion has only a frozen equilibrium geometry (R_0). But this is clearly an oversimplified assumption because the ground vibrational state wavefunction, even under a quantum harmonic oscillator approximation, exhibits a spatial distribution.[9] How does this affect the time evolution of an electronic superposition state?
Hint: Consider the total wavefunction as

$$\psi(r, R, t) = \chi(R)\left[\psi_g(r)e^{-\frac{iE_g t}{\hbar}} + \psi_u(r)e^{-\frac{iE_u t}{\hbar}}\right]$$

The first and second terms on the right-hand side are, respectively, the nuclear and electronic parts of the total wavefunction. See Ref. 8(f) for further hints.

1.6.2 *Vibrational superposition state*

One of the simplest examples of a vibrational superposition state can perhaps be realized from a superposition of the ground and the first

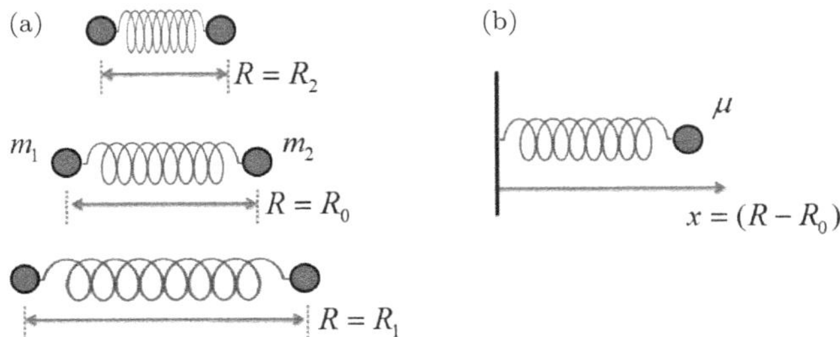

Figure 1.4. (a) Vibrational motion of a diatomic molecule can conveniently be modeled by a vibrating spring with two masses (nuclei, completely ignoring the presence of electrons in the molecule). (b) Reduced mass connected to the wall by a spring (see text for details).

excited state of a diatomic quantum harmonic oscillator. A quantum harmonic oscillator is a good model for a vibrating diatomic molecule (the motion of nuclei in the diatomic species, completely ignoring the presence of electrons in the molecular skeleton). Under this model, a diatomic molecule can be represented by a spring, as shown in Figure 1.4(a). Here, m_1 and m_2 are the masses of two atoms (more specifically, nuclei). The equilibrium bond distance is R_0, and R represents the instantaneous bond length during the vibration. Therefore, if $x(= R - R_0) > 0$, the spring is stretched, and if $x < 0$, the spring is compressed.

Considering the reduced mass $\mu = \frac{m_1 m_2}{m_1 + m_2}$ and based on the relative coordinate $x = (R - R_0)$, the above two-body problem can conveniently be reduced to a one-body problem, as shown in Figure 1.4(b). Details can be found in most of the standard quantum chemistry textbooks.[9] This reduction makes the subsequent mathematical derivation simple. The vibration of the one-body system is governed by $\omega = \sqrt{\frac{k}{\mu}}$, where k is the force constant, representing a measure of the stiffness of the spring (a small value of k implies a weak bond, and a large value of k implies a strong bond), and ω represents the frequency of the vibration (in radian per second).

The TISE for the above-mentioned reduced one-body problem is given by $[-\frac{\hbar^2}{2\mu}(\frac{d^2}{dx^2}) + \frac{1}{2}kx^2]\psi(x) = E\psi(x)$. Here, the first term in the Hamiltonian represents the kinetic energy of the body with reduced mass, and the second term in the Hamiltonian represents

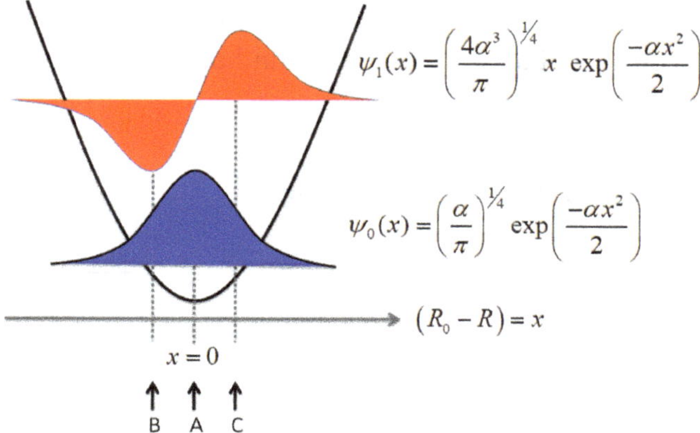

Figure 1.5. Vibrational wavefunctions of the ground and the first excited states of a diatomic molecule under harmonic oscillator approximation. Note that these wavefunctions do not represent the electrons in the diatomic species. These wavefunctions represent vibrational wavefunctions under harmonic oscillator approximation. Three different positions in x-space are labeled by A, B and C, respectively. See text for more details.

the potential experienced by the body with reduced mass as a result of Hooke's law. In this model, the presence of electrons in a molecule is completely ignored. When this second-order differential equation is solved, **vibrational wavefunctions** are obtained, with the vibrational energy levels $E_v = \hbar(\frac{k}{\mu})^{1/2}(v + \frac{1}{2})$, where the vibrational quantum numbers are $v = 0, 1, 2, 3, \ldots$. The wavefunctions corresponding to these E_v are nondegenerate states expressed by **Hermite polynomials.** Details of the solutions and expressions of Hermite polynomials can be found in most of the standard quantum chemistry textbooks.[9] Here, we are only interested in the ground $\psi_0(x)$ and first excited $\psi_1(x)$ vibrational states of the quantum harmonic oscillator. They are given by (as depicted in Figure 1.5)

$$\psi_0(x) = \left(\frac{\alpha}{\pi}\right)^{1/4} \exp\left(\frac{-\alpha x^2}{2}\right) \quad \text{and} \quad \psi_1(x) = \left(\frac{4\alpha^3}{\pi}\right)^{1/4}$$
$$\times\, x \exp\left(\frac{-\alpha x^2}{2}\right), \text{where } \alpha = \sqrt{\frac{k\mu}{\hbar^2}}.$$

Finally, the superposition state created by the ground and the first excited states of a diatomic quantum harmonic oscillator can be written as

$$\begin{aligned}\psi(x,t) &= \psi_0(x)e^{\frac{-iE_0t}{\hbar}} + \psi_1(x)e^{\frac{-iE_1t}{\hbar}} \\ &= \left(\frac{\alpha}{\pi}\right)^{1/4} \exp\left(\frac{-\alpha x^2}{2}\right) e^{\frac{-iE_0t}{\hbar}} + \left(\frac{4\alpha^3}{\pi}\right)^{1/4} x \\ &\quad \times \exp\left(\frac{-\alpha x^2}{2}\right) e^{\frac{-iE_1t}{\hbar}}\end{aligned}$$

The probability density of the vibrational superposition state is then given by

$$\begin{aligned}|\psi(x,t)|^2 &= |\psi_0(x)|^2 + |\psi_1(x)|^2 + 2\psi_0(x)\psi_1(x)\cos\left(\frac{\Delta E_{10}t}{\hbar}\right) \\ &= \left(\frac{\alpha}{\pi}\right)^{1/2} \exp(-\alpha x^2) + \left(\frac{4\alpha^3}{\pi}\right)^{1/2} x^2 \exp(-\alpha x^2) \\ &\quad + 2\left(\frac{\alpha}{\pi}\right)^{1/4} \left(\frac{4\alpha^3}{\pi}\right)^{1/4} \exp\left(-\alpha x^2\right) x \cos\left(\frac{\Delta E_{10}t}{\hbar}\right)\end{aligned}$$

Here, $\frac{\Delta E_{10}}{\hbar} = \frac{\hbar\omega}{\hbar} = \omega$ is the frequency with which the probability density of the vibrational coherent superposition state, created by the superposition of the ground and the first excited vibrational states of the diatomic species, oscillates. Schematically, this oscillation is depicted in Figure 1.6. In the absence of this coherence (assuming that the molecule is at the ground vibrational state), the molecule also oscillates with a frequency of ω; however, experimentally, it is not possible to monitor this oscillation because the probability density (which is experimentally measurable) of the ground vibrational state is time-independent. Therefore, coherence (created by the superposition of the ground and first excited vibrational states) allows one to observe the same oscillation which the molecule exhibits when it is present at the ground vibrational state.

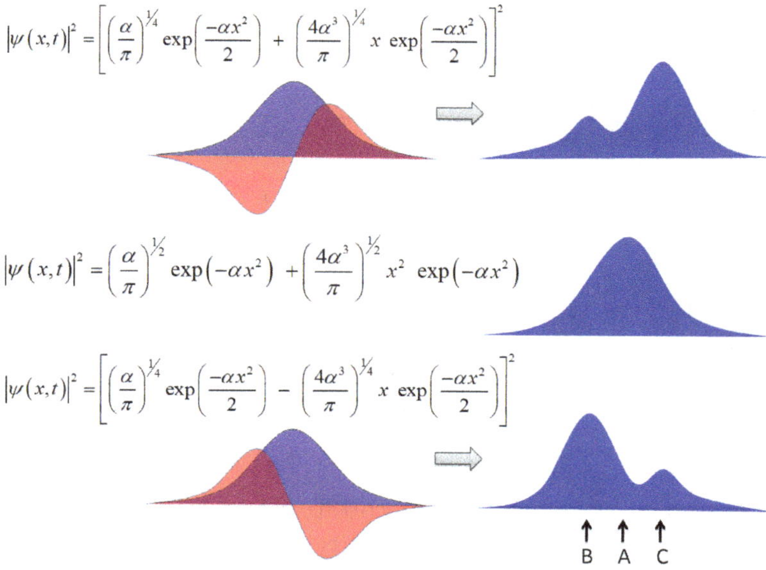

Figure 1.6. Schematic snapshots of the time-dependent density distribution for the vibrational superposition state following its initial preparation. Note that this time-dependent change in probability density represents the motion of the nuclei of the diatomic molecule. Positions A, B and C are correlated with that of Figure 1.5.

Guiding Question

1.4. Consider the superposition of the ground and second excited states of the quantum harmonic oscillator. Show how the time-dependent probability density oscillates for this vibrational superposition state?

Hint: Use the following wavefunctions for the ground and second excited vibrational states of a quantum harmonic oscillator, respectively:

$$\psi_0(x) = \left(\frac{\alpha}{\pi}\right)^{1/4} \exp\left(\frac{-\alpha x^2}{2}\right) \text{ and } \psi_2(x) = \left(\frac{\alpha}{4\pi}\right)^{1/4} (2\alpha x^2 - 1)$$

$$\times \exp\left(\frac{-\alpha x^2}{2}\right), \text{where } \alpha = \sqrt{\frac{k\mu}{\hbar^2}}.$$

To visualize the temporal evolution of the probability density of the above-stated vibrational superposition state, we may consider the values of $+1$, 0 and -1 for the cosine function:

(a) $\cos\left(\frac{\Delta E_{10}t}{\hbar}\right) = +1$ or, at $t = 0$,

$$|\psi(x,t)|^2 = \left(\frac{\alpha}{\pi}\right)^{1/2} \exp(-\alpha x^2) + \left(\frac{4\alpha^3}{\pi}\right)^{1/2} x^2$$
$$\times \exp(-\alpha x^2) + 2\left(\frac{\alpha}{\pi}\right)^{1/4} \left(\frac{4\alpha^3}{\pi}\right)^{1/4} x \exp(-\alpha x^2)$$
$$= \left[\left(\frac{\alpha}{\pi}\right)^{1/4} \exp\left(\frac{-\alpha x^2}{2}\right) + \left(\frac{4\alpha^3}{\pi}\right)^{1/4} x \right.$$
$$\left. \times \exp\left(\frac{-\alpha x^2}{2}\right)\right]^2$$

Thus, at $t = 0$, when the superposition state is just created, the maximum of the probability density appears near position C (as shown in Figure 1.6).

(b) $\cos\left(\frac{\Delta E_{10}t}{\hbar}\right) = 0$ or, at $t = \frac{\pi}{2}\frac{\hbar}{\Delta E_{10}}$,

$$|\psi(x,t)|^2 = \left(\frac{\alpha}{\pi}\right)^{1/2} \exp(-\alpha x^2) + \left(\frac{4\alpha^3}{\pi}\right)^{1/2} x^2 \exp(-\alpha x^2)$$

Thus, at $t = \frac{\pi}{2}\frac{\hbar}{\Delta E_{10}}$ after the superposition state is created, the maximum of the probability density appears near position A (as shown in Figure 1.6).

(c) $\cos\left[\frac{\Delta E_{10}t}{\hbar}\right] = -1$ or, at $t = \pi\frac{\hbar}{\Delta E_{10}}$,

$$|\psi(x,t)|^2 = \left[\left(\frac{\alpha}{\pi}\right)^{1/4} \exp\left(\frac{-\alpha x^2}{2}\right) - \left(\frac{4\alpha^3}{\pi}\right)^{1/4} x \exp\left(\frac{-\alpha x^2}{2}\right)\right]^2$$

Thus, at $t = \pi\frac{\hbar}{\Delta E_{10}}$ after the superposition state is created, the maximum of the probability density appears at position B (as shown in Figure 1.6). Note that Figure 1.6 schematically plots the density distributions. For more rigorous plots, readers are referred to PythonChapter A.

1.7 Optical Analogy to Quantum Superposition

In the context of quantum superposition, one may find a discussion of optical superposition very interesting because of the close resemblance of the two phenomena. Light exhibits both wave and particle natures. When light exhibits a wave nature, it propagates as an electromagnetic wave. For a light wave traveling along the $+z$ direction, its electric field is represented by a plane wave (as presented in Figure 1.7(a)):[10]

$$E(z,t) = E_0 e^{i(\omega_0 t - kz)}$$

Here, E_0 is the maximum electric field amplitude, k is the magnitude of the wave vector (its direction represents the wave propagation direction, which is along the $+z$ direction in the present example), and ω_0 is the angular frequency of the plane wave. Furthermore, the following relationships hold:

$$\omega_0 = 2\pi v_0 \quad \text{and} \quad c = v_0 \lambda_0$$

The term $(\omega_0 t - kz)$ in the above equation features the phase of the plane wave, which represents an angle to manifest a certain linear advancement of the propagating wave. Figure 1.7(b) depicts a close relationship between rotational motion and linear motion. When a wave propagates along the $+z$-direction, the advancement of the wave along the $+z$-direction with respect to a certain frame or point in space can be represented by an angle made on the rotating dial. For example, a 2π-phase advancement in a rotating frame represents a linear advancement of the wave by λ (wavelength).

Similarly, if two plane waves propagating along the same direction are considered, the phase difference between those two plane waves will suggest relative phase advancement of those two waves with respect to one another. Phase is an important physical quantity of a plane wave (just like the phase of a wavefunction in quantum mechanics). It carries information about the velocity of the wave.

Figure 1.7(b) shows that there is a certain time interval after which a particular (constant) phase repeats as the plane wave propagates (pictorially, the same color reappears after a certain time

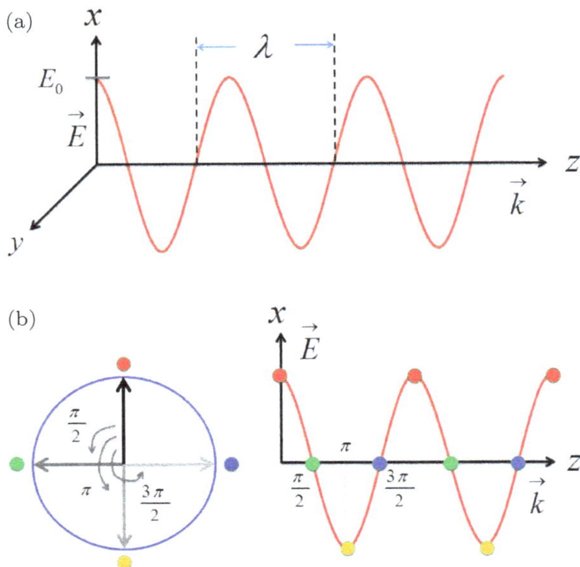

Figure 1.7. (a) A plane wave propagating along $+z$-direction is shown. (b) The meaning of the phase of a plane wave is depicted: A linear advancement of the plane wave is expressed by phase advancement in rotating dial (see text for more details).

interval). For such a constant phase $(\omega_0 t - kz)$, its first derivative with respect to t becomes zero:

$$\frac{d}{dt}(\omega_0 t - kz) = 0, \text{ or } \frac{dz}{dt} = \frac{\omega_0}{k}$$

Here, $\frac{dz}{dt}$ represents the phase velocity, $V_p = \frac{\omega_0}{k}$ ($= c$ in vacuum). Thus, **the phase velocity of a plane wave represents the velocity with which the constant phase front of a plane wave travels.** Note that the phase of a plane wave varies continuously as a function of time; that is why the phase cannot be constant for a plane wave. In order to determine phase velocity, only the velocity of the constant phase front (the velocity with which the same constant phase repeats) is invoked.

The phase of a plane wave not only carries information about the velocity of the wave but also plays an important role in the synthesis of an optical pulse.[10] The idea of an optical pulse comes from the

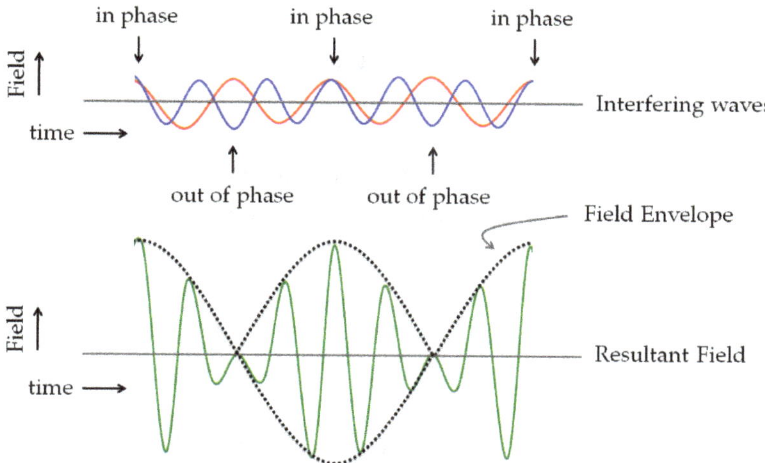

Figure 1.8. Top panel: In-phase and out-of-phase interferences of two plane waves are schematically shown (two different colors represent two slightly different wavelengths or frequencies). Bottom panel: The generation of an optical pulse due to the interference of two plane waves with different frequency components is schematically depicted. This is just a schematic diagram. Refer to PythonChapter A for a more realistic plot of an optical pulse using Python programming.

theory of optical "interference," which states that at a point, the total electric field can be calculated based on the **linear superposition** of electromagnetic plane waves (just like quantum mechanical superposition): $\vec{E}_{\text{total}} = \vec{E}_1 + \vec{E}_2 + \vec{E}_3 + \cdots + \vec{E}_n$, because electric fields are additive.

For a simple representation, only two plane waves of slightly different frequencies may be considered. Furthermore, they are assumed to possess the same maximum field amplitudes and to be traveling along the same direction (say $+z$ direction), as depicted in Figure 1.8 (top panel). These two fields (E_1 and E_2) can be represented by

$$E_1 = E_0 e^{i(\omega_1 t - k_1 z)} \text{ and } E_2 = E_0 e^{i(\omega_2 t - k_2 z)}$$

Note that two different magnitudes of $\vec{k}$-vectors are considered; however, their directions are the same. Let us now define $\omega_{\text{avg}} = \frac{\omega_1+\omega_2}{2}$ and $k_{\text{avg}} = \frac{k_1+k_2}{2}$ as well as $\Delta\omega = \frac{\omega_1-\omega_2}{2}$ and $\Delta k = \frac{k_1-k_2}{2}$. Consequently, one can write, $\omega_1 = \omega_{\text{avg}} + \Delta\omega, \omega_2 = \omega_{\text{avg}} - \Delta\omega,$

$k_1 = k_{\text{avg}} + \Delta k$, and $k_2 = k_{\text{avg}} - \Delta k$. Therefore, the total electric field due to interference can be written as

$$\begin{aligned}
E_{\text{total}} &= E_0 e^{i(\omega_1 t - k_1 z)} + E_0 e^{i(\omega_2 t - k_2 z)} \\
&= E_0 e^{i(\omega_{\text{avg}} t + \Delta\omega t - k_{\text{avg}} z - \Delta k z)} + E_0 e^{i(\omega_{\text{avg}} t - \Delta\omega t - k_{\text{avg}} z + \Delta k z)} \\
&= E_0 e^{i(\omega_{\text{avg}} t - k_{\text{avg}} z)} [e^{i(\Delta\omega t - \Delta k z)} + e^{-i(\Delta\omega t - \Delta k z)}] \\
&= 2E_0 e^{i(\omega_{\text{avg}} t - k_{\text{avg}} z)} \cos(\Delta\omega t - \Delta k z)
\end{aligned}$$

Now, taking the real part, we get the expression for the resultant electric field as

$$E_{\text{total}} = 2E_0 \cos(\omega_{\text{avg}} t - k_{\text{avg}} z) \cos(\Delta\omega t - \Delta k z) \tag{1.11}$$

In order to understand the meaning of the above equation, we take a reductionist approach; let us consider $z = 0$ (turn off z-dependence), and check the time-dependent part only:

$$E_{\text{total}} = 2E_0 \cos(\omega_{\text{avg}} t) \cos(\Delta\omega t)$$

The first term, $\cos(\omega_{\text{avg}} t)$, is a fast-varying component, and the second term, $\cos(\Delta\omega t)$, is a slowly-varying component of the resultant field. These two components are shown in the bottom panel of Figure 1.8 (the black dotted line represents the slowly-varying component, and the green bold line represents the fast-varying component). Therefore, the interference of two plane waves of slightly different frequencies, propagating along the same direction, yields an optical pulse, which is nothing but a product of the rapidly varying and slowly varying cosine waves.

An optical pulse also propagates in time and space with some velocity. **How is the velocity of an optical pulse defined?** The answer lies within equation (1.11) (generated due to the interference of two plane waves of dissimilar frequency components propagating along the same direction). The first component, $\cos(\omega_{\text{avg}} t - k_{\text{avg}} z)$, is the **fast-varying** component, which represents the **carrier wave** of an optical pulse. The frequency of the carrier wave is ω_{avg} (which is the resultant frequency). The second component, $\cos(\Delta\omega t - \Delta k z)$, is the **slowly-varying** component of an optical pulse, which features the **field envelope** of the pulse. Figure 1.8 depicts a carrier wave and the field envelope for an optical pulse. The periods of the envelope component, $\cos(\Delta\omega t - \Delta kz)$, and carrier wave component, $\cos(\omega_{\text{avg}} t - k_{\text{avg}} z)$, differ significantly. However, both components together represent a propagating wave. The velocity of an optical pulse is defined based on both of these components, following the idea of the velocity of a constant phase front:

(a) Phase velocity of an optical pulse: The carrier wave (fast-varying component) of a pulse travels with phase velocity, defined by the velocity of the constant phase front of the fast-varying component. As a result, $V_p = \frac{\omega_{\text{avg}}}{k_{\text{avg}}}$.

(b) Group velocity of an optical pulse: The field envelope (slowly-varying component) of an optical pulse travels with group velocity, defined by the velocity of the constant phase front of the slowly-varying component. As a result, $V_g = \frac{\Delta\omega}{\Delta k}$. Considering an infinitesimal difference in frequency, we can write $V_g = \frac{d\omega}{dk}$.

Thus, the above discussion shows that, similar to a wavepacket, an optical pulse also originates from the effect of superposition. While **a localized electromagnetic wave in time represents an optical pulse, a localized particle wave in position space features a wavepacket**. Both optical pulses and wavepackets have phase velocity and group velocity. Above discussion gives an easily perceptible definition of phase velocity and group velocity for an optical pulse. The same for the wavepacket will be discussed in Chapter 3. An early introduction to these concepts pertaining to an optical pulse is expected to help readers grasp the details of the same for the wavepacket.

Key Points to Remember

- All the physical and chemical properties of a quantum particle are described by its wavefunction, $\psi(x,t)$, which follows the time-dependent Schrödinger equation, $i\hbar\frac{\partial}{\partial t}\psi(x,t) = \hat{H}\psi(x,t)$. One cannot experimentally observe a wavefunction; only the probability density, which is expressed by $\psi^*(x,t)\psi(x,t)$, can be experimentally realized using repeated measurements.
- The motion of a quantum particle is only manifested by the time evolution of the probability density.
- The time-dependent Schrödinger equation is solvable using the variable separation method if the Hamiltonian is not time-dependent.
- Quantum superposition (leading to a phenomenon called coherence) is the origin of quantum dynamics. The general expression for a superposition (or non-stationary) state is given by

$$\psi(x,t) = \sum_{n=0}^{\infty} a_n \psi_n(x) e^{-i\frac{E_n t}{\hbar}} \text{ (for discrete energy states)}$$

$$\text{or } \psi(x,t) = \int_0^{\infty} a_E \psi_E(x) e^{-\frac{iEt}{\hbar}} dE \text{ (for continuous energy states)}$$

- The characteristic timescales of motion of a quantum system are given by $\tau = \frac{h}{\Delta E}$, where ΔE represents the energy gap between two states involved in superposition:

	ΔE	τ
Vibrational motion of nuclei	100 meV	41 femtoseconds
Valence electrons in atoms	13 eV	318 attoseconds
Inner shell electrons in atoms	1 keV	4 attoseconds

Notes, References, and Further Reading

1. Theory-minded curious readers are referred to (a) E. Schrödinger, *Collected Papers on Wave Mechanics*, 3rd (Augmented) English Edition. Chelsea Publishing Company, New York (1982); (b) J. S. Briggs and J. M. Rost, On the derivation of the time-dependent equation of Schrödinger. *Found. Phys.* 31, 693–712 (2001).
2. A general discussion on the wavefunction and its statistical or probabilistic interpretations can be found in most of the standard quantum chemistry books. The same has been briefly documented in Chapter 2 of this book from a time-dependent perspective. Curious new physical chemistry students are further referred to the following four lovely text books: (a) I. N. Levine, Chapter 7: Theorems of quantum mechanics. *Quantum Chemistry*, 7th edn. Pearson Education, Inc., New Jersey (2014); (b) D. J. Griffiths, Chapter 1: The wave function. *Introduction to Quantum Mechanics*, 2nd edn. Pearson

Education, Inc., New Jersey (2005); (c) P. Atkins and R. Freidnman, Chapter 1: The foundations of quantum mechanics. *Molecular Quantum Mechanics*, 5th edn. Oxford University Press, New York (2011); and (d) D. A. McQuarrie, Chapter 4: The postulates and general principles of quantum mechanics. *Quantum Chemistry*, 2nd edn. University Science Books, California (2008).

3. In classical mechanics, the conservative force is defined as the force for which the work done by the moving particle between two points is independent of the path taken by the particle between the two points.

Thus, the energy of the particle, subjected to a conservative force, depends only on its position, not on the path taken by the particle to reach that position. This energy is called **potential energy** — an energy that depends on the position of the particle. For an infinitesimal displacement dx, if the potential energy change is dV, under the conservative force, Energy = Force × Displacement, or $dV = -F \times dx$, or $F = -\frac{dV}{dx}$. Here, the negative sign comes from the fact that potential energy increases when displacement increases, but the force is active on the opposite direction to reduce the potential energy. Curious readers are further referred to H. Goldstein, C. Poole and J. Safko, *Classical Mechanics*, 3rd edn. Pearson, Edinburgh (2001).

4. A general discussion on the variable separation method can be found in most of the standard quantum chemistry textbooks. To name a few lovely books: (a) I. N. Levine, Chapter 1: The Schrödinger equation. *Quantum Chemistry*, 7th edn. Pearson Education, Inc., New Jersey (2014); (b) D. A. McQuarrie, Chapter 4: The postulates and general principles of quantum mechanics. *Quantum Chemistry*, 2nd edn. University Science Books (2008).

5. One of the oldest quantum chemistry textbooks where superposition state has been mentioned in the context of quantum dynamics is perhaps H. Eyring, J. Walter and G. E. KimBall, Chapter XVII: The quantum mechanical theory of reaction rates. *Quantum Chemistry*. John Wiley and Sons, Inc., New York (1944). Curious readers are also referred to two relatively new books: (a) D. J. Tannor, Chapter 1: The time-dependent Schrödinger equation. *Introduction to Quantum Mechanics a Time-Dependent Approach*. University Science Books, Sausalito (CA) (2007); (b) D. A. McQuarrie, Chapter 4: The postulates and general principles of quantum mechanics. *Quantum Chemistry*, 2nd edn. University Science Books, California (2008).

6. On the contrary, for multi-electron molecules, electronic wavefunction is presented by the antisymmetrized product of all molecular orbitals. For more

details, see A. Szabo and N. S. Ostlund, *Modern Quantum Chemistry: Introduction to Advanced Electronic Structure Theory*, 1st edn. Dover Publications Inc., New York (1996). For a general discussion on the molecular orbital theory and LCAO method, refer to (a) M. Huheey, *Inorganic Chemistry: Principle of Structure and Reactivity*. Pearson, New Delhi (2006) and (b) E. V. Anslyn and D. A. Dougherty, *Modern Physical Organic Chemistry*. University Science Books, California (2005).

7. M. Beyer and F. Merkt, Structure and dynamics of H_2^+ near the dissociation threshold: A combined experimental and computational investigation. *J. Mol. Spectrosc.* 330, 147 (2016).
8. Electronic superposition leads to the attosecond phenomenon. Readers are referred to the following few interesting recent articles in this context: (a) F. Remacle and R. D. Levine, An electronic time scale in chemistry. *Proc. Natl. Acad. Sci. (USA)*, 103, 6793–6798 (2006); and (b) S. Chandra and A. Bhattacharya, Attochemistry of ionized halogen, chalcogen, pnicogen and tetrel bonded clusters. *J. Phys. Chem. A*, 120, 10057–10071 (2016); (c) H. J. Wörner, C. A. Arrell, N. Banerji, A. Cannizzo, M. Chergui, A. K. Das, P. Hamm, U. Keller, P. M. Kraus, E. Liberatore, P. Lopez-Tarifa, M. Lucchini, M. Meuwly, C. Milne, J.-E. Moser, U. Rothlisberger, G. Smolentsev, J. Teuscher, J. A. van Bokhoven, and O. Wenger, Charge migration and charge transfer in molecular systems. *Struct. Dyn.* 4, 061508 (2017); (d) E. Goulielmakis , Z.-H. Loh, A. Wirth, R. Santra, N. Rohringer, V. S Yakovlev, S. Zherebtsov, T. Pfeifer, A. M Azzeer, M. F. Kling, S. R. Leone, F. Krausz, Real-time observation of valence electron motion. *Nature*, 466, 739–743 (2010); (e) F. Calegari, D. Ayuso, A. Trabattoni, L. Belshaw, S. De Camillis, S. Anumula, F. Frassetto, L. Poletto, A. Palacios, P. Decleva, J. B. Greenwood, F. Martín, M. Nisoli, Ultrafast electron dynamics in phenylalanine initiated by attosecond pulses. *Science*, 346, 336–339 (2014); (f) S. Bag, S. Chandra, J. Ghosh, A. Bera, E. R. Bernstein and A. Bhattacharya, The attochemistry of chemical bonding. *Int. Rev. Phys. Chem.* 40, 405–455 (2021).
9. See, for example, D. A. McQuarrie, Chapter 5: The harmonic oscillator and vibrational spectroscopy. *Quantum Chemistry*, 2nd edn. University Science Books, California (2008).
10. A. Bhattacharya, Chapter 2: Mathematical representation of ultrafast pulse. *Ultrafast Optics and Spectroscopy in Physical Chemistry*. World Scientific, Singapore (2018).

Exercises

1.1. If ψ_m and ψ_n, represented by $\psi_m(x,t) = \psi_m(x)e^{-\frac{iE_m t}{\hbar}}$ and $\psi_n(x,t) = \psi_n(x)e^{-\frac{iE_n t}{\hbar}}$, respectively, are solutions to the TDSE, show that the following wavefunction also satisfies the TDSE:

$$\psi(x,t) = a_m\psi_m(x)e^{-\frac{E_m t}{\hbar}} + a_n\psi_n(x)e^{-\frac{iE_n t}{\hbar}}$$

1.2. For a certain one-dimensional particle of mass m, the wavefunction is given by $\psi(x,t) = Ae^{-\frac{iamx}{\hbar}}e^{-iat}$, where A and a are real positive constants. Find the potential energy for which the above wavefunction satisfies the TDSE.

1.3. For a certain one-dimensional particle of mass m, the wavefunction is given by $\psi(x,t) = Ae^{ikx}e^{-\frac{iEt}{\hbar}}$, where $k = \sqrt{\frac{2mE}{\hbar^2}}$ and E is the total energy of the particle. Find the potential energy for which the above wavefunction satisfies the TDSE.

1.4. Assume that $\psi(x,t)$ is a superposition state constructed using a linear combination of the first two lowest lying states of a particle in a one-dimensional box. Calculate the probability density. Pictorially depict how the probability density periodically moves from one side of the box to the other side of the box as a function of time.

1.5. The TDSE, given by $i\hbar\frac{\partial}{\partial t}\psi(x,t) = [-\frac{\hbar^2}{2m}(\frac{\partial^2}{\partial x^2}) + V(x)]\psi(x,t)$, represents a single particle of mass m experiencing a time-independent potential. If, suddenly, an additional time-independent constant potential V_0 starts acting on the particle, what changes do you expect in the wavefunction and the probability density?

1.6. The π electron in the butadiene molecule can be approximately modeled as an electron in a one-dimensional box, given the C–C single bond length of 1.454 Å and the C=C double bond length of 1.338 Å. Using this model, find the ground state energy and the first excited state energy of the π electron in butadiene. If a superposition state is constructed by a linear combination of the ground and first excited states of the π electron, find the timescale over which the π electron density moves from one side to the other side of the butadiene molecule.

1.7. Assume that $\psi(x,t)$ is a superposition state constructed using a linear combination of the ground state and the second excited state of a particle in a one-dimensional box. Calculate the probability density. Pictorially depict how the probability density periodically moves from one side of the box to the other side of the box as a function of time?

PythonChapter A

First Session with Python Programming

Highlights: *Running Python Programs, Simple Arithmetic Computations, Computations with Standard Mathematical Functions, Performing Repetitive Task using While and For Loops, SciPy (Scientific Python), Matplotlib.PyPlot (Python Graphing Tools).*

A.1 Introduction

Quantum mechanics is very often taught at both the undergraduate and graduate levels, with almost the entire focus on systems having known analytical solutions. A particle in a one-dimensional box, a simple harmonic oscillator, and the hydrogen atom are very popular examples. On the other hand, most of the practical problems in quantum mechanics, even at the very preliminary research level, do not have analytical solutions. They must be solved numerically. This mismatch poses a vexing problem, particularly when it comes to encouraging students both in experimental and theoretical physical chemistry laboratories to think freely about any quantum mechanical problem. Furthermore, even for problems with known analytical solutions, often, an inability to build a physically perceptible mental picture at the end of a rigorous mathematical derivation also leads to a state of being ill-equipped for free-thinking of quantum mechanical problems. As the proverb goes, "a picture is worth a thousand words," graphs are the best way of portraying the physical concept hidden in a mathematical expression. Therefore, to make the study of quantum mechanics practically useful and to make the learning process comfortable, the blackboard (or whiteboard) presentation of quantum mechanics must be supplemented by computer-programming-based graphing tools.[1] This is true for both the time-independent and time-dependent versions of quantum mechanics.

The aim of the PythonChapters given in this book is at introducing numerical techniques, which are commonly required to find solutions to very frequently encountered quantum mechanical problems, such as the solution to the Schrödinger equation for an arbitrary potential and the simulation of quantum dynamics under the influence of a time-independent potential. Furthermore, the PythonChapters are prepared to provide a set of graphing tools to promote easy construction of mental pictures for the analytical or numerical solutions.

No prior programming knowledge is expected from any enthusiastic student who wants to go over the PythonChapters in this book without an expert's help. In the PythonChapters, first, the core concepts of programming for scientific computing are systematically introduced using the modern programming language Python,[2–4] which has been extensively developed with the scientific community in mind. The Python programming language is selected for many favorable reasons: it is very easy to learn, it is becoming increasingly popular with the modern scientific community for its rich scientific libraries, and it is freely available. After introducing the core concepts of scientific computing, several examples based on quantum mechanical problems are discussed with a hands-on approach.

In the present PythonChapter, first, a quick introduction to Python programming is provided, and then, a set of Python tools for constructing two-dimensional graphs are introduced. Students can comfortably go through the notes and instructions if they have a Windows-based computer, where Python 3, at version 3.8.0 (which was released on October 10, 2019), is already installed (one can download this version from https://www.python.org/downloads/release/python-380/). The installation of Python 3.8.0 creates a directory (folder) "Python38-32" in the "C" drive (C:\Program Files (x86)\Python 38-32).

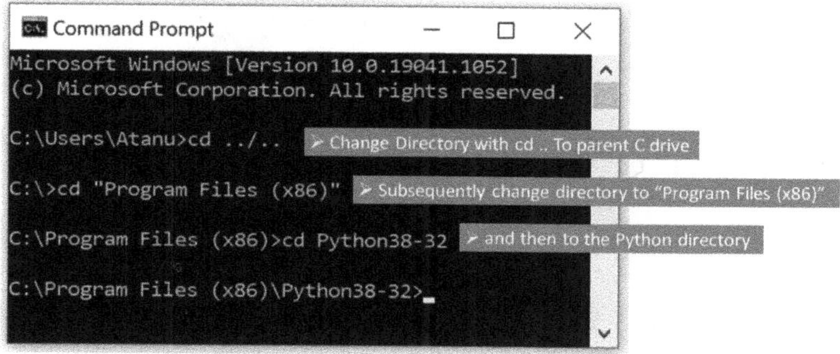

Figure A.1. Steps are depicted to reach the Python directory where the Python programs are executed. Python code can be simply written on Notepad and saved as a text file with a **.py** extension.

To run a Python program, one may first save the Python code (say, using Notepad) with the extension **.py** in the Python directory (C:\Program Files (x86)\Python38-32). Then, after launching Command Prompt, one may change the directory to "Python38-32" (follow the steps given in Figure A.1), and finally, one may run the file using the construct **python *filename*** followed by pressing the **Enter** key (keyboard). As Python is an interpreted language, once a Python interpreter is installed (say, Python 3.8.0), the Python script can be executed directly. There is no need for a compiler.

A.2 Simple Computations

A.2.1 *Arithmetic computation*

A simple arithmetic computation (addition, subtraction, multiplication, division, or exponentiation) can be performed, and the result can be easily printed using Python's built-in functionality **print()**:

(a) Directly using numerical values:

```
print(((10*2)/5)**2)

16.0
```

> ➢ One can use brackets () to group different arithmetic operations to preserve the order of operations and make the code easier to read.

Python's five built-in standard arithmetic operators (programming constructs to perform predefined operations) are addition ($+$), subtraction (−), multiplication (*), division (/), and exponentiation (**).

(b) Using expression and variables:

```
a=10
b=2
c=5
d=2
y=(((a*b)/c)**d)
print(y)

16.0
```

> ➢ Python programming is case-sensitive. Instead of print() with the small letter p, if one types Print() with the capital letter P, one gets an error message (NameError: name 'Print' is not defined) because Python does not recognize Print() with the capital letter P. Similarly, in Python, a and A are two different variables.

In general, variable names can contain any lower- or upper-case letter, number, or underscore, e.g., v_0, $v1$.

(c) Using inputs taken from a user:

```
a=int(input("What is the value of a?"))
b=int(input("What is the value of b?"))
c=int(input("What is the value of c?"))
d=int(input("What is the value of d?"))
y=(((a*b)/c)**d)
print(y)
```

16.0 (if a = 10, b = 2, c = 5, and d = 2 are used)

Any input given (by the user using the computer's keyboard) through Python's built-in functionality **input()** is stored as a string (which is defined as a collection of characters or text). A string cannot be used in numerical computations. One always has to convert a (string) input to an integer [or floating point] number using Python's built-in functionality **int()** [or **float()**] for subsequent arithmetic computations with the input. In the above example, two required functionalities, **int()** and **input()**, are clubbed together. Alternatively, one can rewrite as follows to get the same answer.

```
a=input("What is the value of a?")
b=input("What is the value of b?")
c=input("What is the value of c?")
d=input("What is the value of d?")
a=int(a)
b=int(b)
c=int(c)
d=int(d)
y=(((a*b)/c)**d)
print(y)
# I am a physical chemist and this is my first Python program.
```

> ➢ int(), float(), print(), input() are built-in functionalities of Python to pass the data through the parentheses ().

Note that in the last line of the above example, an informative comment with the # character is inserted. Everything written after this character (#) is ignored when the Python program is executed. Hashtag (#)-lines are frequently used in Python programs to insert informative comments because good comment lines become very useful for any program to understand without difficulty.

(d) Formatted final printing:

One may use formatted print() to print more informative text along with the numerical value.

```
a=10
b=2
c=5
d=2
y=(((a*b)/c)**d)
print("Final Result=%d"%(y))

Final Result=16.0
```

> ➢ **Logical line:** A single statement line of Python.
> **Physical line:** A single line of the text.
> In this example, four logical lines are written on four different physical lines to define variables, *a*, *b*, *c*, and *d*. Instead, all four logical lines can be written on a single physical line using semicolons:
> a = 10; b = 2; c = 5; d = 2

Here, the formatted print statement **print("…..%d" % ())** prints a string (everything enclosed within the quote). This quoted string has a slot with a **%d** sign where the program can place the value of **y**, which is kept inside the parenthesis separated from the quoted string by another **%** sign. **%d** defines an integer format for the result. Instead of %d, one can use **%e** for compact scientific notation or **%f** for a floating point number.

A.2.2 *Computation with standard mathematical functions*

By default, Python does not make standard mathematical functions (such as **sin, cos, tan, log, exp, ln, sqrt, e**, and **pi**) available.

We have to **import** a specific Python module, such as **math**, **numpy** (numerical module), or **scipy** (scientific computing module), to access these mathematical functions. The **scipy** module of Python provides not only standard mathematical functions but also numerous sophisticated numerical tools frequently used in scientific computing, such as different array operations, numerical integration, and differential equation solvers. We will be extensively using the **scipy** module throughout this book, and this is why, here, we use the **scipy** module to import standard mathematical functions.

To compute the square root of a number:

```
from scipy import sqrt              #Importing a mathematical function from scipy module (library)
a=input("Enter the number ")  #input() function stores everything as string
a=int(a)                                   #Converting string to integer for numerical computation
y=sqrt(a)
print(y)
```

Two lines are new and important here: the first line, which defines the module to be imported by the construct **from scipy import sqrt**, and the fourth line, where the mathematical function to compute the square root is defined by **sqrt()**.

A.3 Repetitive Computations

Let us consider a formula to convert a to b and then print a and b using Python's built-in functionality **print(a,b)**. We can repeat the statements as many times as required. The following is an example.

```
a=10
b=a*2
print(a,b)
a=20
b=a*2
print(a,b)
a=30
b=a*2
print(a,b)
```

```
10 20
20 40
30 60
```

Instead of repeating identical statements, repetitive tasks can be executed using two variants of Python loops: **while loop** and **for loop**.

A.3.1 *The while loop*

A while loop is used to repeat a statement as long as **a condition is true**. The repetitive conversion of a to b, which is mentioned above, can be accomplished using the while loop given in the following.

```
A=10              #Start value
da=10             #Increment
while a<=30:      #Loop heading with condition
        b=a*2             #1st statement inside the loop
        print(a,b)        #2nd statement inside the loop
        a=a+da            #3rd statement inside the loop
```

```
10 20
20 40
30 60
```

Note that the statement block (here, the last three lines) is executed in each pass of the while loop. This block is **indented** with a tab (a space can also be used). The colon **(the : sign)** at the end of the while loop heading line is essential and indicates the beginning of the loop. The condition in the while loop heading line is a Boolean expression (true or false). Python's built-in comparison operators (they compare the values and return either True or False) are $==$, $!=$, $<$, $>$, $<=$, and $>=$. More than one condition can be used in the while loop heading line (e.g., $a < 10$ and $a > 5$).

<= indicates is less than or equal to,
== indicates is equal to,
!= indicates is not equal to,
> indicates is greater than,
< indicates is less than.

A.3.2 *The for loop*

A **for loop** can also be used to repeat a certain task. Repetitive conversions of a to b, which are mentioned above, can also be accomplished using the following **for loop**.

```
L=[10,20,30]          #Creating a list
for a in L:           #For Loop heading ends with colon (the : sign)
        b=a*2               #1st statement inside the loop
        print(a,b)          #2nd statement inside the loop
```

```
10 20
20 40
30 60
```

A list of numbers can be created using the list functionality **L=[10,20,30]**. In the list, numbers are placed within a square bracket ([]), each separated by commas. In the current example, we have three elements in the list. The index of the list elements starts with 0; thus, in the above list, the zeroth element is 10, the first element is 20, and the second element is 30. Python can take any name for the list. So, instead of **L=[10,20,30]**, we could also name it as **M=[10,20,30]**.

The **for loop** heading, **for a in L**, creates a loop over all elements in the list L. In each pass of the loop, the variable **a** refers to an element in the list, starting with the zeroth element (10), followed by the first element, and so on. Note that the entire statement block is **indented** (just like a while loop).

While constructing a list, instead of writing every element in the list, one may automate the list construction using the **arange(start,stop,step)** functionality of the **scipy** module of Python. Note that the **arange(start,stop,step)** ***does not include*** **stop** in the sequence. Therefore, an automated version of the above program would be (using the **arange** functionality of **scipy**):

```
#Importing arange functionality from the scipy library
from scipy import arange
L=arange(10,40,10)
for a in L:
        b=2*a
        print(a,b)
```

```
10 20
20 40
30 60
```

As the **arange(start,stop,step)** functionality ***does not include*** **stop** in the sequence, arange(10 20 30), returns a list [10 20 30], excluding 40.

A.4 Simple Graphing

The **matplotlib** module of Python is a plotting library which is a powerful (yet easy to learn) tool for constructing graphs of mathematical functions. Following simple steps along with the **plot()** functionality of the **matplotlib.pyplot** submodule can be used to create an X–Y plot.

First, create a list of *X* values using the arange functionality of scipy, and then get the *Y* (function) value for each *X* value; thereafter, plot the function using the plot() functionality of matplotlib.pyplot, and finally, display the plot using the show() functionality of matplotlib.pyplot.

Example: Plot a sine function over the range 0–5.

```
#Importing required functionalities from libraries
from scipy import sin,arange
from matplotlib.pyplot import plot,show
#Creating X values (or X-grid)
X=arange(0,5.0,0.1)
#Creating corresponding Y values
Y=sin(X)
#Plotting X-Y function
plot(X,Y)
#Displaying X-Y function
show()
```

Here, a number of points are worth noting. As discussed already, the **arange(start,stop,step)** functionality (this functionality should be read as *a-range*, not as *arrange*) of **scipy** returns a list of evenly spaced numbers within the given range (more specifically, it returns an array, but for now, we may skip this detail). As mentioned before, the range defined by the **arange** functionality includes **start** but excludes **stop**. For example, arange(1,5,1) returns a list [1 2 3 4] in which 5 (**stop**) is not included.

When the entire list, prepared by the **arange** functionality, is used as a variable for a mathematical function, we get back another list with the corresponding function's values. In the above example, since X is a list of numbers, sin(X) returns another list containing sin(Xi) for each value Xi in the array X.

Finally note that **plot()** functionality does not directly display the graph: **show()** functionality of the **matplotlib.pyplot** submodule helps in displaying all graphs. Furthermore, Python remains running while displaying the graph using the **show()** functionality. The above program returns a graph, as shown in the following. Clearly, this graph requires further formatting for better presentation.

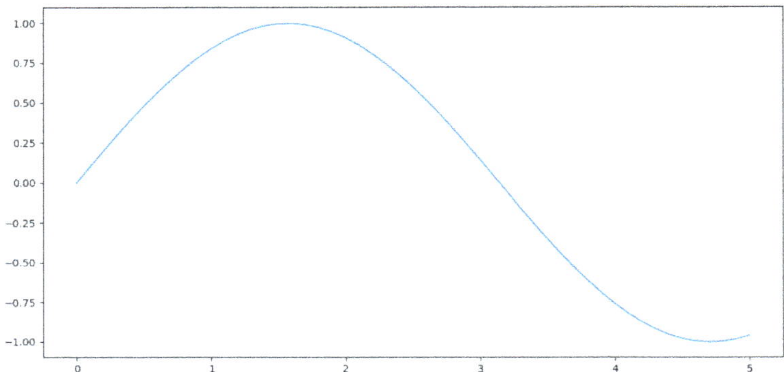

Example: Format the graph and the plot.

```
from scipy import sin,arange
#Importing additional functionalities from library for formatting the graph
from matplotlib.pyplot import plot,show,xlabel,ylabel,tick_params,title
X=arange(0,5,0.1)
Y=sin(X)
#Plotting X-Y function with required formatting of the line
plot(X,Y,color='red',linestyle='--',linewidth=3)
#Formatting the graph
xlabel('X',fontsize=15)
ylabel('Sin(X)',fontsize=15)
tick_params(direction='in',bottom='bool',top='bool',
left='bool',right='bool',width=2,length=5,labelsize=15)
title('Sine Function Plot',fontsize=15,loc='left')
#Displaying graph
show()
```

(a) To control the appearance of the plot:

The linestyle, color of the line, and linewidth of the plot can be conveniently formatted using different in-built features of the **plot()** functionality of the **matplotlib.pyplot** submodule. In the construct **plot(X,Y,color='red',linestyle='–',linewidth=3)** used above, the first and second items are, respectively, lists of X and

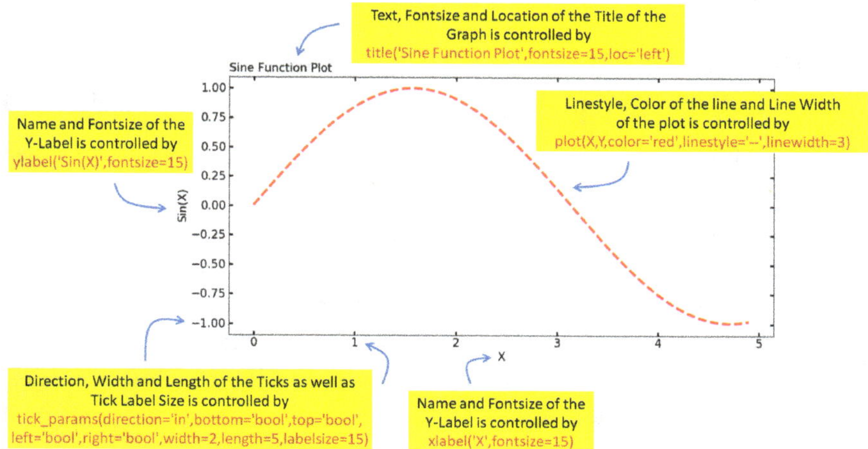

Figure A.2. Details of the graph and plot formatting are given.

Y values. The third item specifies the color of the line (plot) using the construct **color='red'**. Other colors (such as blue or black) can be selected as desired. The fourth item controls the linestyle using the construct **linestyle='–'**. Different characters are available in the **matplotlib.pyplot** submodule to specify the linestyle: solid line, dashed line, and dotted line are represented, respectively, by **'-'**, **'–'**, and **':'**. Finally, line width is specified by the construct **linewidth='3'**. A reasonable value of the line width can be selected for the desired plot.

(b) To control the appearance of the overall graph:

(1) The **xlabel()** and **ylabel()** functionalities of the **matplotlib.pyplot** submodule set the labels on the X and Y axes, respectively:

```
xlabel('X',fontsize=15)
ylabel('Sin(X)',fontsize=15)
```

A reasonable value of the font size can be selected for the desired graph.

(2) The **tick_params()** functionality of the **matplotlib.pyplot** module is used to control the appearance of the ticks and tick labels:

```
tick_params(direction='in',bottom='bool',top='bool',
left='bool',right='bool',width=2,length=5,labelsize=15)
```

The item **direction='in'** puts ticks inside the axes. Four items, **bottom='bool',top='bool',left='bool',right='bool'**, all together place axes and ticks in all four directions. The items **width=2,length=5** control the width and length of the ticks, respectively. Finally, the item **labelsize=15** controls the font size of the tick labels.

(3) The **title()** functionality of **matplotlib.pyplot** specifies a title for the graph:

```
title('Sine Function Plot',fontsize=15,loc='left')
```

First, the string item **'Sine Function Plot'** refers to the actual title. The font size of the title is specified by **fontsize=15**. The **loc='left'** argument specifies the **left location** of the title. Right or center location can also be selected based on convenience.

A.5 Examples

(1) Computing exponential function using a while loop

Aim of this exercise: A real exponential function can be defined by the infinite power series

$$e^x = 1 + x + \frac{x^2}{2!} + \frac{x^3}{3!} + \frac{x^4}{4!} + \cdots \infty$$

However, one can truncate the infinite sum at some high power value (say $N = 25$), and compute an approximate value of the function.

```
from scipy.special import factorial #Factorial functionality imported from scipy.special
x=int(input("Integer"))
N=int(input("Truncation Point"))
k=1
sum=1
while k<=N:
        term=((x**k)/factorial(k))
        sum=sum+term                #Sum will be updated in each cycle
        k=k+1                       #k will be updated in each cycle
print("exp(%f)=%f" % (x,sum))       #Formatted print option is used
```

exp(2.000000)=7.389056 if x = 2 and N = 25 are selected.

Note that the factorial of a number can be computed using the **factorial()** functionality of the special function submodule of scipy (**scipy.special**).

Self-Study

Select the truncation point to be very low and monitor the result. How does it compare with the above result?

(2) Displaying an optical pulse as a sum of many cosine waves

Aim of this exercise: To construct an optical pulse as a superposition of electromagnetic plane waves of the form given in Chapter 1.

```
#Importing Libraries
from scipy import cos,arange
from matplotlib.pyplot import plot,show,xlabel,ylabel,tick_params,title
#Defining X-grid
X=arange(-35,35,0.1)
#Defining 11 plane waves
Y1=0.1*cos(X)
Y2=0.2*cos(1.1*X)
Y3=0.3*cos(1.2*X)
Y4=0.4*cos(1.3*X)
Y5=0.5*cos(1.4*X)
Y6=0.6*cos(1.5*X)
Y7=0.5*cos(1.6*X)
Y8=0.4*cos(1.7*X)
Y9=0.3*cos(1.8*X)
Y10=0.2*cos(1.9*X)
Y11=0.1*cos(2*X)
#Defining the pulse as a sum of plane waves
Y= Y1+Y2+Y3+Y4+Y5+Y6+Y7+Y8+Y9+Y10+Y11
#Plotting X-Y function
plot(X,Y,color='black',linestyle='-',linewidth=3)
#Formatting the graph
xlabel('X',fontsize=15)
ylabel('Amplitude',fontsize=15)
tick_params(direction='in',bottom='bool',top='bool',
left='bool',right='bool',width=2,length=5,labelsize=15)
title('Optical Pulse: Finite Sum of Plane Waves',fontsize=15,loc='left')
#Displaying the graph
show()
```

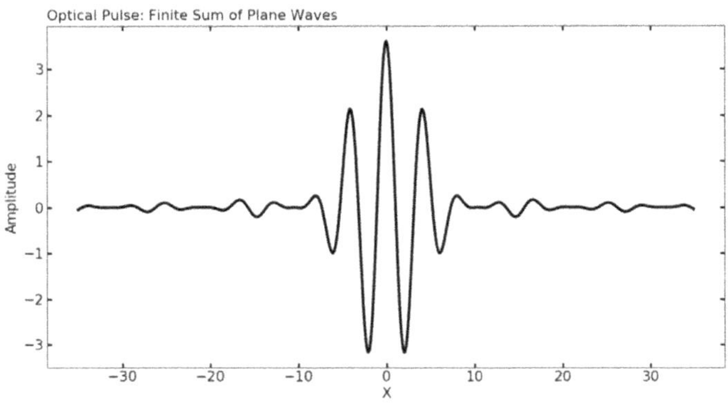

Self-Study

(1) Study the phase velocity of each plane wave and both the phase and group velocities of the resulting pulse. The definitions of phase velocity and group velocity are given in Chapter 1.
(2) Repeat the above exercise by selecting all amplitudes to be 1. Describe your observation.
(3) Repeat the above exercise by decreasing the number of plane waves in the sum: Take the sum up to $Y6$. Describe your observation. Do you find the resulting sum to be periodic?

Key Points to Remember

- Standard arithmetic computation (addition, subtraction, multiplication, and exponentiation) can be performed using Python's built-in **print()** functionality and the corresponding arithmetic operations.
- Python's built-in **input()** functionality always stores strings.
- Computations using standard mathematical functions, such as cos, sin, exp, ln and log can be performed by importing the respective function from the **scipy** module.
- Both **while loop** and for **for loop** headings must end with a colon sign, and the statements in the loop must be indented.
- The graph-plotting functionality **plot()** must be accompanied by the **show()** functionality to display a graph. Both functionalities must be imported from the **matplotlib.pyplot** submodule.
- A list of numbers can be easily prepared using scipy's **arange(start, stop,step)** functionality.

Notes, References, and Further Reading

1. A lovely textbook depicting several pictures of different quantum mechanical problems is S. Brandt and H. D. Dahmen, *The Picture Book of Quantum Mechanics*, 4th edn. Springer, New York (2012).
2. Explore the very useful online Python resource https://www.python.org/ and its SciPy documentation https://docs.scipy.org/doc/ for updated details of available functionalities, methods, arguments, and syntax.
3. Readers are referred to two lovely textbooks focusing on general scientific computations with the Python programming language: (a) C. Hill, *Learning Scientific Programming with Python*, 2nd edn. Cambridge University Press, UK (2020); (b) H. P. Langtangen, *A Primer on Scientific Programming with Python*, 4th edn. Springer, Berlin (2014).
4. A lovely textbook entirely focusing on computational quantum mechanics using Python-based numerical methods: J. Izaac and J. Wang, *Computational Quantum Mechanics*. Springer, Switzerland (2018).

Exercises

A.1. Write a program which takes a value of x from the user and prints the value of the function $\frac{\sin(x)}{x}$. If the user provides $x = 0$, the program should produce an error message and immediately should ask the user to provide a new value of x.

A.2. Create a list of squares of the integer numbers between 10 and 20, including the start and end points.

A.3. Plot a Gaussian function using the matplotlib.pyplot submodule.

A.4. Plot the probability density distribution of a superposition state created by a linear combination of the ground and first excited vibrational states (under a simple harmonic oscillator approximation) for three different times (corresponding to the cosine values 1, 0, and -1).

A.5. Write a Python program to calculate the energy (in Joule) of a hydrogen atom in any Bohr orbit defined by the user-provided quantum number $n(> 0)$.

A.6. Write a Python program to convert a temperature value given in Celsius to Fahrenheit.

A.7. Write a Python program to find the roots of a quadratic polynomial by taking coefficients and constant terms as input.

A.8. Plot the function $f(x) = k_1 \sin x + k_2 \cos x$.

A.9. For two user-input integers k and n, write a Python program that prints the results of all common arithmetic operations between these two integers.

A.10. Plot a cosine function using the matplotlib.pyplot submodule.

A.11. Write a Python code which takes three values a, b, and c from the user and prints the largest one.

A.12. Plot the function $f(x) = x \exp(x)$.

A.13. Create a list of cubes of the numbers between 1 and 20, including the start and end points.

A.14. Write a Python program which takes somebody's date of birth as input from the user and prints the current age.

A.15. Write a Python program to convert a binary number to a decimal number.

A.16. Plot the function $f(x) = \frac{\sin(x)}{x}$.

A.17. Write a Python program to sort a set of numbers in ascending order.

A.18. Plot the Maxwell–Boltzmann speed distribution function $f(x) = Av^2 \exp(-Bv^2)$, where A and B are constants and v is speed.

A.19. Plot the probability density distribution of a superposition state created by a linear combination of the ground and first excited states of particle in a one-dimensional box for three different times (corresponding to the cosine values 1, 0, and −1).
A.20. Plot the function $V = 2x^2 + 3x^3$ using the matplotlib.pyplot submodule.
A.21. Write a Python code which takes a number from the user and tells you whether the number is a prime number.
A.22. Plot the function $\frac{x^2}{a^2} + \frac{y^2}{b^2} = 1$ for $a = 4$ and $b = 3$.
A.23. Print a random number from 1 to 10.
A.24. Write a Python code to find the factorial of a positive integer.
A.25. Plot the function $\sin(2x) + \cos(3x)$.

Chapter 2

Introduction to Quantum Dynamics with a Classical Mechanical Flavor

Highlights: *Meaning of Probability Density, Time Dependence of Probability Density, Time Dependence of Normalization Constant, Expectation Value and its Time Dependence, Ehrenfest Theorem, Equation of Continuity, Bohmian Mechanics.*

2.1 Introduction: A Closer Look at the Probability Density

In quantum mechanics and quantum dynamics, perhaps the most important introductory question is the following: "How can a delocalized mathematical function (wavefunction) represent a tiny little particle which is supposed to be localized (according to our classical or general consensus)?" Note that quantum mechanics only provides a probabilistic answer to this question in terms of the probability density (or amplitude) $\rho(x,t)$ for finding the particle at point x and time t. This point is further illuminated in the following, for reasons to be revealed soon.

A quantum "particle" is defined by its space- and time-dependent wavefunction, $\psi(x,t)$. **Unless clearly stated, hereinafter, always assume that the wavefunction $\psi(x,t)$ represents a non-stationary state of the particle, and in that case, its probability density remains to be time-dependent (manifesting a quantum dynamics).** The wavefunction $\psi(x,t)$ itself

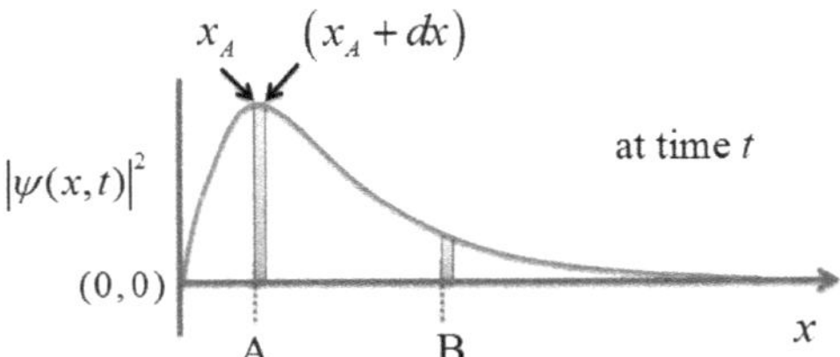

Figure 2.1. A typical example of probability density distribution of a particle in one dimension.

can be a complex function; however, the probability density $\rho(x,t) = |\psi(x,t)|^2 = \psi^*(x,t) \cdot \psi(x,t)$ (where $\psi^*(x,t)$ is a complex conjugate of $\psi(x,t)$) must always be real. Figure 2.1 illustrates a typical example of the probability density $\rho(x,t)$ distribution of a particle at a given time t. As $\rho(x,t)dx = |\psi(x,t)|^2 dx$ represents the area under the curve $|\psi(x,t)|^2$ for an interval of dx, one can easily conclude that the particle is statistically more likely to be found near the point A than near the point B on the x-axis (because the shaded area near point A is larger than that near point B). But *what does this probability density $\rho(x,t)$ distribution at a given time imply for a quantum particle with regard to the experimental observation?*

To address the above question, let us first consider a hydrogen atom which has one electron, and imagine that the theory (quantum mechanics) gives us the probability density distribution of that electron, as depicted in Figure 2.2(a) at a given time t, assuming that the nucleus of the hydrogen atom always remains anchored at the $(0,0)$ position.[1] Therefore, Figure 2.2(a) schematically depicts the probability amplitude or density of the electron at a distance r from the nucleus in the hydrogen atom at a time t.

Next, imagine that we have an experimental device which can produce identical, fresh hydrogen atoms (one at a time) if we push a button on that device. Furthermore, assume that in this thought experiment, one can very precisely determine the position of the electron in a hydrogen atom with respect to the corresponding nuclear position. If we could repeat the position measurement process at the given time t with freshly prepared identical hydrogen atoms, what would be the expected experimental results?

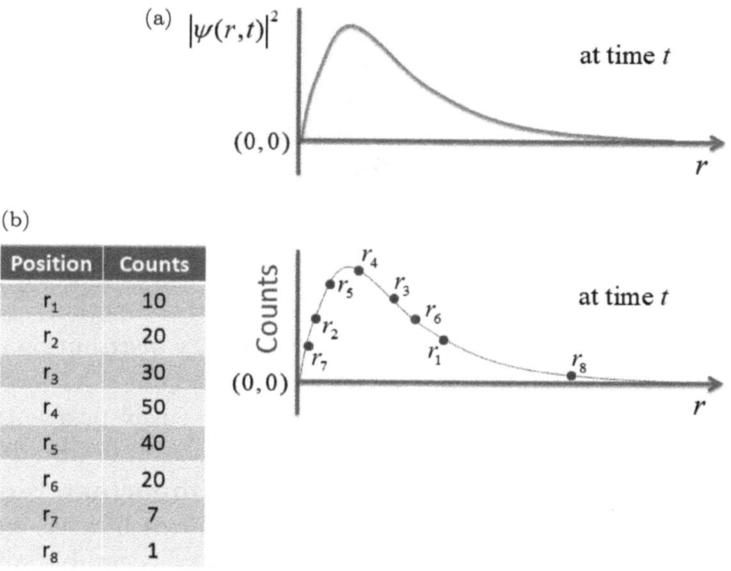

Position	Counts
r_1	10
r_2	20
r_3	30
r_4	50
r_5	40
r_6	20
r_7	7
r_8	1

Figure 2.2. An illustrative example: Assume that Figure (a) represents the theoretically predicted probability density distribution of the electron in a hydrogen atom, considering that the nucleus of the hydrogen atom is anchored at the $(0, 0)$ position, and Figure (b) represents the possible outcome of repeated experimental measurements to determine the position of the electron with respect to the position of the nucleus in the hydrogen atom. For the repeated measurements, identically prepared fresh hydrogen atoms are used. See the text for more details.

In the first experiment, the device would produce a fresh hydrogen atom, and the position measurement process would reveal a specific position for the electron at, say, r_1. In the second experiment, the device would produce another fresh identical hydrogen atom, and the measurement process would reveal another specific position, say, r_2. In the second measurement, it is not necessary that we would find the electron at the same r_1 position (the outcome of the first experiment). If the measurement process is repeated several times, each time with one freshly prepared identical hydrogen atom, we would finally get a table depicting the counts (i.e., how many times the same value is obtained) as a function of different positions, as exemplified in Figure 2.2(b). The same is plotted in the adjacent graph. Quite interestingly, we would notice that the r_4 position possesses the highest counts. Therefore, repeated experimental measurements should finally render the same distribution which is predicted by quantum mechanics if an infinitely large collection of identically prepared fresh hydrogen atom samples is used. Therefore, **the probability density distribution for a particle directly connects quantum mechanical prediction with the expected experimental outcome of repeated measurements.** This is called the **standard interpretation** of the probability density distribution (also called the **Born interpretation** of the wavefunction).[2]

In quantum dynamics, the probability density distribution $\rho(x, t)$ for a particle in position space varies as a function of t; however, the total probability of finding the particle over the entire position space must always remain 1 at all times because we have only **one** dynamically evolving quantum particle. This is an **important requirement for exploring quantum dynamics** and is fulfilled only by considering a **normalized wavefunction** as an acceptable solution to the time-dependent Schrödinger equation (TDSE) for the given quantum mechanical system.

Let us pause here and ask two important questions relevant to the time-dependent behavior of a wavefunction:

(a) *How do we normalize a wavefunction at $t = 0$?*
(b) *How do we prove that, as argued above, if a wavefunction is normalized at $t = 0$, it remains normalized for all later times?*

Normalizing a wavefunction: One can normalize a wavefunction using the normalization condition

$$\int_{-\infty}^{+\infty} |\psi(x,t)|^2 dx = 1 \tag{2.1}$$

Take an example. Assume that the wavefunction of a particle at $t = 0$ has the following Gaussian form:

$$\psi(x, 0) = Ae^{-ax^2}, \text{ where } A \text{ and } a \text{ are real and positive constants.}$$

Then, the probability density is given by

$$\rho(x, 0) = A^2 e^{-2ax^2}$$

Now, employing the normalization condition, we get

$$\int_{-\infty}^{+\infty} A^2 e^{-2ax^2} dx = 1$$

or $A^2\sqrt{\dfrac{\pi}{2a}} = 1$, using the standard Gaussian integral

$$\int_{-\infty}^{+\infty} e^{-ax^2+ibx+ic} dx = \sqrt{\frac{\pi}{a}} e^{\left(\frac{-b^2}{4a}+ic\right)}$$

or $A = \left(\dfrac{2a}{\pi}\right)^{1/4}$, which is a normalization constant.

Thus, the normalized Gaussian wavefunction at $t = 0$ has following form:

$$\psi(x,0) = \left(\frac{2a}{\pi}\right)^{1/4} e^{-ax^2}$$

The above exercise reveals that even if the wavefunction of a particle is not normalized (in that case, the integral in equation (2.1) gives some **finite positive constant**), it can be normalized by using the normalization condition (equation (2.1)). Here, note that **if a wavefunction cannot be normalized, it is not acceptable as a solution to the TDSE because the statistical interpretation (which is the only physically realizable interpretation) of the wavefunction fails.** In fact, both $\psi(x,0)$ and $\psi(x,t)$ are considered physically acceptable wavefunctions (a physically acceptable wavefunction is also called a well-behaved wavefunction) in quantum dynamics (and in quantum mechanics in general) if and only if

(a) $\psi^*\psi$ must be single-valued,
(b) they are not infinite over a finite range,
(c) they are continuous everywhere,
(d) they possess continuous first derivatives, and
(e) they are normalizable.

Time-dependence of normalization constant: Our next concern pertaining to the time-dependent behavior of a wavefunction is that even if a wavefunction is normalized at $t = 0$, does it remain normalized for all later times? This question can be rephrased as, "Can the normalization constant be a function of time?"

To find the answer, consider the time derivative of the probability density:

$$\frac{\partial}{\partial t}|\psi(x,t)|^2 = \frac{\partial}{\partial t}[\psi^*(x,t)\cdot\psi(x,t)] = \frac{\partial\psi^*(x,t)}{\partial t}\psi(x,t) + \psi^*(x,t)\frac{\partial\psi(x,t)}{\partial t} \tag{2.2}$$

From the TDSE, on the other hand, we get

$$i\hbar\frac{\partial}{\partial t}\psi(x,t) = \left[-\frac{\hbar^2}{2m}\frac{\partial^2}{\partial x^2} + V(x)\right]\psi(x,t)$$

$$\text{or}\quad \frac{\partial\psi(x,t)}{\partial t} = \frac{i\hbar}{2m}\frac{\partial^2\psi(x,t)}{\partial x^2} - \frac{iV}{\hbar}\psi(x,t) \tag{2.3}$$

Taking the complex conjugate of the above equation, we get

$$\frac{\partial\psi^*(x,t)}{\partial t} = -\frac{i\hbar}{2m}\frac{\partial^2\psi^*(x,t)}{\partial x^2} + \frac{iV}{\hbar}\psi^*(x,t) \tag{2.4}$$

(considering that V is real)

Inserting equations (2.3) and (2.4) into equation (2.2), we get

$$\begin{aligned}\frac{\partial}{\partial t}|\psi(x,t)|^2 &= \left[-\frac{i\hbar}{2m}\frac{\partial^2\psi^*(x,t)}{\partial x^2} + \frac{iV}{\hbar}\psi^*(x,t)\right]\psi(x,t) \\ &\quad + \psi^*(x,t)\left[\frac{i\hbar}{2m}\frac{\partial^2\psi(x,t)}{\partial x^2} - \frac{iV}{\hbar}\psi(x,t)\right] \\ &= \frac{i\hbar}{2m}\left[\psi^*(x,t)\frac{\partial^2\psi(x,t)}{\partial x^2} - \frac{\partial^2\psi^*(x,t)}{\partial x^2}\psi(x,t)\right] \\ &= \frac{i\hbar}{2m}\frac{\partial}{\partial x}\left[\psi^*(x,t)\frac{\partial\psi(x,t)}{\partial x} - \frac{\partial\psi^*(x,t)}{\partial x}\psi(x,t)\right]\end{aligned}$$

Integrating both sides in the $[-\infty, +\infty]$ limit of x, we get

$$\int_{-\infty}^{+\infty} \frac{\partial}{\partial t}|\psi(x,t)|^2 dx$$

$$= \frac{i\hbar}{2m}\left[\psi^*(x,t)\frac{\partial\psi(x,t)}{\partial x} - \frac{\partial\psi^*(x,t)}{\partial x}\psi(x,t)\right]_{-\infty}^{+\infty} = 0$$

$$\text{or } \frac{d}{dt}\left[\int_{-\infty}^{+\infty} |\psi(x,t)|^2 dx\right]^2 = 0 \tag{2.5}$$

Note that we have made use of the fact that both ψ and ψ^* are zero at $x = +\infty$ or $x = -\infty$ (an important characteristic of a well-behaved wavefunction). Note also that a total derivative $\frac{d}{dt}$ is used for the integral $\int_{-\infty}^{+\infty} |\psi(x,t)|^2 dx$, which is a function only of t, and a partial derivative $\frac{\partial}{\partial t}$ is used for the integrand $|\psi(x,t)|^2$, which is a function of both x and t.

Equation (2.5) illuminates an important fact that once $\psi(x,t)$ at $t = 0$ is normalized, it remains normalized at any later time. In other words, the TDSE does not allow the total probability density to change over time. This is a remarkable property of the TDSE that it automatically preserves the normalization of the wavefunction (note that, here, we are dealing with **only a time-independent Hamiltonian**). Thus, the requirement for exploring quantum dynamics is automatically fulfilled if one uses a normalized wavefunction to solve the TDSE.

Guiding Questions

2.1. In the variable separation method (discussed in Chapter 1), we have seen that E is a constant. Prove that E must be a *real* constant for a normalizable wavefunction.
Hint: Use $\frac{d}{dt}\left[\int_{-\infty}^{+\infty} |\psi(x,t)|^2 dx\right] = 0$.

2.2. If we assume that $\frac{d}{dt}\left[\int_{-\infty}^{+\infty} |\psi(x,t)|^2 dx\right] = A^2|\psi(x)|^2 e^{\frac{2\varepsilon t}{\hbar}}$ is true for a system, find the meaning of this time-dependent total probability density.

(*Continued*)

(Continued)

2.3. Write down the complex conjugate of the following equations:

$$i\hbar\frac{\partial}{\partial t}\psi(x,t) = \left[-\frac{\hbar^2}{2m}\frac{\partial^2}{\partial x^2} + V(x)\right]\psi(x,t) \quad \text{and} \quad \hat{H}\psi(x) = E\psi(x)$$

2.4. Determine whether the following wavefunctions at $t = 0$ are physically acceptable solution to the TDSE:
(a) $\psi(x,0) = Ae^{-x}$ over the interval $(0,+\infty)$, (b) $\psi(x,0) = Ae^{-x}$ over the interval $(-\infty,+\infty)$, (c) $\psi(x,0) = Ae^{ikx}$ over the interval $(0,+\infty)$, (d) $\psi(x,0) = Ae^{ikx}$ over the interval $(-\infty,+\infty)$.

The above analysis of the conservation of total probability density is constructed based on the following two important postulates[3] of quantum mechanics in the context of the motion of a quantum particle.

Postulate 1 and its consequences: *The state of a quantum particle is completely defined by its (position- and time-dependent) wavefunction, $\psi(x,t)$. At time $t, \psi^*(x,t)\cdot\psi(x,t)dx = \rho(x,t)dx$ gives the probability for finding the particle in the dx interval between the x and $(x+dx)$ positions. Only a well-behaved wavefunction represents a physically realizable state of the particle. A well-behaved wavefunction must be normalizable, and its first derivative must be continuous and finite. Once a wavefunction is normalized, it remains normalized at any later time if the TDSE is used to explore the quantum dynamics.*

Postulate 2 and its consequences: *The wavefunction of a particle evolves in time according to the TDSE: $i\hbar\frac{\partial}{\partial t}\psi(x,t) = \hat{H}\psi(x,t)$. We have already made use of this postulate in Chapter 1 and have seen some of the consequences, which render two important concepts — stationary and superposition states. A stationary state represents a state wherein the probability density remains independent of time, and a superposition state represents a state wherein the probability density exhibits a time-dependent change (giving birth to coherence).*

Important Note

In quantum mechanics, the wavefunction of a particle can be represented both in position space and in momentum space. Based on mathematical and numerical convenience, one representation is selected to solve a particular quantum mechanical problem. In this chapter, we have represented the wavefunction only in position space (as a function of its position coordinate, x). In Chapter 3 and PythonChapter C, for convenience, we will also represent the wavefunction in momentum space (as a function of the momentum coordinate, k). One can conveniently change the representation using a well-known mathematical transformation, called the Fourier transform:

(a) Fourier transform of position-space wavefunction to momentum-space wavefunction:

$$\phi(k,t) = \frac{1}{\sqrt{2\pi}} \int_{-\infty}^{+\infty} \psi(x,t) e^{-ikx} dx$$

(b) Inverse Fourier transform of momentum-space wavefunction to position-space wavefunction:

$$\psi(x,t) = \frac{1}{\sqrt{2\pi}} \int_{-\infty}^{+\infty} \phi(k,t) e^{ikx} dk$$

Details are given in Chapter 3 and PythonChapter C.

2.2 Quest for Classical Mechanical Flavor in Quantum Dynamics

Based on what we have discussed above, a generic view of the possible one-dimensional motion of a quantum particle can now be constructed, as schematically exemplified in Figure 2.3(a). The probability density distribution changes as a function of time when the quantum particle moves in position space (a space defined by the position coordinate x). Each distribution (quantum mechanically predicted at a given time) can be experimentally obtained by performing many (say, Avogadro's number) repeated measurements of the position of the particle at that given time. A rigorous mathematical derivation and an in-depth discussion of the motion of a quantum particle will be presented in Chapter 3. Here, just a brief

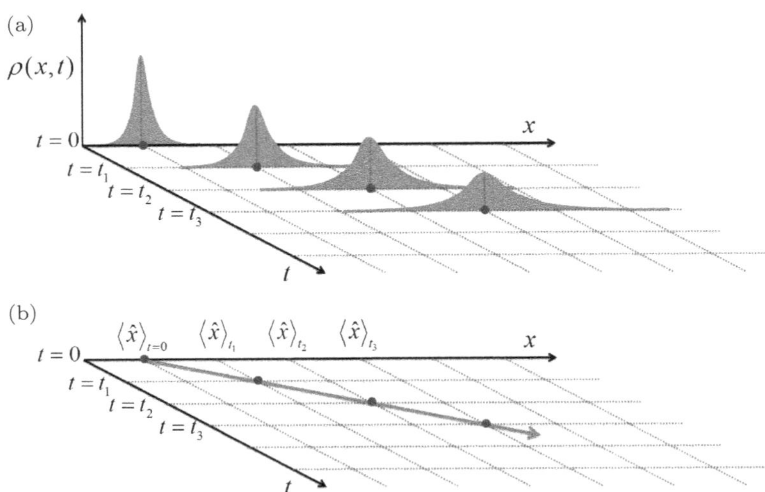

Figure 2.3. (a) A generic view of possible one-dimensional motion of a quantum particle is schematically presented. A Gaussian distribution is assumed for illustration purpose only. A quantum particle can adopt any distribution depending on the potential it experiences. (b) The average of each probability density distribution (called the expectation value, as depicted by the filled circle) of position shows a trajectory (point-by-point information) very similar to a classical mechanical trajectory.

idea is featured to obtain a classical mechanical flavor in quantum dynamics.

In classical mechanics (and in our daily lives), the motion of a (classical) particle (with a mass) is very conveniently described using a **trajectory (flight path)**, which is the path that the particle follows in position space as a function of time. As an example, the trajectory (dotted line) of a classical particle starting from point A and reaching point B through certain intermediate points is depicted in Figure 2.4. In our daily experience, we feel very uncomfortable when we do not get this local (point-by-point) information of the motion of a particle. Discomfort is further magnified in quantum dynamics because quantum mechanics gives only information of possible positions through its probability density distribution. This is why an equivalent **quantum trajectory** (point-by-point information of the motion of a quantum particle) is not possible to obtain directly

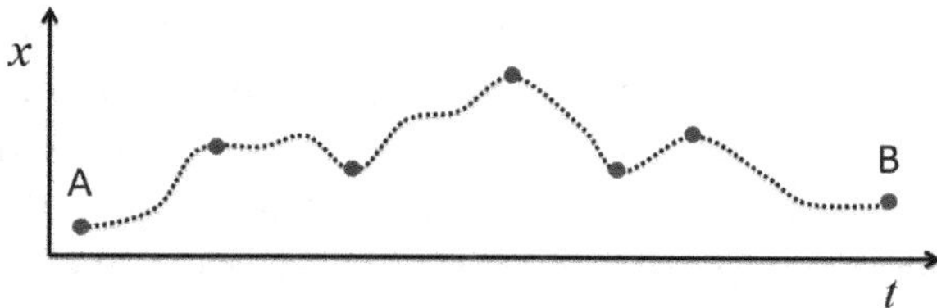

Figure 2.4. A typical example of a one-dimensional trajectory of a classical particle is shown.

using the standard interpretation (Born interpretation or probabilistic interpretation) of quantum mechanics. However, three theorems, namely the Ehrenfest theorem, the equation of continuity, and the Bohmian mechanics, provide a classical flavor in quantum dynamics. The rest of this chapter is dedicated to discussing these theories to help readers visualize quantum dynamics with trajectories (more specifically, with a classical mechanical flavor).[4]

2.3 Expectation Value

Although the wavefunction and the associated probability density distribution of a particle in position space at a given time are always global (delocalized) in nature, one can very easily determine the average position of the particle at a given time from the normalized wavefunction using the following expression:

$$\langle \hat{x} \rangle (t) = \int_{-\infty}^{+\infty} \psi^*(x,t)\hat{x}\psi(x,t)dx \tag{2.6}$$

The average position obtained using the above equation is called the **expectation value of position** at the given time t. In fact, for a given normalized wavefunction, **the expectation value of a physical or dynamical quantity provides a way to compute the average of repeated experimental measurements (which is equivalent to the average of the probability density distribution)**.

Note that a wavefunction, as pointed out earlier, does not represent a point in position space; rather, it is always spread out in position space, unraveling the fact that the probability density for the particle is finite at different points in position space. This is why a wavefunction is global (or delocalized) in the same sense that the English language is a global language. The probability of finding an English-speaking person is finite almost everywhere around the globe. On the contrary, the expectation value of position represents a point in position space, and its time evolution, therefore, provides an easily perceptible (with a classical mechanical flavor) trajectory in quantum dynamics (as schematically shown in Figure 2.3(b)). Before we go over the time evolution of the expectation value, note that the above argument originates from the third postulate of quantum mechanics.

Postulate 3 and its consequences: *If a quantum mechanical system is described by a normalized wavefunction $\psi(x,t)$, the average value of an observable corresponding to the operator $\langle\hat{A}\rangle(t)$ at a given time (t) is given by $\langle\hat{A}\rangle(t) = \int_{-\infty}^{+\infty} \psi^*(x,t)\hat{A}\psi(x,t)dx$. Similarly, $\langle\hat{A}^2\rangle(t) = \int_{-\infty}^{+\infty} \psi^*(x,t)\hat{A}^2\psi(x,t)dx$, and finally, the variance of the probability density distribution or amplitude at a given time t is given by $\sigma_A^2(t) = \langle\hat{A}^2\rangle(t) - \langle\hat{A}\rangle^2(t)$.*

In this chapter, no further discussion will be developed revolving around two important physical quantities, $\langle\hat{A}^2\rangle$ and σ_A^2. In Chapter 3, we will make use of them to understand Heisenberg's uncertainty principle and the correspondence between position-space and momentum-space representations of the wavefunction.

Recapitulation

Note that, in this chapter, we are going over the general postulates of quantum mechanics from a time-dependent point of view in the interest of developing the subject of ***time-dependent quantum mechanics***. The postulates of quantum mechanics and their consequences, based on which different quantum descriptions of physical or chemical systems are formulated, construct the foundation of quantum theory. Readers are referred to a lovely textbook *Quantum Mechanics* by Cohen-Tannoudji *et al.*, wherein several postulates of quantum mechanics are discussed in adequate detail.[5]

(*Continued*)

(*Continued*)

We have already gone over three postulates of quantum mechanics from a time-dependent point of view. In this recapitulation box, a brief discussion is presented based on the fourth postulate of quantum mechanics to give a formal introduction to the quantum mechanical operators. Our familiar $\hat{H}$ is an operator corresponding to the total energy (classical observable) which appears in the TDSE.

Postulate 4 and its consequences: *For every observable in classical mechanics, there corresponds a linear and Hermitian operator in quantum mechanics. Table 2.1 lists some examples of the quantum mechanical operators and their classical observables.* A formal definition of the Hermitian operator is given in Chapter 4.

Table 2.1. Classical mechanical observables and corresponding quantum mechanical operators.

Quantum Mechanical Operator (one-dimensional)			
Name	**Symbol**	**Operation**	**Classical Observable**
Hamiltonian	$\hat{H}$	$\left[-\frac{\hbar^2}{2m}\left(\frac{\partial^2}{\partial x^2}\right)+V(x)\right]$	Total Energy, E
Potential Energy	$\hat{V}$	Multiply by $V(x)$	Potential Energy, $V(x)$
Kinetic Energy	$\hat{K}_x$	$-\frac{\hbar^2}{2m}\left(\frac{\partial^2}{\partial x^2}\right)$	Kinetic Energy, K_x
Momentum	$\hat{P}_x$	$-i\hbar\left(\frac{\partial}{\partial x}\right)$	Momentum, P_x

2.4 Ehrenfest Theorem: Time Evolution of Expectation Value

As pointed out earlier, the expectation value of the position, which is calculated from a global function (wavefunction), represents a point in position space. How does the expectation value of position $\langle\hat{x}\rangle(t)$ change as time progresses? This question is addressed by the Ehrenfest theorem,[6] which provides a relationship between the expectation values, such as the expectation value of position

$\langle \hat{x} \rangle(t)$ and the expectation value of momentum $\langle \hat{P} \rangle(t)$, to unravel the underlying connection between quantum and classical dynamics.

The expectation value of x at given time t is expressed as

$$\langle \hat{x} \rangle(t) = \int_{-\infty}^{+\infty} \psi^*(x,t)\hat{x}\psi(x,t)dx$$

Taking the first derivative of the above equation with respect to time, we get

$$\frac{d\langle \hat{x} \rangle}{dt} = \int_{-\infty}^{+\infty} \frac{\partial \psi^*(x,t)}{\partial t}\hat{x}\psi(x,t)dx + \int_{-\infty}^{+\infty} \psi^*(x,t)\hat{x}\frac{\partial \psi(x,t)}{\partial t}dx \quad (2.7)$$

Note that on the left-hand side, we have the total derivative $\frac{d}{dt}$ and on the right-hand side, we have the partial derivative $\frac{\partial}{\partial t}$. Now, from the TDSE, we know

$$i\hbar \frac{\partial}{\partial t}\psi(x,t) = \hat{H}\psi(x,t)$$

$$\text{or} \quad \frac{\partial}{\partial t}\psi(x,t) = \frac{1}{i\hbar}\hat{H}\psi(x,t) \quad (2.8)$$

and taking the complex conjugate of the TDSE, we have

$$\frac{\partial}{\partial t}\psi^*(x,t) = -\frac{1}{i\hbar}[\hat{H}\psi(x,t)]^* \quad (2.9)$$

Inserting equations (2.8) and (2.9) into equation (2.7), we get

$$\frac{d\langle \hat{x} \rangle}{dt} = \int_{-\infty}^{+\infty} -\frac{1}{i\hbar}[\hat{H}\psi(x,t)]^*\hat{x}\psi(x,t)dx + \int_{-\infty}^{+\infty} \psi^*(x,t)\hat{x}\frac{1}{i\hbar}[\hat{H}\psi(x,t)]dx \quad (2.10)$$

As $\hat{H}$ is a Hermitian operator, one can write $\int_{-\infty}^{+\infty} f^*\hat{H}g dx = \int_{-\infty}^{+\infty} g(\hat{H}f)^* dx$. The properties of a Hermitian operator will be discussed in detail in Chapter 4. For now, we just accept it without question and use the final result. Carefully note that the order of operation does not matter when a Hermitian operator is involved. This operator can act either from the right or from the left.

Thereby, we can write

$$-\frac{1}{i\hbar}\int_{-\infty}^{+\infty}[\hat{H}\psi(x,t)]^{*}\hat{x}\psi(x,t)dx = -\frac{1}{i\hbar}\int_{-\infty}^{+\infty}\psi(x,t)^{*}\hat{H}\hat{x}\psi(x,t)dx \tag{2.11}$$

Inserting equation (2.11) into equation (2.10), we get

$$\frac{d\langle\hat{x}\rangle}{dt} = \frac{1}{i\hbar}\left[\int_{-\infty}^{+\infty}\psi^{*}(x,t)\hat{x}\hat{H}\psi(x,t)dx - \int_{-\infty}^{+\infty}\psi(x,t)^{*}\hat{H}\hat{x}\psi(x,t)dx\right]$$

$$\text{or } \frac{d\langle\hat{x}\rangle}{dt} = \frac{i}{\hbar}\left[\int_{-\infty}^{+\infty}\psi^{*}(x,t)(\hat{H}\hat{x}-\hat{x}\hat{H})\psi(x,t)dx\right] \tag{2.12}$$

As $\hat{H}$ is expressed as $[-\frac{\hbar^2}{2m}\frac{\partial^2}{\partial x^2}+V(x)]$ (considering time-independent Hamiltonian), we can write

$$(\hat{H}\hat{x}-\hat{x}\hat{H})\psi(x,t) = \left[-\frac{\hbar^2}{2m}\frac{\partial^2}{\partial x^2}+V\right]x\psi(x,t) - x\left[-\frac{\hbar^2}{2m}\frac{\partial^2}{\partial x^2}+V\right]\psi(x,t)$$

As $\frac{\partial^2}{\partial x^2}(x\psi) = \frac{\partial}{\partial x}\left(x\frac{\partial\psi}{\partial x}+\psi\right) = x\frac{\partial^2\psi}{\partial x^2}+2\frac{\partial\psi}{\partial x}$, one can write

$$(\hat{H}\hat{x}-\hat{x}\hat{H})\psi(x,t) = -\frac{\hbar^2}{2m}x\frac{\partial^2\psi}{\partial x^2} - \frac{\hbar^2}{2m}2\frac{\partial\psi}{\partial x} + Vx\psi + \frac{\hbar^2}{2m}x\frac{\partial^2\psi}{\partial x^2} - Vx\psi = -\frac{\hbar^2}{m}\frac{\partial\psi}{\partial x}$$

Note that $\hat{V}$ and $\hat{x}$ are multiplication operators. As a result, their order of operation does not matter. So, one can easily write, $xV\psi = Vx\psi$. More details about the properties of quantum mechanical operators from a linear algebra point of view are given in Chapter 4.

The above equation can be further rewritten in terms of the momentum operator (consider one dimension). As $\hat{P}_x\psi = -i\hbar\frac{\partial\psi}{\partial x}$,

$\frac{\partial\psi}{\partial x} = -\frac{1}{i\hbar}\hat{P}_x\psi$. Therefore, one can write

$$(\hat{H}\hat{x} - \hat{x}\hat{H})\psi(x,t) = -\frac{\hbar^2}{m}\left(-\frac{1}{i\hbar}\hat{P}_x\right)\psi(x,t) = -\frac{i\hbar}{m}\hat{P}_x\psi(x,t) \quad (2.13)$$

Thus, inserting equation (2.13) into equation (2.12), we get

$$\frac{d\langle\hat{x}\rangle}{dt} = \frac{i}{\hbar}\left[\int_{-\infty}^{+\infty}\psi^*(x,t)\left(-\frac{i\hbar}{m}\right)\hat{P}_x\psi(x,t)dx\right]$$

$$= \frac{1}{m}\left[\int_{-\infty}^{+\infty}\psi^*(x,t)\hat{P}_x\psi(x,t)dx\right] = \frac{\langle\hat{P}_x\rangle}{m}$$

$$\text{or} \quad m\frac{d\langle\hat{x}\rangle}{dt} = \langle\hat{P}_x\rangle \quad (2.14)$$

Here, the expectation value of momentum at a given time t is expressed as $\langle\hat{P}_x\rangle(t) = \int_{-\infty}^{+\infty}\psi^*(x,t)\hat{P}_x\psi(x,t)dx$. Note that the above equation represents the quantum mechanical version of the classical mechanical description of linear momentum, $(m\frac{dx}{dt} = P)$.

Guiding Question

2.5. Prove that, in general, if an operator $\hat{A}$ is a Hermitian operator and does not depend on time, the time evolution of its expectation value can be described by

$$\frac{d\langle\hat{A}\rangle(t)}{dt} = \frac{i}{\hbar}\left[\int_{-\infty}^{+\infty}\psi^*(x,t)(\hat{H}\hat{A} - \hat{A}\hat{H})\psi(x,t)dx\right]$$

Next, to obtain another interesting expression, consider the time derivative of the expectation value of momentum (in one dimension):

$$\frac{d\langle\hat{P}_x\rangle}{dt} = \frac{i}{\hbar}\left[\int_{-\infty}^{+\infty}\psi^*(x,t)(\hat{H}\hat{P}_x - \hat{P}_x\hat{H})\psi(x,t)dx\right] \quad (2.15)$$

Given that $\hat{H} = [-\frac{\hbar^2}{2m}\frac{\partial^2}{\partial x^2} + V(x)]$ and $\hat{P}_x = -i\hbar\frac{\partial}{\partial x}$, we can write (check it yourself)

$$[\hat{H}\hat{P}_x - \hat{P}_x\hat{H}]\psi = i\hbar\frac{dV}{dx}\psi \tag{2.16}$$

Thus, inserting equation (2.16) into equation (2.15), we get

$$\frac{d\langle\hat{P}_x\rangle}{dt} = \frac{i}{\hbar}\left[\int_{-\infty}^{+\infty}\psi^*(x,t)i\hbar\left(\frac{d\hat{V}}{dx}\right)\psi(x,t)dx\right]$$

$$= -\int_{-\infty}^{+\infty}\psi^*(x,t)\left(\frac{d\hat{V}}{dx}\right)\psi(x,t)dx$$

$$\text{or} \quad \frac{d\langle\hat{P}_x\rangle}{dt} = -\left\langle\frac{d\hat{V}}{dx}\right\rangle \tag{2.17}$$

Here, the expectation value of $(\frac{d\hat{V}}{dx})$ at a given time t is expressed as $\langle\frac{d\hat{V}}{dx}\rangle(t) = \int_{-\infty}^{+\infty}\psi^*(x,t)(\frac{d\hat{V}}{dx})\psi(x,t)dx$. As in classical mechanics, the rate of change of momentum gives force; equation (2.17) features the sense of classical mechanical force which depends on the negative gradient of the potential energy, $(F = -\frac{dV}{dx})$.

Both equations (2.14) and (2.17) represent the Ehrenfest theorem, which reveals that the time evolution of the expectation values obtained from the respective normalized wavefunction follows classical mechanics. Therefore, a trajectory can be easily constructed from the time evolution of the expectation values for the visualization of quantum dynamics with a classical mechanical flavor. This is schematically depicted in Figure 2.3(b). Note that the expectation value of position does not always need to follow a linear trajectory. It may follow a nonlinear trajectory, depending on how the probability density distribution of the particle changes as a function of time. Figure 2.3(b) represents just a schematic illustration of the concept behind the Ehrenfest theorem. A detailed discussion on the time evolution of the expectation value of the position of a free particle will be presented in Chapter 3.

2.5 The Equation of Continuity: Hydrodynamic Formulation of TDSE

We have understood that a practically realizable interpretation of a wavefunction is obtained only by using the concept of probability density, $\rho(x,t) = |\psi(x,t)|^2 = \psi^*(x,t) \cdot \psi(x,t)$; $|\psi(x,t)|^2$ gives the probability density or amplitude for the particle at position x and time t. In quantum dynamics, as noted earlier, the probability density changes as a function of time. This is schematically shown again in Figure 2.5(a). One may conveniently represent the time-dependent change of the probability density at a certain position x (which is called local probability density and is illustrated using the vertical bar at position x in Figure 2.5(a)) as the flow of the probability density through that point (considering an analogy equivalent to a fluid flowing through a point), as schematically depicted in Figure 2.5(b).

The flow rate of the probability density through a point is called the **probability current**, which is defined by the rate at which the probability flows through the point x at time t. The probability

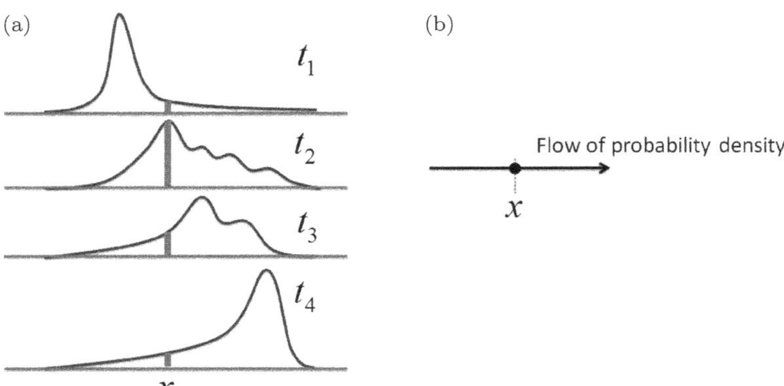

Figure 2.5. (a) A schematic illustration to represent the temporal change in the probability density of a particle as a function of time. (b) The time-dependent change in the probability density at a certain position x can be represented by the flow of the probability density through that point.

current at point x and at time t is given by

$$J(x,t) = \frac{i\hbar}{2m}\left[\psi(x,t)\frac{\partial\psi^*(x,t)}{\partial x} - \psi^*(x,t)\frac{\partial\psi(x,t)}{\partial x}\right] \tag{2.18}$$

The derivation of the above expression is not given here;[7] we directly make use of this expression for the following discussion.

Using the expression of the probability current, as given in equation (2.18), an alternative form of the TDSE (called the **equation of continuity**) is obtained by first multiplying the TDSE by ψ^* from the left and then by subtracting its complex conjugate equation from the resulting equation.

We begin with the TDSE:

$$i\hbar\frac{\partial\psi(x,t)}{\partial t} = \hat{H}\psi(x,t) \quad \text{or} \quad i\hbar\psi^*(x,t)\frac{\partial\psi(x,t)}{\partial t} = \psi^*(x,t)\hat{H}\psi(x,t) \tag{2.19}$$

and its complex conjugate equation:

$$-i\hbar\psi(x,t)\frac{\partial\psi^*(x,t)}{\partial t} = \psi(x,t)\hat{H}\psi^*(x,t) \tag{2.20}$$

Here, $\hat{H}$ is assumed to be a real Hermitian operator, which is expressed as

$$\hat{H} = \left[-\frac{\hbar^2}{2m}\frac{\partial^2}{\partial x^2} + V(x)\right]$$

Subtracting equation (2.20) from equation (2.19), we get

$$i\hbar\psi^*(x,t)\frac{\partial\psi(x,t)}{\partial t} + i\hbar\psi(x,t)\frac{\partial\psi^*(x,t)}{\partial t}$$
$$= \psi^*(x,t)\hat{H}\psi(x,t) - \psi(x,t)\hat{H}\psi^*(x,t)$$

or

$$\psi^*(x,t)\frac{\partial\psi(x,t)}{\partial x} + \psi(x,t)\frac{\partial\psi^*(x,t)}{\partial x}$$
$$= \frac{1}{i\hbar}\left[\psi^*(x,t)\hat{H}\psi(x,t) - \psi(x,t)\hat{H}\psi^*(x,t)\right]$$

$$= \frac{1}{i\hbar}\left[\psi^*(x,t)\left(-\frac{\hbar^2}{2m}\frac{\partial^2}{\partial x^2} + V(x)\right)\psi(x,t)\right.$$
$$\left. -\psi(x,t)\left(-\frac{\hbar^2}{2m}\frac{\partial^2}{\partial x^2} + V(x)\right)\psi^*(x,t)\right]$$
$$= \frac{i\hbar}{2m}\left[\psi^*(x,t)\frac{\partial^2\psi(x,t)}{\partial x^2} - \psi(x,t)\frac{\partial^2\psi^*(x,t)}{\partial x^2}\right]$$

because V is just a multiplication operator; therefore, $\psi^*(x,t)V(x)\psi(x,t) = \psi(x,t)V(x)\psi^*(x,t)$

$$\text{or} \quad \frac{\partial}{\partial t}[\psi^*(x,t)\psi(x,t)]$$
$$= -\frac{i\hbar}{2m}\left[\psi(x,t)\frac{\partial^2\psi^*(x,t)}{\partial x^2} - \psi^*(x,t)\frac{\partial^2\psi(x,t)}{\partial x^2}\right]$$
$$\text{or} \quad \frac{\partial}{\partial t}[\psi^*(x,t)\psi(x,t)]$$
$$= -\frac{\partial}{\partial x}\left[\frac{i\hbar}{2m}\left(\psi(x,t)\frac{\partial\psi^*(x,t)}{\partial x} - \psi^*(x,t)\frac{\partial\psi(x,t)}{\partial x}\right)\right] = -\frac{\partial}{\partial x}J(x,t)$$
$$\text{or} \quad \frac{\partial\rho(x,t)}{\partial t} = -\frac{\partial}{\partial x}J(x,t) \tag{2.21}$$

(using equation (2.18))

The above equation is called the **equation of continuity**, which features the local conservation of the probability. Note that the global conservation of probability is given by $\int_{-\infty}^{-\infty}\rho(x,t)dx = 1$.

Although the equation of continuity is just another mathematical form of the TDSE, there exists an important difference between the equation of continuity and the TDSE in the sense of their interpretation. The equation of continuity directly deals with the probability density, which is an experimentally realizable (and observable) quantity. On the other hand, the TDSE deals with the wavefunction, which is not a directly experimentally realizable (or observable) quantity.

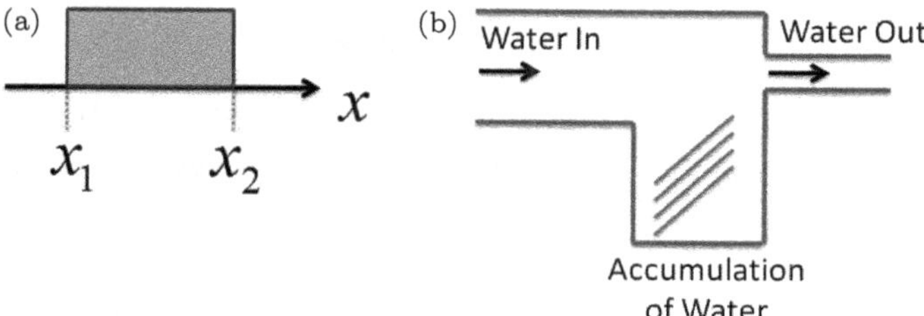

Figure 2.6. (a) A localized space in one dimension between x_1 and x_2 is selected to calculate the time-dependent change of the local probability. (b) The hydrodynamic flow of liquid is schematically shown to demonstrate that liquid starts to accumulate if the current-in is higher than the current-out.

To further understand the meaning of the equation of continuity, let us consider a localized space in one dimension between x_1 and x_2, as shown in Figure 2.6(a). The total probability of finding the particle in the $[x_1, x_2]$ space is given by

$$\int_{x_1}^{x_2} |\psi(x,t)|^2 dx = P_{x_1 x_2}(t)$$

According to the equation of continuity, the time-dependent change of this local probability can be expressed as

$$\begin{aligned}\frac{dP_{x_1 x_2}}{dt} &= \int_{x_1}^{x_2} \frac{\partial |\psi(x,t)|^2}{\partial t} dx = -\int_{x_1}^{x_2} \frac{\partial J(x,t)}{\partial x} dx = -[J(x,t)]_{x_1}^{x_2} \\ &= J(x_1,t) - J(x_2,t)\end{aligned}$$

Therefore, the time-dependent change in the local probability of finding the particle in the $[x_1, x_2]$ space can be expressed by the difference in probability currents between the two points. If $J(x_1,t) > J(x_2,t)$, then $\frac{dP_{x_x x_2}}{dt}$ is positive, i.e., the probability of finding the particle will increase in the $[x_1, x_2]$ space as time progresses because more probability is flowing into the $[x_1, x_2]$ space than that is flowing out of the $[x_1, x_2]$ space. This picture is equivalent to the hydrodynamic flow of fluid, as demonstrated in Figure 2.6(b), in which fluid starts accumulating if current-in is higher than current-out. This is why equation (2.21) is called the hydrodynamic formulation of the TDSE.

Guiding Questions

2.6. Show that (a) $\frac{d}{dt}\left[\int_{-\infty}^{+\infty}\psi^*\psi dx\right] = 0$ and (b) $\frac{d}{dt}\left[\int_{-\infty}^{+\infty}\psi_1^*\psi_2 dx\right] = 0$.

2.7. Find the probability current J for the free-particle wavefunction

$$\psi(x,t) = A\,e^{-i(\frac{Et}{\hbar}-kx)}$$

This wavefunction for a free particle will be introduced in Chapter 3. Which direction does the probability current follow (along the increasing x-direction or along the decreasing x-direction)?

2.8. In Chapter 3, we will realize that the free-particle wavefunction of the form $\psi(x,t) = A\,e^{-i(\frac{Et}{\hbar}-kx)}$ is not a normalizable wavefunction. In other words, one cannot normalize the wavefunction using the normalization condition $\int_{-\infty}^{+\infty}|\psi(x,t)|^2dx = 1$. Test it yourself. Further details of the criteria for normalizable wavefunctions will be given in Chapter 4. The wavefunction $\psi(x,t) = A\,e^{-i(\frac{Et}{\hbar}-kx)}$ can be normalized to unit probability current. Deduce the normalized free-particle wavefunction if it is normalized to unit probability current.

2.6 Bohmian Mechanics: Quantum Trajectories

In the above section, we have gone over a hydrodynamic formulation of the TDSE. We have learned that the temporal change in the probability density of a particle at a point in space depends on the negative gradient of the probability current at that point in space. This renders the local conservation of the probability. We now examine another formulation of the TDSE, often regarded as Bohmian mechanics.[8] In this formulation, we will rewrite the TDSE as two equations: one for the probability density and another for a phase function whose spatial derivative gives the local (Bohmian) velocity.

In Chapter 1, we have understood that the variable separation method (which is developed under the assumption that $\hat{H}$ does not depend on time) renders the following form as a solution to the TDSE:

$$\psi_n(x,t) = \psi_n(x)e^{-i\frac{E_n t}{\hbar}} \tag{2.22}$$

where $\psi_n(x)$ represents the nth stationary state wavefunction of the particle and $e^{-i\frac{E_n t}{\hbar}}$ is the corresponding time-dependent phase factor. It is now obvious that the probability density of the particle whose wavefunction is represented by equation (2.22) does not change as a function of time. Only a superposition state of the particle features the time-dependent change of the probability density. By now, we have already developed a clear consensus that a superposition state can be expressed as

$$\psi(x,t) = \sum_{n=0}^{\infty} \overbrace{a_n \; \psi_n(x)}^{\text{call it amplitude}} \; \overbrace{e^{-i\frac{E_n t}{\hbar}}}^{\text{Temporal Phase}} \tag{2.23}$$

For the following discussion, instead of taking a mathematically complicated form for the superposition state, as given in equation (2.23), one can propose a convenient trial wavefunction for the superposition state as follows:

$$\psi(x,t) = A(x,t)e^{\frac{i}{\hbar}S(x,t)} \tag{2.24}$$

Here, $A(x,t)$ and $S(x,t)$ are, respectively, the unknown **amplitude and phase functions. They are assumed to be real functions.** Note that the probability density of the quantum particle represented by the wavefunction $\psi(x,t) = A(x,t)e^{\frac{i}{\hbar}S(x,t)}$ depends on time: $\rho(x,t) = |A(x,t)|^2$. As a consequence, arguably, equation (2.24) represents a superposition state of the particle.

The motivation behind considering the form given in equation (2.24) as a trial wavefunction for a superposition state comes from its close resemblance to the form of the stationary state wavefunction (as given in equation (2.22)). The major difference between the expressions given in equations (2.22) and (2.24), however, is that a stationary state wavefunction has only a position-space-dependent amplitude, $\psi(x)$, and only a time-dependent (complex) phase factor, $e^{-i\frac{Et}{\hbar}}$, whereas, a superposition state wavefunction (as expressed by equation (2.24)) exhibits both position- and time-dependent amplitude and phase. Arguably, many other trial forms for the superposition state of a quantum particle, such as $\psi(x,t) = A(x,t)e^{-iS(x,t)}$ and $\psi(x,t) = A(x,t)e^{-\frac{i}{\hbar}S(x,t)}$, could also be considered; however, we will only consider the form given in equation (2.24) for the development of Bohmian mechanics.

Insert the trial wavefunction given in equation (2.24) into the TDSE to unravel important information hidden inside the $S(x,t)$ function. Calculate the time and space derivatives separately:

$$\frac{\partial \psi}{\partial t} = \frac{\partial A}{\partial t} e^{\frac{i}{\hbar}S(x,t)} + A\frac{i}{\hbar}\frac{\partial S}{\partial t} e^{\frac{i}{\hbar}S(x,t)} \tag{2.25}$$

and $\dfrac{\partial \psi}{\partial x} = \dfrac{\partial A}{\partial x} e^{\frac{i}{\hbar}S(x,t)} + A\dfrac{i}{\hbar}\dfrac{\partial S}{\partial x} e^{\frac{i}{\hbar}S(x,t)}$

$$\text{or } \frac{\partial^2 \psi}{\partial x^2} = \frac{\partial^2 A}{\partial x^2} e^{\frac{i}{\hbar}S(x,t)} + 2\frac{\partial A}{\partial x}\frac{i}{\hbar}\frac{\partial S}{\partial x} e^{\frac{i}{\hbar}S(x,t)} + A\frac{i}{\hbar}\frac{\partial^2 S}{\partial x^2} e^{\frac{i}{\hbar}S(x,t)} - \frac{A}{\hbar^2}\left(\frac{\partial S}{\partial x}\right)^2 e^{\frac{i}{\hbar}S(x,t)} \tag{2.26}$$

Inserting equations (2.25) and (2.26) into the TDSE yields

$$\left[i\hbar\frac{\partial A}{\partial t} e^{\frac{i}{\hbar}S(x,t)} - A\frac{\partial S}{\partial t} e^{\frac{i}{\hbar}S(x,t)}\right] = \left[-\frac{\hbar^2}{2m}\frac{\partial^2 A}{\partial x^2} e^{\frac{i}{\hbar}S(x,t)} - \frac{i\hbar}{m}\frac{\partial A}{\partial x}\frac{\partial S}{\partial x} e^{\frac{i}{\hbar}S(x,t)} - \frac{i\hbar}{2m}A\frac{\partial^2 S}{\partial x^2} e^{\frac{i}{\hbar}S(x,t)} + \frac{A}{2m}\left(\frac{\partial S}{\partial x}\right)^2 e^{\frac{i}{\hbar}S(x,t)}\right] + VAe^{\frac{i}{\hbar}S(x,t)}$$

Now, the above equation can be resolved into its real and imaginary parts by equating separately the real and the imaginary parts on both sides.

For the real part:

$$-A\frac{\partial S}{\partial t} = -\frac{\hbar^2}{2m}\frac{\partial^2 A}{\partial x^2} + \frac{A}{2m}\left(\frac{\partial S}{\partial x}\right)^2 + VA$$

$$\text{or} \quad \frac{\partial S}{\partial t} + \frac{1}{2m}\left(\frac{\partial S}{\partial x}\right)^2 + V = \frac{\hbar^2}{2mA}\frac{\partial^2 A}{\partial x^2} \tag{2.27}$$

For the imaginary part:

$$\hbar\frac{\partial A}{\partial t}=-\frac{\hbar}{m}\frac{\partial A}{\partial x}\frac{\partial S}{\partial x}-\frac{\hbar}{2m}A\frac{\partial^2 S}{\partial x^2}$$

$$\text{or}\quad \frac{\partial A}{\partial t}+\frac{1}{m}\frac{\partial A}{\partial x}\frac{\partial S}{\partial x}+\frac{A}{2m}\frac{\partial^2 S}{\partial x^2}=0 \tag{2.28}$$

For the reason to be revealed later, multiply equation (2.28) by $2A$ and obtain

$$2A\frac{\partial A}{\partial t}+\frac{2A}{m}\frac{\partial A}{\partial x}\frac{\partial S}{\partial x}+\frac{A^2}{m}\frac{\partial^2 S}{\partial x^2}=0 \tag{2.29}$$

Both the real and imaginary parts carry useful perspectives which construct the Bohmian mechanics; however, for brevity, we focus only on the imaginary part (equation (2.29)).

The probability density for a particle represented by the $\psi(x,t)=A(x,t)e^{\frac{i}{\hbar}S(x,t)}$ trial wavefunction is given by $\rho(x,t)=A^2$, as A is assumed to be a real function. This infers, as also noted earlier, that the amplitude function used in the trial wavefunction is related to the probability density of the particle. To reach a useful conclusion, let us first get the time and space derivatives of the probability density:

$$\frac{\partial \rho}{\partial t}=2A\frac{\partial A}{\partial t} \tag{2.30}$$

and

$$\frac{\partial \rho}{\partial x}=2A\frac{\partial A}{\partial x} \tag{2.31}$$

Inserting equations (2.30) and (2.31) into equation (2.29), we get

$$\frac{\partial \rho}{\partial t}+\frac{1}{m}\frac{\partial \rho}{\partial x}\frac{\partial S}{\partial x}+\frac{\rho}{m}\frac{\partial^2 S}{\partial x^2}=0$$

$$\text{or}\quad \frac{\partial \rho}{\partial t}+\frac{\partial}{\partial x}\left[\frac{\rho}{m}\frac{\partial S}{\partial x}\right]=0$$

$$\text{or}\quad \frac{\partial \rho}{\partial t}=-\frac{\partial}{\partial x}\left[\frac{\rho}{m}\frac{\partial S}{\partial x}\right] \tag{2.32}$$

We may now compare this equation with the hydrodynamic formulation of the TDSE obtained earlier in terms of the probability current: $\frac{\partial \rho}{\partial t} = -\frac{\partial}{\partial x} J$, in which J represents the probability current. This comparison renders another definition of the probability current in Bohmian mechanics:

$$J(x,t) = \left[\frac{\rho(x,t)}{m} \frac{\partial S(x,t)}{\partial x} \right] \tag{2.33}$$

Here, ρ is a dimensionless function because it features probability. On the other hand, J represents the local velocity v (in one dimension, fluid current is represented by its velocity). As a result, one can easily identify that $\frac{\partial S(x,t)}{\partial x}$ (which is the space derivative of the phase function) represents the local momentum of the particle, mv. In other words, the local velocity of the particle is given by

$$v = \frac{1}{m} \left(\frac{\partial S}{\partial x} \right) \tag{2.34}$$

The above mathematical exercise points to an important fact that if the state of a moving quantum particle is represented by the form $\psi(x,t) = A(x,t) e^{\frac{i}{\hbar} S(x,t)}$, the amplitude function (A) provides the probability density of the particle at any given time and space, and the phase function (S) gives the local velocity v of the particle, $v = \frac{1}{m}\left(\frac{\partial S}{\partial x}\right)$. Thus, Bohmian mechanics provides a **quantum trajectory** for a moving particle. But what is the meaning of the local velocity of the particle in Bohmian mechanics? Before we answer this question, let us revisit a pictorial representation of the standard quantum dynamics of a particle.

Figure 2.7(a) schematically depicts the standard interpretation of quantum dynamics. The probability density distribution function, $|\psi(x,t)|^2$, of the quantum particle changes as time progresses after the time-evolution process is initiated at $t = 0$. As a result, the expectation value of the position of the quantum particle evolves as a function of time. The expectation value of position at a particular time represents the mean position of many repeated measurements of position at that particular time.

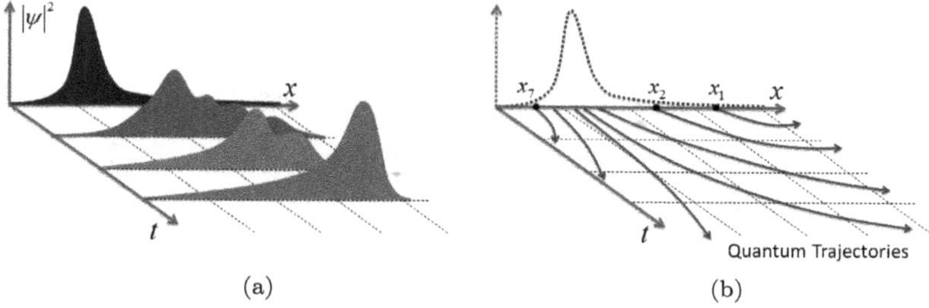

(a) (b)

Figure 2.7. (a) Standard picture of the quantum dynamics of a particle: Probability density of the particle changes as function of time. (b) The Bohmian picture of the quantum dynamics of a particle: If a quantum particle sets off from a possible initial position with local velocity $[\frac{1}{m}(\frac{\partial S(x,t)}{\partial x})]$, it follows a quantum trajectory. The initial position can be defined based on the initial probability density distribution of the particle.

On the other hand, Bohmian mechanics represents the quantum dynamics using **quantum trajectories**. The initial (at $t = 0$) probability density distribution provides information about the possible initial positions of the particle. As an example, seven representative initial positions, namely $x_1, x_2, \ldots, x_7$, are depicted in Figure 2.7(b). Bohmian mechanics describes how the particle would move if it were to set off from each possible initial position with Bohmian velocity or local velocity expressed by $[\frac{1}{m}(\frac{\partial S(x,t)}{\partial x})]$. Thus, one may construct a trajectory for the moving particle associated with each possible initial position. These trajectories are called quantum trajectories.

Guiding Question

2.9. Attempt this exercise after going through Chapter 3 where, we will realize that a free-particle (e.g., Gaussian) wavepacket can be expressed using the following functional form:

$$\psi(x,t) = \left(\frac{2a}{\pi}\right)^{1/4} e^{\left[-\alpha_t(x-x_t)^2+\frac{i}{\hbar}P_0(x-x_t)+\frac{i}{\hbar}\gamma_t\right]}$$

$$= \left(\frac{2a}{\pi}\right)^{1/4} e^{-\alpha_t(x-x_t)^2} e^{\frac{i}{\hbar}[P_0(x-x_t)+\gamma_t]}$$

(Continued)

(Continued)

where $a = \frac{2\ln 2}{\Delta x_0^2}$, Δx_0 is the initial FWHM of the Gaussian wavepacket, and

$$\alpha_t = \frac{a}{\left(1+\frac{2i\hbar t a}{m}\right)}, x_t = \frac{\hbar k_0 t}{m}, \gamma_t = \frac{P_0^2}{2m}t + \frac{i\hbar}{2}\ln\left(1+\frac{2i\hbar t a}{m}\right) \text{ and } P_0 = k_0\hbar$$

The above form of the free-particle Gaussian wavepacket cannot be directly compared with the trial wavefunction used to derive the Bohmian mechanics: $\psi(x,t) = A(x,t)e^{\frac{i}{\hbar}S(x,t)}$, where both $A(x,t)$ and $S(x,t)$ are assumed to be real functions of space and time.
As α_t and γ_t are both complex functions, we have to further modify the equation for the Gaussian wavepacket to make both the phase and amplitude functions real.

(a) Show (using mathematical derivation) that the free-particle Gaussian wavepacket can be rewritten in the following form to make the phase function $S(x,t)$ to be real:

$$\psi(x,t) = \left(\frac{2a}{\pi}\right)^{1/4}\left(1+\frac{4\hbar^2t^2a^2}{m^2}\right)^{-1/4} e^{\frac{-a(x-x_t)^2}{\left(1+\frac{4\hbar^2t^2a^2}{m^2}\right)}}$$
$$\times e^{\frac{i}{\hbar}\left[\frac{2\hbar^2 t a^2}{m\left(1+\frac{4\hbar^2t^2a^2}{m^2}\right)}(x-x_t)^2 + k_0\hbar(x-x_t) - \frac{\hbar}{2}\tan^{-1}\left(\frac{2\hbar t a}{m}\right)\right]}$$

(b) If the real $S(x,t)$ function of the free particle is expressed as

$$S(x,t) = \frac{2\hbar^2 t a^2}{m\left(1+\frac{4\hbar^2t^2a^2}{m^2}\right)}(x-x_t)^2 + k_0\hbar(x-x_t) - \frac{\hbar}{2}\tan^{-1}\left(\frac{2\hbar t a}{m}\right)$$

find the Bohmian velocity of the particle and, finally, the Bohmian trajectory of the particle.
If $\frac{4\hbar^2a^2}{m^2} = b$, $\frac{k_0\hbar}{m} = u$, $(x - ut) = y$, and $y_0 = x_0$ (initial position at $t = 0$) are assumed, the Bohmian trajectory (quantum trajectory) of the particle adopts the following form:

$$x = ut + x_0\sqrt{1+bt^2}$$

The above equation represents the Bohmian trajectory of a free particle. To specifically denote the trajectory for the nth particle belonging to the initial Gaussian distribution, we may rewrite the equation as

$$x_n = ut + x_{0,n}\sqrt{1+bt^2}$$

where u is the initial group velocity of the wavepacket and $x_{0,n}$ is the initial position of the nth particle belonging to the initial Gaussian distribution. Immeasurable thanks will be always due to Prof. Alok Kumar Pan (IIT Hyderabad) for illuminating discussion on this exercise.

Key Points to Remember

- The standard interpretation of quantum mechanics represents quantum dynamics as a time-evolving probability density.
- Three theorems, namely Ehrenfest theorem, the hydrodynamic formulation of the TDSE, and Bohmian mechanics, provide the sense of classical trajectory in quantum dynamics.
- Ehrenfest theorem shows that the expectation value of position (which is nothing but the average of the probability density distribution of position) follows classical mechanics: $m\frac{d\langle\hat{x}\rangle}{dt} = \langle\hat{P}_x\rangle$.
- The hydrodynamic formulation, which leads to the equation of continuity, shows that the time-dependent change in the probability density at a particular time and space depends on the negative gradient of probability current at that time and space: $\frac{\partial\rho(x,t)}{\partial t} = -\frac{\partial}{\partial x}J(x,t)$. This picture is equivalent to the hydrodynamic flow of a fluid.
- Bohmian mechanics, on the other hand, gives quantum trajectories. They describe how a quantum particle would move with Bohmian velocity $\left[\frac{1}{m}\left(\frac{\partial S(x,t)}{\partial x}\right)\right]$ if it were to set off from an initial position, which can be predicted from the initial probability density distribution of the particle.

Notes, References, and Further Reading

1. Note that the present thought experiment is for simple illustration purpose only. First of all, the electron in a hydrogen atom should not be viewed as a tiny particle that orbits around the nucleus (which depicts the classical Bohr atomic model); rather, its probability density has a distribution in position space. A quantum mechanical solution to the hydrogen atom problem (in which the electron experiences a coulombic potential) unravels the fact that the ground state of the electron exhibits a $1s$ probability density distribution. Excited states, such as $2s$ and $2p$, have different probability density distributions. A free electron which does not experience any potential, on the other hand, features a probability density distribution different from that of the electron in a hydrogen atom.
2. The Born interpretation provides a statistical interpretation of the wavefunction (and quantum mechanics). Readers may note that the basic philosophy behind the interpretation of wavefunction and quantum mechanics is still a controversial subject. One illuminating book which gives a lovely historical perspective on the interpretation of quantum mechanics is M. Jammer, *The Philosophy of Quantum Mechanics.* John Wiley & Sons, Inc., USA (1974). See also standard textbooks (a) I. N. Levine, Chapter 7: Theorems of quantum mechanics. *Quantum Chemistry,* 7th edn. Pearson Education, Inc., New Jersey (2014); (b) D. J. Griffiths, Chapter 1: The wave function. *Introduction to Quantum Mechanics*, 2nd edn. Pearson Education, Inc., New Jersey (2005); (c) P. Atkins and R. Friednman, Chapter 1: The foundations of quantum mechanics. *Molecular Quantum Mechanics*, 5th edn. Oxford University Press, New York (2011); and (d) D. A. McQuarrie, Chapter 4: The postulates and

general principles of quantum mechanics. *Quantum Chemistry*, 2nd edn. University Science Books, California (2008).

3. Quantum mechanical postulates are statements that are assumed to be true and to be starting points for further reasoning and arguments. For example, one very commonly used postulate of quantum mechanics is that the observables of classical mechanics are represented in quantum mechanics by operators. The ultimate test of the fidelity of the postulates must be obtained by comparing the results with experimental data. Depending on the context of the presentation, textbook writers make use of a certain set of postulates which are sufficient for the development of context. This is why a number of postulates often differ from textbook to textbook. However, there is a fairly elementary set of postulates that suffices for the presentation of interest in chemistry. Here, in this textbook, a set of postulates is presented, keeping the time-dependent viewpoint of quantum mechanics in mind.
4. To realize the classical–quantum correspondence, readers are further referred to David J. Tannor, Chapter 4: Correspondence between classical and quantum dynamics. *Introduction to Quantum Mechanics: A Time-Dependent Perspective*. University Science Books, California (2007).
5. C. Cohen-Tannoudji, B. Diu and F. Laloe, Chapter III: The postulates of quantum mechanics. *Quantum Mechanics*, Vol. 1 (English Translation). Hermann and John Wiley and Sons, New York (1977).
6. Ehrenfest's original work was not published in English (P. Ehrenfest, *Bemerkung über die angenäherte Gültigkeit der klassischen Mechanik innerhalb der Quantenmechanik. Z. Phys.* 45, 455 (1927)). Derivation is given in David J. Tannor, Chapter 4: Correspondence between classical and quantum dynamics. *Introduction to Quantum Mechanics: A Time-Dependent Perspective*. University Science Books, California (2007).
7. In fluid dynamics, volumetric flow rate (or volume flow rate or volume velocity) is the volume of fluid passing per unit time. The SI unit of volumetric flow rate is $\mathrm{m}^3\,\mathrm{s}^{-1}(Av$ volume per second).

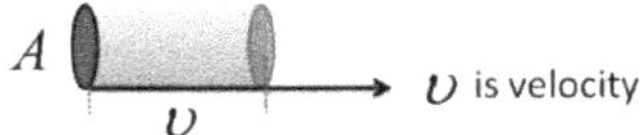

In one dimension, it is simply velocity. So, $J(x, t)$ simply represents velocity.

8. For further details, see P. R. Holland, *The Quantum Theory of Motion*. Cambridge, New York (1993). (b) D. Bohm, A suggested interpretation of the quantum theory in terms of "hidden" variables. I. *Phys. Rev.* 85, 166 (1952); (c) D. Bohm, A suggested interpretation of the quantum theory in terms of "hidden" variables. II. *Phys. Rev.* 85, 180 (1952).

Exercises

2.1. The time-independent wavefunction $\psi(x)$ can always be taken to be real. Even if $\psi(x)$ is complex, one can express $\phi(x) = \psi(x) + \psi^*(x)$ to obtain a real solution, where $\psi(x)$ and $\psi^*(x)$ both satisfy the TISE. Prove this argument.

2.2. Prove that if $\psi(x,t)$ is a solution to the TDSE, $A\psi(x,t)$ is also a solution to the TDSE.
(*Important note:* $\psi(x,t)$ is necessarily complex due to the presence of the phase factor; however, $\psi(x)$ and E are always real).

2.3. Determine whether the following wavefunctions at $t = 0$ are physically acceptable solutions to the TDSE:
(a) $\psi(x,0) = \sin^{-1}(x)$ over the interval $(-1,+1)$, (b) $\psi(x,0) = Ae^{-|x|}$ over the interval $(-\infty,+\infty)$.

2.4. In quantum dynamics, how is a quantum trajectory different from the trajectory created by the expectation value of position?

2.5. What is the difference between the TDSE and the "equation of continuity"?

2.6. Show that the probability current vanishes if the wavefunction is real.

2.7. For a particle in a one-dimensional box of length L, the time-dependent wavefunction of the nth state of the particle is given by $\psi_n(x,t) = \sqrt{\frac{2}{L}}\sin(\frac{\pi x n}{L})e^{-\frac{iE_n t}{\hbar}}$. Show that the expectation value of position is not time-dependent.

2.8. For the ground, $\psi_0(x)$, and first excited, $\psi_1(x)$, vibrational states of the quantum harmonic oscillator, find the expectation value of position. The respective wavefunctions are given by $\psi_0(x) = (\frac{\alpha}{\pi})^{1/4}\exp(\frac{-\alpha x^2}{2})$ and $\psi_1(x) = (\frac{4\alpha^3}{\pi})^{1/4}x\exp(\frac{-\alpha x^2}{2})$, where $\alpha = \sqrt{\frac{k\mu}{\hbar^2}}$.

PythonChapter B

Grid Representation of Wavefunction, Computing Norm, and Finding Expectation Value

Highlights: *Position Grid Representation of Wavefunction, Probability Density, Numerically Normalize a Wavefunction and Compute Expectation Value of Position.*

B.1 Introduction

The wavefunction and the operator are the two key constituents of quantum mechanics. The wavefunction represents the state of a system, and when the operator acts on the wavefunction, we get the average or expectation value of the observables. Therefore, the first step toward obtaining a numerical solution to any quantum mechanical problem is to present the wavefunction and the operator in an appropriate computer programming data structure.

Almost all currently available numerical methods for solving quantum mechanical problems make use of the grid representation of the wavefunction and the operator. A wavefunction, by its nature, is continuous (stated by a postulate of quantum mechanics) in position space; however, in the position grid representation, a continuous wavefunction is expressed on a set of position grid points. The process of discretization of the wavefunction on the position grid is described in the following.

First, a certain range of the x-coordinate (say $x_{\text{min}} \leq x \leq x_{\text{max}}$) is divided by a suitably small step size Δx to produce a uniform discrete x-grid, as shown in Figure B.1. This means that we construct

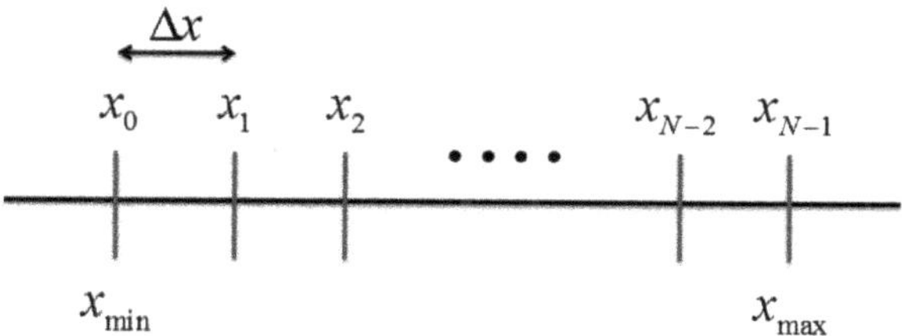

Figure B.1. Discretized x-grid or x-axis or x-coordinate is shown.

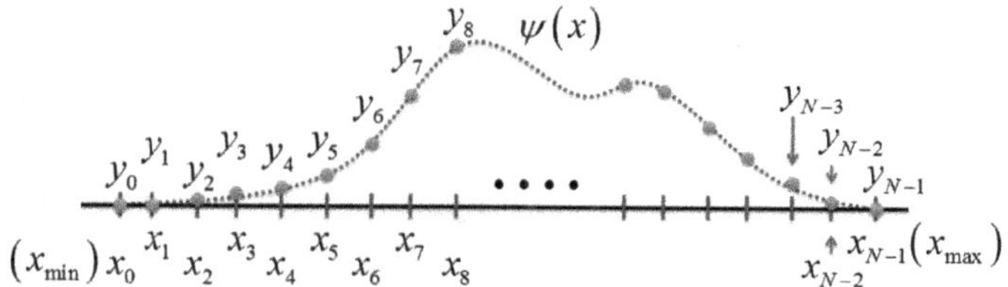

Figure B.2. Discretized wavefunction $\psi(x)$ is represented on the x-grid.

N discrete values of x within the $x_{\min} \le x \le x_{\max}$ range, each separated by the step size Δx. Therefore, we have

$$x_{N-1} = x_0 + (N-1)\Delta x$$

and by rearranging, we get

$$N = 1 + \frac{x_{N-1} - x_0}{\Delta x} \tag{B.1}$$

Here, x_0 is $x_{\min}$, and x_{N-1} is $x_{\max}$.

Finally, a continuous wavefunction $\psi(x)$ is represented on the x-grid, as shown in Figure B.2. As a result, we get the discretized values of the wavefunction as follows:

$$\begin{aligned}\psi(x_0) &= y_0\\ \psi(x_1) &= y_1\\ \psi(x_2) &= y_2\\ &\vdots\\ \psi(x_{N-1}) &= y_{N-1}\end{aligned}$$

The above-discretized wavefunction is conveniently represented by a column matrix:

$$\psi(x) = \begin{pmatrix} y_0 \\ y_1 \\ y_2 \\ . \\ . \\ . \\ y_{N-1} \end{pmatrix}$$

> ➢ In linear algebra, a column matrix is called a vector consisting of a single column of certain number of elements.

Once a wavefunction is discretized on the x-grid, the following two important questions arise: (a) *How do we normalize a wavefunction in position grid representation?* and (b) *How do we determine the expectation value of position for a given discretized wavefunction?* These two questions are addressed systematically in this chapter.

B.2 Grid Representation of Wavefunction

B.2.1 *Creating the x-grid*

In PythonChapter A, we have already realized that the **arange(start,stop,step)** functionality of **scipy** returns a list of evenly spaced discrete values of x within the given range. To be technically correct, we have also briefly mentioned that the **arange** functionality returns an **array**. Both arrays and lists represent a collection of numbers; however, the primary difference between a **list** and an **array** is that Python's built-in functionality **list** cannot be used for linear algebra routines implemented in **scipy**, while an **array** created by the **arange** functionality of **scipy** can be used for linear algebra routines (the usefulness of linear algebra numerical routines will be evident soon).

In general, the data structure which is used to represent a mathematical matrix in computer programming is called an **array**. An array can be N-dimensional, but a two-dimensional array is called a matrix, and a one-dimensional array is called a vector. Python does not have intrinsic (built-in) functionality to deal with arrays (and different array operations or, more broadly, linear algebra routines). For that, one needs to import specific functionalities from **scipy**.

In PythonChapter A, we have used a one-dimensional array of discrete x-values to create an x–y plot using the **arange** functionality of **scipy**. Here, we use the same functionality to create the x-grid.

```
#Importing the libraries
from scipy import arange
#Creating the x-grid
xmin=-100
xmax=100.1
dx=0.1
x=arange(xmin,xmax,dx)
#Printing the length of the array and the array itself
N=len(x)
print(N)
print(x)
```

```
2001
[-100. -99.9 -99.8 ... 99.8 99.9 100. ]
```

The above program creates an x-grid in which $x_{\text{min}} = -100$, $x_{\text{max}} = +100$, and step $\Delta x = 0.1$. Note that Python's built-in functionality **len()** returns the length (number of items) of an array object. Therefore, in the above example, **len(x)** returns the number of elements present in the x-array. After executing the above program, we find that the x-grid contains 2,001 elements. One may analytically calculate the same using equation (B.1) for the given range and step:

$$N = 1 + \frac{x_{\text{max}} - x_{\text{min}}}{\Delta x} = 1 + \frac{100 - (-100)}{0.1} = 2001$$

Note that $x_{\text{max}} = 100.1$ (instead of 100) is used because, as mentioned in PythonChapter A, the **arange(start,stop,step)** functionality of **scipy** does not include **stop** in the sequence.

B.2.2 *Discretizing the wavefunction*

As an illustrative example, let us assume that a Gaussian profile $\psi(x) = e^{-x^2}$ represents the wavefunction of a quantum system. This wavefunction can be easily represented on the above-created x-grid using the following command lines.

```
#Importing the libraries
from scipy import arange,exp
from matplotlib.pyplot import plot,show
#Creating the x-grid
xmin=-100
xmax=100.1
dx=0.1
x=arange(xmin,xmax,dx)
#Discretizing the wavefunction
psi=exp(-(x)**2)
#Plotting the wavefunction
plot(x,psi)
show()
```

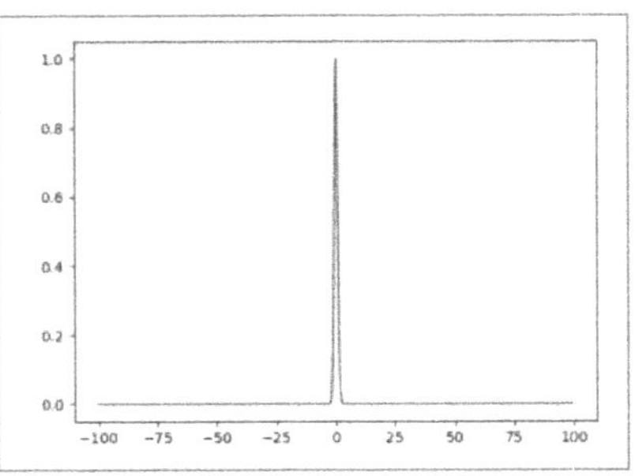

➢ A quick plot of the function gives us an idea of nature of the function.

As we have noticed in PythonChapter A for the x–y graph plot, when an entire **array** prepared by the **arange** functionality of **scipy** is used as a variable for a mathematical function, we produce another **array** containing function values corresponding to each element of the array variables (in this case, it is the x-grid).

B.3 Normalizing Discretized Wavefunction

The Gaussian form $\psi(x) = e^{-x^2}$ does not represent a normalized wavefunction because

$$\int_{-\infty}^{+\infty} |\psi(x)|^2 dx = \int_{-\infty}^{+\infty} e^{-2x^2} dx = \sqrt{\frac{\pi}{2}} \neq 1$$

However, it can be normalized by using a simple mathematical trick. Let us first define a physical quantity, called the "norm" of a wavefunction $\psi(x)$:

$$\text{norm} = \left(\int_{-\infty}^{+\infty} |\psi(x)|^2 dx \right)^{\frac{1}{2}}$$

Further details of the above definition of the norm of a wavefunction are given in Chapter 4. Here, we make use of this definition directly. A normalized wavefunction is obtained simply by dividing the wavefunction by its norm:

$$\psi_{\text{normalized}}(x) = \frac{\psi(x)}{\left(\int_{-\infty}^{+\infty} |\psi(x)|^2 dx\right)^{\frac{1}{2}}} \tag{B.2}$$

Analytically, one can find the "norm" of the given Gaussian wavefunction as follows:

$$\text{Norm} = \left(\int_{-\infty}^{+\infty} |\psi(x)|^2 dx\right)^{\frac{1}{2}} = \left(\int_{-\infty}^{+\infty} e^{-2x^2} dx\right)^{\frac{1}{2}} = \left(\frac{\pi}{2}\right)^{\frac{1}{4}}$$

Consequently, the analytical form of the normalized Gaussian wavefunction is given by

$$\psi_{\text{normalized}}(x) = \left(\frac{2}{\pi}\right)^{\frac{1}{4}} e^{-x^2} \tag{B.3}$$

Our task here, however, is to normalize the discretized Gaussian wavefunction numerically. The first step in this task is to compute the integral given in equation (B.2) numerically.

There are several ways one can perform numerical integration using Python; however, here, we make use of a highly optimized integration package of **scipy**. The **scipy.integrate** submodule of **scipy** provides a number of integration techniques, including **Simpson's rule**.[1] The **simps(Y,X)** functionality of the **scipy.integrate** submodule integrates the Y-array numerically, while the X-array provides the points at which Y is sampled to employ the **Simpson's rule** of numerical integration.[2]

In our present problem, the square of the absolute value of the wavefunction, presented by $|\psi(x)|^2$, which is called the probability density function, features the Y-array, which is to be integrated. The probability density function $|\psi(x)|^2$ is sampled at the points present on the x-grid. Furthermore, the integration limit is automatically set by the x-grid limit, which is $[-100,+100]$ for the present problem.

Python's built-in functionality **abs()** returns the absolute value of a given number. If the number is a complex number $(a+ib)$, **abs()** returns its magnitude, $\sqrt{a^2+b^2}$. If the number is a real number, **abs()** removes the negative sign of the number. As $\psi(x)$ on the

x-grid is represented by an array, abs($\psi(x)$) returns an array with the absolute value of each element of the $\psi(x)$ array.

Thus, with the above realization, one can now use Simpson's rule to calculate the norm of the wavefunction $\psi(x) = e^{-x^2}$.

```
#Importing the libraries
from scipy import arange,exp,sqrt
from scipy.integrate import simps
#Creating the x-grid
xmin=-100
xmax=100.1
dx=0.1
x=arange(xmin,xmax,dx)
#Discretizing the wavefunction
psi=exp(-(x)**2)
#Calculating the norm
prob_density_psi=abs(psi)**2
norm=sqrt(simps(prob_density_psi,x))
print(norm)
```

```
1.1195151349202477
```

Finally, one can normalize the wavefunction by dividing the discretized wavefunction by its norm.

```
#Importing the libraries
from scipy import arange,exp,sqrt
from scipy.integrate import simps
#Creating the x-grid
xmin=-100
xmax=100.1
dx=0.1
x=arange(xmin,xmax,dx)
#Discretizing the wavefunction
psi=exp(-(x)**2)
#Calculating the norm
prob_density_psi=abs(psi)**2
norm=sqrt(simps(prob_density_psi,x))
#Normalize the wavefunction
psiNorm=psi/norm
#Checking whether we have really normalized the wavefunction
prob_density_psiNorm=abs(psiNorm)**2
norm_1=sqrt(simps(prob_density_psiNorm,x))
print(norm_1)
```

```
0.9999999999999999 (which can be considered 1.0)
```

The above exercise proves that we have normalized the wavefunction numerically.

Self-Study

e^{-x^2} and $e^{-(x-a)^2}$ are two Gaussian profiles centered at $x = 0$ and $x = a$, respectively. Write a Python program to find out the norms of both profiles.

B.4 Determining Expectation Value of Position Using Discretized Wavefunction

For a given normalized wavefunction, the expectation value of position is given by

$$\langle \hat{x} \rangle = \int_{-\infty}^{+\infty} \psi^*(x)\hat{x}\psi(x)dx = \int_{-\infty}^{+\infty} \psi^*(x)x\psi(x)dx$$

Since x is just a multiplication factor, one can change its position in the integrand without any constraints. Therefore, one may write

$$\langle \hat{x} \rangle = \int_{-\infty}^{+\infty} \psi^*(x)\psi(x)xdx = \int_{-\infty}^{+\infty} |\psi(x)|^2 xdx \tag{B.4}$$

The above expression infers that the integrand is a product of two functions: probability density distribution function, $|\psi(x)|^2$ and position function, x. In the position grid representation, each discretized function is represented by its characteristic one-dimensional array (column matrix), and the product of the two functions is nothing but the element-wise multiplication of two arrays. In Chapter 4, we will learn that this product is nothing but the **scalar product of two arrays (which is different from the vector product of two arrays).** In this chapter, we directly make use of the terminology, scalar product, without delving into details of its construct:

$$|\psi(x)|^2 x = \begin{pmatrix} |\psi(x_0)|^2 x_0 \\ |\psi(x_1)|^2 x_1 \\ |\psi(x_2)|^2 x_2 \\ \vdots \\ |\psi(x_{N-1})|^2 x_{N-1} \end{pmatrix}$$

In Python programming, the **scalar product of two arrays** is obtained simply by **element-wise multiplication** using Python's built-in * (multiplication) operator.

The first step in evaluating the expectation value of position is to normalize the wavefunction. Then, we numerically evaluate the integral given in equation (B.4).

```
#Importing the libraries
from scipy import arange,exp,sqrt
from scipy.integrate import simps
#Creating the x-grid
xmin=-100
xmax=100.1
dx=0.1
x=arange(xmin,xmax,dx)
#Discretizing the wavefunction
psi=exp(-(x)**2)
#Calculating the norm
prob_density_psi=abs(psi)**2
norm=sqrt(simps(prob_density_psi,x))
#Normalize the wavefunction
psiNorm=psi/norm
#Finding expectation value of position
prob_density_psiNorm=abs(psiNorm)**2
integrand=(prob_density_psiNorm)*(x)
expectV=simps(integrand,x)
print(expectV)
```

```
-6.938893903907228e-18 (which can be considered 0.0)
```

The above exercise numerically proves that the expectation value of position associated with the Gaussian function e^{-x^2} is $\langle \hat{x} \rangle = 0$.

Self-Study

e^{-x^2} and $e^{-(x-a)^2}$ are two Gaussian profiles centered at $x = 0$ and $x = a$, respectively. Write a Python program to find the expectation value of position for both functions. Note the observation if one selects $a = 2$, $a = 5$, and $a = 10$. Compare the numerical results with analytical results (one can evaluate using equation (B.4) analytically).

Key Points to Remember

- To represent a one-dimensional wavefunction, first, an x-grid is prepared using the **arange(start,stop,step)** functionality of **scipy**, then the function value is evaluated at each grid point.
- When a list of numbers is used as a variable for a function, we get another list of the corresponding function's values. We make use of this realization to discretize a wavefunction.
- The **simps(Y,X)** functionality of the **scipy.integrate** submodule integrates the Y-array using the Simpson's rule of numerical integration.
- Python's built-in functionality **abs()** returns the absolute value of a number.

Notes, References, and Further Reading

1. Navigate the SciPy.Integrate documentation available at https://docs.scipy.org/doc/scipy/tutorial/integrate.html for updated details of the Python implementation of several integration techniques.
2. For further details of numerical integration methods using Python, see J. Izaac and J. Wang, Chapter 6: Numerical integration. *Computational Quantum Mechanics.* Springer, Switzerland (2018).

Exercises

B.1. The wavefunction of the nth state of a particle in a one-dimensional box of length L is given by $\psi_n(x) = \sqrt{\frac{2}{L}}\sin(\frac{\pi x n}{L})$. Numerically check whether this wavefunction is normalized. Plot the normalized wavefunction. Numerically find the expectation value of position for the first three low-lying states of the particle.

B.2. Find the expectation value of position numerically for the ground and first excited vibrational states of the quantum harmonic oscillator. The respective wavefunctions are given by $\psi_0(x) = (\frac{\alpha}{\pi})^{1/4}\exp(\frac{-\alpha x^2}{2})$ and $\psi_1(x) = (\frac{4\alpha^3}{\pi})^{1/4} x\exp(\frac{-\alpha x^2}{2})$, where $\alpha = \sqrt{\frac{k\mu}{\hbar^2}}$

B.3. Write a Python code to normalize a Gaussian function $e^{-(x-3)^2}$. Find the expectation value of position.

B.4. Normalize the wavefunction $e^{-i\theta}$ within the limit $0 \leq \theta \leq 2\pi$.

B.5. Write a Python program to find $\langle x^2 \rangle$ for the wavefunction $\sin^2(\frac{\pi x}{a})$ within the limit $0 \leq x \leq a$. Present your argument.

B.6. Compute the normalization constant for the wavefunction $\psi(x,t) = e^{-(x-2)^2} e^{i\frac{x}{2}}$. Present your argument.

B.7. Find the expectation value of position for the wavefunction $\psi(x) = \cos(2x + 3x^2)$. Present your argument.

B.8. Find the expectation value of position for the $1s$ orbital of the hydrogen atom after normalizing the respective function. Compare the value with the Bohr radius of the $1s$ orbital.

B.9. Plot a Lorentzian function, normalize the function, and find the expectation value of position.

B.10. Plot the third excited state of a particle in a one-dimensional box, and find the expectation value of position.

B.11. Plot the wavefunction for a Slater-type orbital, normalize it, and find the expectation value of position.

Chapter 3

Time-Dependent Quantum Mechanics of Translational Motion

Highlights: *Free Particle, Plane Wave Solution, Paradoxical Consequences of Plane Wave Solution, Concept of Wavepacket, Phase and Group Velocity of Wavepacket, Rigorous Treatment of Wavepacket, Properties and General Form of Gaussian Wavepacket, Wavepacket Dynamics under the Influence of a Linear Potential.*

3.1 Free Particle: Plane Wave Solution

One of the simplest forms of motion is perhaps a free particle traveling in one-dimensional space (translational motion). The potential experienced by the particle is zero everywhere. Classically, this particle moves at a constant velocity $\left(V_{\text{cl}} = \frac{P\,(\text{momentum})}{m\,(\text{mass})}\right)$, but in quantum mechanics, this problem needs to be described using the time-dependent Schrödinger equation (TDSE). The TDSE governing the dynamics of this system can be written as (a free particle experiences zero potential)

$$i\hbar\frac{\partial}{\partial t}\psi(x,t) = \left[-\frac{\hbar^2}{2m}\left(\frac{\partial^2}{\partial x^2}\right)\right]\psi(x,t) \tag{3.1}$$

Here, $\psi(x,t)$ is the wavefunction of the particle (representing the **matter wave or particle wave**). Since the Hamiltonian operator is time-independent, as we did before, we may separate the variables (space and time) and obtain a solution of the following form:

$$\psi(x,t) = \psi(x)\, e^{-\frac{iEt}{\hbar}} \tag{3.2}$$

where E represents the kinetic energy of the free particle.

Inserting equation (3.2) into equation (3.1), we obtain the time-independent Schrödinger equation (TISE):

$$\left[-\frac{\hbar^2}{2m}\left(\frac{d^2}{dx^2}\right)\right]\psi(x) = E\psi(x) \tag{3.3}$$

One can propose that a solution to this differential equation can be given by[1]

$$\psi(x) = A\, e^{+ikx} \quad \text{or} \quad \psi(x) = B\, e^{-ikx} \tag{3.4}$$

Here, A and B are constants (representing the amplitude) of the respective wavefunction. In addition, by inserting any one form given in equation (3.4) into equation (3.3), we get, $k = \sqrt{\frac{2mE}{\hbar^2}}$. Here, m and E are mass and kinetic energy of the free particle, respectively.

Further theoretical analysis reveals that the two wavefunctions given in equation (3.4) correspond to the motion of a free particle with the same magnitude of linear momentum but in the opposite direction (as depicted in Figures 3.1(a) and 3.1(b)). This can be proved easily by calculating the respective linear momentum (when

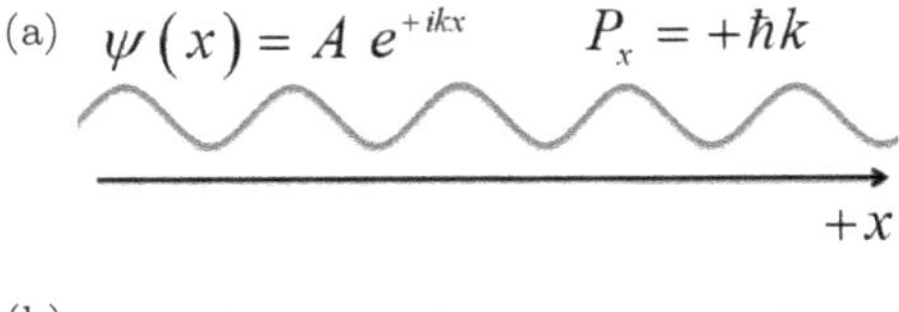

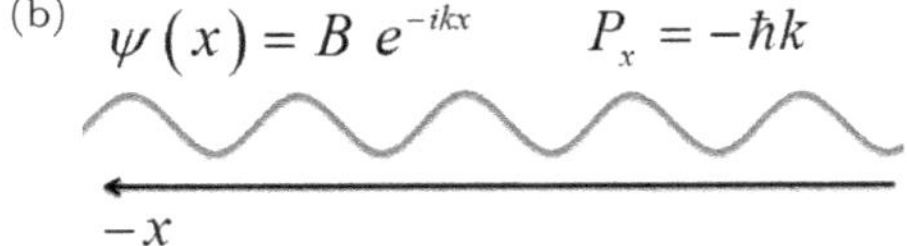

Figure 3.1. (a) The plane wave solution which represents a free particle moving toward the increasing x-direction (with linear momentum $P_x = +\hbar k$). (b) The plane wave solution which represents a free particle moving toward the decreasing x-direction (with linear momentum $P_x = -\hbar k$). Note that, for the reasons to be revealed soon, the plane wave solution does not truly represent a free particle, as it leads to several paradoxical consequences. See the text for more details.

the linear momentum operator acts on the wavefunction of a particle, we obtain the linear momentum of that particle):

$$\hat{P}_x(A\,e^{+ikx}) = -i\hbar\frac{d}{dx}(Ae^{+ikx}) = \hbar k(Ae^{+ikx})$$

$$\text{and}\quad \hat{P}_x(Ae^{-ikx}) = -i\hbar\frac{d}{dx}(Ae^{-ikx}) = -\hbar k(Ae^{-ikx})$$

For our next analysis, we consider a free particle moving only toward the increasing x-direction (with linear momentum $P_x = +\hbar k$). Note that a similar derivation for a free particle moving toward the decreasing x-direction leads to very similar conclusions.

Based on the above TISE solution, the time-dependent wavefunction of the free particle moving toward increasing x-direction can be expressed by

$$\psi(x,t) = Ae^{+ikx}\,e^{-\frac{iEt}{\hbar}} = Ae^{i\left(kx-\frac{Et}{\hbar}\right)} \quad \text{where} \quad E = \frac{\hbar^2k^2}{2m} \tag{3.5}$$

The above expression is nothing but an equation of a plane wave with the total phase

$$\varphi_{\text{total}} = \left(kx - \frac{Et}{\hbar}\right)$$

Therefore, the phase velocity of this free-particle wave (phase velocity is defined by the velocity of the constant phase front of a plane wave)[2] is given by

$$\frac{d\varphi_{\text{total}}}{dt} = \left(k\frac{dx}{dt} - \frac{E}{\hbar}\right) = 0 \quad \text{(for the constant phase front)}$$

$$\text{or}\quad kV_p = \frac{E}{\hbar} \text{ where phase velocity } V_P = \frac{dx}{dt}$$

or the phase velocity of the free-particle wave expressed by the wavefunction $\psi(x,t) = Ae^{i\left(kx-\frac{Et}{\hbar}\right)}$ is given by $V_P = \frac{E}{\hbar k} = \frac{\hbar k}{2m}$ $\left(\text{using the relationship } E = \frac{\hbar^2k^2}{2m}\right)$.

3.2 Paradoxical Consequences of the Plane Wave Solution

Multiple features of the above theoretical analysis appear to be very strange (paradoxical). They are discussed below:

(1) The plane wave solution of a free particle expressed by $\psi(x,t) = Ae^{i\left(kx-\frac{Et}{\hbar}\right)}$ exhibits the fact that the particle is moving along the $+x$-direction with a momentum of $+\hbar k$ and with the phase velocity $\frac{\hbar k}{2m}$; however, its probability density is independent of time and space: $\psi^*(x,t)\psi(x,t) = |A|^2$, where A is a positive real constant. *How can a time- and space-independent probability density represent a particle which is moving with a certain momentum and velocity?*

(2) We have already developed a clear consensus that only a normalizable wavefunction is an acceptable solution to the TDSE. The plane wave form of the free-particle wavefunction $\psi(x,t) = Ae^{i\left(kx-\frac{Et}{\hbar}\right)}$ is not normalizable within the limit $[-\infty, +\infty]$ in the x-space.[3] *How can a non-normalizable wavefunction represent a quantum particle which is freely moving?*

(3) The plane wave form of the free particle expressed by $\psi(x,t) = Ae^{i\left(kx-\frac{Et}{\hbar}\right)}$ gives us a velocity (more specifically, the phase velocity, which is defined by the velocity of the constant phase front) of the particle as $V_p = \frac{\hbar k}{2m}$. On the other hand, the velocity of the particle can also be calculated from its momentum (this velocity is called classical velocity because it is related to classical momentum): $mV_{\rm cl} = \hbar k$ or $V_{\rm cl} = \frac{\hbar k}{m}$. *How can the same particle have two different velocities?*

Conclusions: *Therefore, it is clear that the plane wave form of the wavefunction of a free particle does not provide a physically realizable interpretation of its quantum state and quantum dynamics. It cannot be accepted as a solution to the TDSE for a free particle. The above analysis teaches us that an analytical or numerical method may provide a solution to a quantum mechanical problem; however, our task in quantum mechanics is to further analyze the solution to judge whether the solution is physically acceptable.*

3.3 Free-Particle Wavepacket

To eliminate paradoxical consequences of the plane wave solution of a free particle, the general solution to the TDSE of the free particle should be presented as a linear combination of the plane wave solutions. In that case, the wavefunction which finally describes the free particle should be represented as a superposition of plane waves with slightly different (kinetic) energies. **Such a superposition is called a wavepacket, which is nothing but a localization of a matter wave in position space** (as schematically depicted in Figures 3.2(a) and 3.2(b), taking an example of the interference of two plane waves).

To develop a physically transparent model of the motion of a wavepacket, it suffices to consider a wavepacket composed of two time-dependent wavefunctions of the form $\psi(x,t) = Ae^{+ikx}e^{-\frac{iEt}{\hbar}}$ with slightly different energies $E_0 \pm \Delta E$ and so wavenumbers

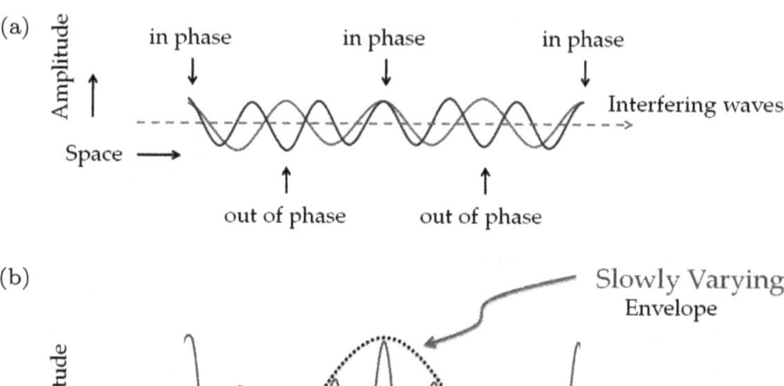

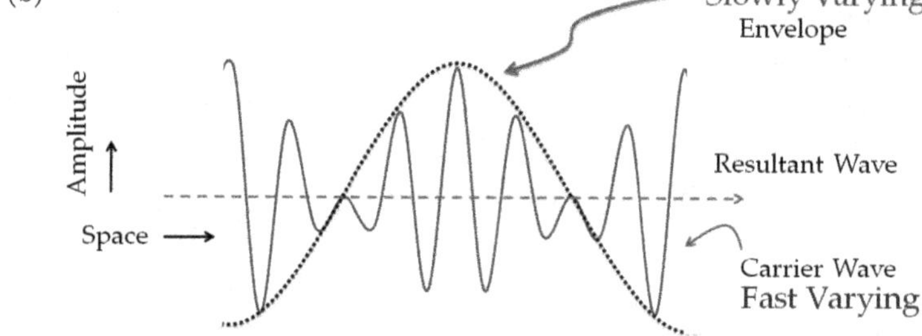

Figure 3.2. Spatial interference of two different plane waves (a), propagating along the same direction, results in a wavepacket (b). A wavepacket can be decomposed into a carrier wave (solid line) and an envelope (dotted line). The dashed horizontal line represents the direction along which the wavepacket propagates. Carefully compare this figure with Figure 1.8 in Chapter 1.

$(k_0 \pm \Delta k)$:

$$\psi_{k_1} = Ae^{+ik_1x}e^{-\frac{iE_1t}{\hbar}} \quad \text{and} \quad \psi_{k_2} = Ae^{+ik_2x}\, e^{-\frac{iE_2t}{\hbar}}$$

To keep the analytical derivation simple, we consider the same amplitude for both functions. Then, the wavepacket can be represented by

$$\psi_{w.p.}(x,t) = Ae^{+ik_1x}e^{-\frac{iE_1t}{\hbar}} + Ae^{+ik_2x}\, e^{-\frac{iE_2t}{\hbar}} \tag{3.6}$$

Here, $k_1 = (k_0 + \Delta k)$, $k_2 = (k_0 - \Delta k)$ and $E_1 = (E_0 + \Delta E)$, $E_2 = (E_0 - \Delta E)$, where $k_0 \left(= \frac{k_1+k_2}{2}\right)$ represents the average wave number, $E_0 \left(= \frac{E_1+E_2}{2}\right)$ represents the average kinetic energy, $\Delta k \left(= \frac{k_1-k_2}{2}\right)$ represents the difference between the wavenumbers, and $\Delta E \left(= \frac{E_1-E_2}{2}\right)$ represents the difference between kinetic energies. Inserting these relations into equation (3.6), we obtain

$$\begin{aligned}\psi_{w.p.}(x,t) &= Ae^{+i(k_0+\Delta k)x}e^{-\frac{i(E_0+\Delta E)t}{\hbar}} + Ae^{+i(k_0-\Delta k)x}e^{-\frac{i(E_0-\Delta E)t}{\hbar}} \\ &= Ae^{+ik_0x}e^{-\frac{iE_0t}{\hbar}}\left[e^{i\left(\Delta kx - \frac{\Delta Et}{\hbar}\right)} + e^{-i\left(\Delta kx - \frac{\Delta Et}{\hbar}\right)}\right]\end{aligned}$$

$$\text{or} \quad \psi_{w.p.}(x,t) = 2Ae^{+i\left(k_0x - \frac{E_0t}{\hbar}\right)}\cos\left(\Delta kx - \frac{\Delta Et}{\hbar}\right) \tag{3.7}$$

The first term in equation (3.7) represents a fast-varying component of the wavepacket (the solid line, as depicted in Figure 3.2(b)), and the second term features a slowly-varying component of the wavepacket (the dotted line, as depicted in Figure 3.2(b)), because the magnitudes of both k_0 and E_0 are much larger than those of Δk and ΔE, respectively. Both components, however, represent a plane wave. Therefore, the motion of a wavepacket can be viewed as two plane wave components (slowly varying and fast varying) traveling simultaneously.

The first component, $e^{i\left(k_0x - \frac{E_0t}{\hbar}\right)}$ or its real form $\cos\left(k_0x - \frac{E_0t}{\hbar}\right)$, which represents the fast-varying component of the wavepacket, is called the **carrier wave** of the wavepacket. The second component of the wavepacket, $\cos\left(\Delta kx - \frac{\Delta Et}{\hbar}\right)$, is called the envelope of the wavepacket. The velocity of the wavepacket is defined based on these two components of the wavepacket.

3.3.1 *Phase velocity of the wavepacket*

The phase velocity of the wavepacket is nothing but the velocity of the constant phase front of the fast-varying component of the wavepacket. This is obtained by

$$\varphi_{\text{fast}} = \left(k_0 x - \frac{E_0 t}{\hbar}\right)$$

$$\text{or} \quad \frac{d\varphi_{\text{fast}}}{dt} = 0 \quad \text{(for the constant phase front)}$$

$$\text{or} \quad k_0 \frac{dx}{dt} - \frac{E_0}{\hbar} = 0$$

$$\text{or} \quad V_p = \frac{E_0}{\hbar k_0}$$

As the average kinetic energy is expressed by $E_0 = \frac{\hbar^2 k_0^2}{2m}$,

$$V_p = \frac{\hbar k_0}{2m} \tag{3.8}$$

3.3.2 *Group velocity of the wavepacket*

The group velocity of the wavepacket is nothing but the velocity of the constant phase front of the slowly-varying component of the wavepacket. This is obtained by

$$\varphi_{\text{slow}} = \left(\Delta k x - \frac{\Delta E t}{\hbar}\right)$$

$$\text{or} \quad \frac{d\varphi_{\text{slow}}}{dt} = 0 \text{ (for the constant phase front)}$$

$$\text{or} \quad \Delta k \frac{dx}{dt} - \frac{\Delta E}{\hbar} = 0$$

$$\text{or} \quad V_g = \frac{\Delta E}{\hbar \Delta k}$$

Considering an infinitesimal difference both in energy and wavenumber of the particle waves interfering to produce the wavepacket, one may write

$$V_g = \frac{1}{\hbar} \left[\frac{dE}{dk} \right]_{E=E_0} \tag{3.9}$$

As the general expression of the kinetic energy is $E = \frac{\hbar^2 k^2}{2m}$, one may write $\frac{dE}{dk} = \frac{\hbar^2 k}{m}$.

$$\text{As a result, } V_g = \frac{\hbar k_0}{m} \tag{3.10}$$

Here, note that the group velocity of the wavepacket exactly represents the classical velocity of the particle (we have noted earlier that the classical velocity of the particle is $V_{\text{cl}} = \frac{\hbar k}{m}$).

It is now instructive to closely inspect the different features of the simplest form of the wavepacket representing a free particle: $\psi_{w.p.}(x,t) = 2Ae^{+i\left(k_0 x - \frac{E_0 t}{\hbar}\right)} \cos\left(\Delta k x - \frac{\Delta E t}{\hbar}\right)$:

(1) Unlike the plane wave solution, the probability density of the particle represented by the above wavepacket form is clearly time- and space-dependent: $\psi^*_{w.p.}\psi_{w.p.} = 4|A|^2 \cos^2\left(\Delta k x - \frac{\Delta E t}{\hbar}\right)$. Therefore, the natural expectation that the probability density of a freely moving particle should depend on time and space is now fulfilled.

(2) The group velocity of the wavepacket represents the classical velocity of the free particle. Thus, the origin of a quantum particle with two different velocities is well-understood.

(3) One can also easily prove that the above wavepacket form of the free particle is normalizable within the limit $[-\infty, +\infty]$ in the x-space (check it yourself). Therefore, the requirement that only a normalizable wavefunction is an acceptable solution to the TDSE is naturally fulfilled.

Conclusions: The above analysis, although based on a wavepacket created by only two time-dependent plane waves, clearly shows that an acceptable picture of a free particle in quantum mechanics is the wavepacket. When a superposition of only two plane waves is involved in constructing the wavepacket, as schematically shown in Figures 3.2(a) and 3.2(b), we obtain a train of wavepackets. Figure 3.2(b) schematically depicts that on both sides of the central wavepacket, there is a rise in other wavepackets. However, if an infinite number of plane waves interfere, we obtain a single (isolated yet traveling) wavepacket in the entire position space. Based on this realization, one can now construct an acceptable picture of the free quantum particle, which is schematically depicted in Figure 3.3, in comparison with the corresponding classical picture.

3.4 Rigorous Mathematical Definition of Free-Particle Wavepacket

After getting familiarized with the above simple (yet physically transparent) model of the free-particle wavepacket, we may now consider the superposition of all possible plane waves (propagating along the same increasing $+x$-direction) with their characteristic time- and energy-dependent phases:

$$\psi_{w.p.}(x,t) = \sum_k A(k) e^{ikx} e^{-i\frac{E_k}{\hbar}t} \quad \text{(summing over all possible } k\text{)}$$

Classical Mechanical View of Particle

m

$+x$

Quantum Mechanical View of Particle: Wavepacket

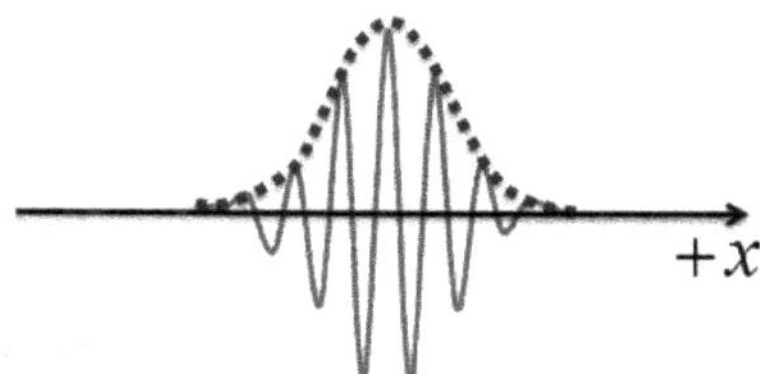

Figure 3.3. Classical versus quantum mechanical view of a particle.

Here, $A(k)$ is the k-dependent expansion coefficient (representing the contribution of the corresponding plane wave to the construction of the wavepacket). Note that in the mathematical derivation of the wavepacket presented earlier based on two plane waves, the expansion coefficient, $A(k)$, was taken to be independent of k to keep the derivation simple. As a free particle's motion is unconfined, the (kinetic) energy or k of the particle is not quantized (not discrete). Therefore, k may adopt any value or can continuously vary. This confirms that an integral form of the superposition (instead of a discrete summation as given above) is quite acceptable as the free-particle wavepacket (this supports the fact that a free particle has a continuous spectrum):

$$\psi_{w.p.}(x,t) = \int_{-\infty}^{+\infty} A(k)e^{ikx}e^{-i\frac{E_k}{\hbar}t}dk \tag{3.11}$$

As a result of the interference (both constructive and destructive) among the component plane waves, at any instant of time, the wavepacket exhibits a large amplitude in a small region of position space and an amplitude close to zero elsewhere. However, the region of constructive interference changes with time because of the presence of the time-dependent phase factor, and as a result, the center of the wavepacket moves (as schematically depicted in Figure 3.4).

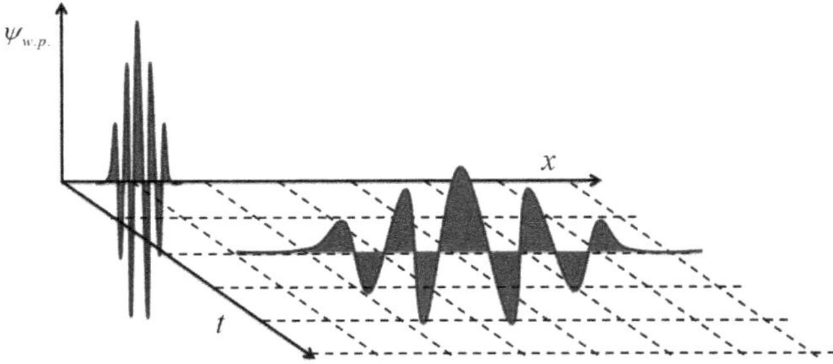

Figure 3.4. Nature of a free-particle wavepacket at two different times during its propagation is schematically depicted. A rigorous mathematical representation of the nature of the free-particle wavepacket propagation is given in the text. For reasons to be revealed soon, note that as the center of the wavepacket moves, it spreads out in space, and consequently, its amplitude decreases.

Note that, for reasons to be revealed soon, the wavepacket also spreads out in position space as its center moves. For the rigorous analytical derivation of free-particle wavepacket motion presented in the following, we closely follow Tannor's lovely textbook.[4]

At $t = 0$ (initial time, which represents the moment just before the onset of wavepacket propagation), we may write (from equation (3.11)

$$\psi_{w.p.}(x,0) = \int_{-\infty}^{+\infty} A(k)e^{ikx}dk$$

Now, multiply both sides of the above equation by $e^{-ik'x}$ and then integrate over x (from $-\infty$ to ∞) to obtain

$$\begin{aligned}\int_{-\infty}^{+\infty} \psi_{w.p.}(x,0)e^{-ik'x}dx &= \int_{-\infty}^{+\infty}\left[\int_{-\infty}^{+\infty} A(k)e^{ikx}e^{-ik'x}dk\right]dx \\ &= \int_{-\infty}^{+\infty} A(k)\left[\int_{-\infty}^{+\infty} e^{ikx}e^{-ik'x}dx\right]dk\end{aligned} \tag{3.12}$$

Next, make use of an important representation of the Dirac delta function[5]:

$$\int_{-\infty}^{+\infty} e^{ikx}e^{-ik'x}dx = 2\pi\,\delta\left(k-k'\right)$$

Consequently, the expression given in equation (3.12) can be rewritten as

$$\int_{-\infty}^{+\infty} \psi_{w.p.}(x,0)\,e^{-ik'x}dx = 2\pi\int_{-\infty}^{+\infty} A(k)\delta\left(k-k'\right)dk$$

When a delta function appears inside an integral as one component of a product with a function, the value of the integral becomes simply the value of the integrand at the position of the delta function. For example, $\int_{-\infty}^{+\infty} f(x)\delta(x - x_0)\, dx = f(x_0)$. Therefore, we may write

$$\int_{-\infty}^{+\infty} \psi_{w.p.}(x,0)e^{-ik'x}dx = 2\pi A(k')$$

$$\text{or} \quad A(k') = \frac{1}{2\pi}\int_{-\infty}^{+\infty} \psi_{w.p.}(x,0)e^{-ik'x}dx$$

If we represent k' as k (this is acceptable because $e^{-ik'x}$ was arbitrarily chosen), we can simply write the above equation as

$$A(k) = \frac{1}{2\pi}\int_{-\infty}^{+\infty} \psi_{w.p.}(x,0)e^{-ikx}dx \tag{3.13}$$

Thus, equation (3.13) gives an expression for $A(k)$ which can readily be used in equation (3.11) to obtain $\psi_{w.p.}(x,t)$. This points to an important fact that the exploration of any quantum dynamical problem is possible when the normalized initial wavefunction, $\psi(x,0)$, is known (this is the characteristic requirement of any initial value problem). As a result, the expansion coefficient $A(k)$ can be determined from the known initial wavefunction (using equation (3.13)), and finally, equation (3.11) helps us find $\psi_{w.p.}(x,t)$, which reveals the dynamical evolution of the wavepacket (free particle) at any later time t. In the following illuminating example, we apprehend that a Gaussian wavepacket serves as a very good model for $\psi(x,0)$. This model gives birth to the **Gaussian wavepacket dynamics**.

A Gaussian Function and a Gaussian Wavepacket

A Gaussian function: The function $A_0e^{-ax^2}$, where A_0 and a are real positive constants, represents a Gaussian function. The center of this function does not move in position space because the $A_0e^{-ax^2}$ function remains anchored at $x = 0$ (property of a Gaussian function), as shown in the following figure.

(Continued)

(*Continued*)

Similarly, $A_0e^{-a(x-x_0)^2}$ also represents another Gaussian function which is anchored at $x = x_0$.

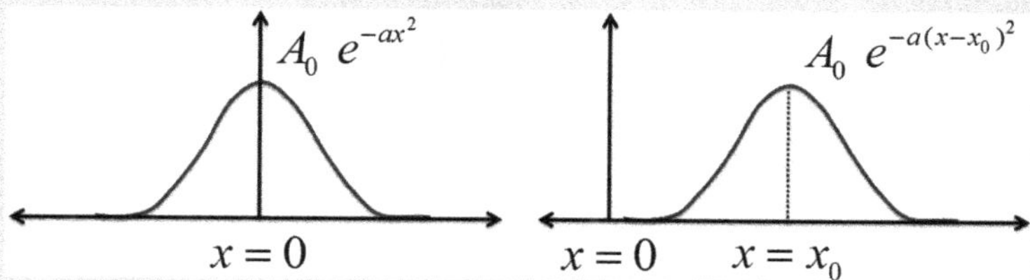

A Gaussian wavepacket: We have already realized that a wavepacket can be viewed as a spatial localization of a particle wave. The manifestation of such a spatial localization of a particle wave in space can be realized by a group of plane waves interfering. The relative phase of the plane waves interfering determines the width (or spread) of the wavepacket. This is depicted in the following figure (taking an example of six plane waves interfering).

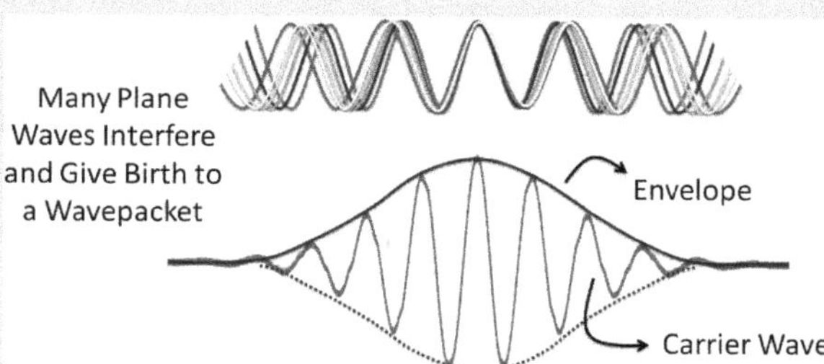

The interference of an infinite number of plane waves yields an isolated wavepacket, which can be (approximately) represented as

$$\psi_{w.p.} = a(x)e^{ik_0x}$$

in which the envelop of the wavepacket can be approximately expressed by an appropriate envelope function $a(x)$ and the carrier wave is represented by the oscillatory term e^{ik_0x} associated with the average wavenumber. Often, a Gaussian function represents a good approximation for $a(x)$. This is why a Gaussian wavepacket, as expressed by $\psi_{w.p.}(x) = A_0\,e^{-ax^2}\,e^{ik_0x}$, is often used to represent a particle in quantum mechanics. Note that the Gaussian wavepacket, presented by the expression $\psi_{w.p.}(x) = A_0\,e^{-ax^2}\,e^{ik_0x}$, features a wavepacket frozen at $t = 0$: Consequently, a more appropriate expression for the same is $\psi_{w.p.}(x, 0) = A_0\,e^{-ax^2}\,e^{ik_0x}$.

3.5 Illuminating Example: Gaussian Wavepacket Dynamics of a Free Particle

Let us assume that the initial (at $t = 0$) wavefunction of a free particle is represented by a Gaussian wavepacket:

$$\psi_{w.p.}(x,0) = A_0 e^{-ax^2} e^{ik_0 x},$$
$$\text{where } k_0 \text{ is a real constant and so are } A_0 \text{ and } a \quad (3.14)$$

As mentioned before, our first task is to normalize $\psi_{w.p.}(x,0)$ to find out $A(k)$ to finally obtain $\psi_{w.p.}(x,t)$.

Math Guide

Standard Gaussian integrals:

For complex forms : $\int_{-\infty}^{+\infty} e^{-ax^2+ibx+ic} dx = \sqrt{\frac{\pi}{a}}\, e^{\frac{-b^2}{4a}} e^{ic}$ where $a > 0$

For real forms : $\int_{-\infty}^{+\infty} e^{-ax^2+bx+c}\, dx = \sqrt{\frac{\pi}{a}} e^{\left(\frac{b^2}{4a}+c\right)}$

The normalized initial Gaussian wavepacket is expressed as (check it yourself)

$$\psi_{w.p.}(x,0) = \left(\frac{2a}{\pi}\right)^{1/4} e^{-ax^2} e^{ik_0 x} \quad (3.15)$$

Finally, the analytical expression for the free-particle wavepacket at any given time t is obtained:

$$\psi_{w.p.}(x,t) = \int_{-\infty}^{+\infty} A(k) e^{i\left(kx - \frac{E_k t}{\hbar}\right)}\, dk \text{ (from equation (3.11))}$$

$$\text{where} \quad A(k) = \frac{1}{2\pi} \int_{-\infty}^{+\infty} \psi_{w.p.}(x,0)\, e^{-ikx} dx$$

Inserting equation (3.15) into the above expression, we obtain

$$A(k) = \frac{1}{2\pi}\left(\frac{2a}{\pi}\right)^{1/4}\int_{-\infty}^{+\infty} e^{-ax^2+i(-k+k_0)x}dx$$

$$\text{or}\quad A(k) = \frac{1}{2\pi}\left(\frac{2a}{\pi}\right)^{1/4}\sqrt{\frac{\pi}{a}}e^{-\frac{(k-k_0)^2}{4a}} \tag{3.16}$$

As a result, inserting equation (3.16) and relation $E_k = \frac{\hbar^2k^2}{2m}$ into equation (3.11), we get

$$\psi_{w.p.}(x,t) = \frac{1}{2\pi}\left(\frac{2a}{\pi}\right)^{1/4}\sqrt{\frac{\pi}{a}}\int_{-\infty}^{+\infty} e^{-\frac{(k-k_0)^2}{4a}+ikx-i\frac{\hbar k^2t}{2m}}dk \tag{3.17}$$

Here, the exponent can be rearranged as follows:

$$-\frac{(k-k_0)^2}{4a}+ikx-i\frac{\hbar k^2t}{2m}$$
$$= \left[-k^2\left(\frac{1}{4a}+\frac{i\hbar t}{2m}\right)+k\left(ix+\frac{k_0}{2a}\right)-\frac{k_0^2}{4a}\right] \tag{3.18}$$

Inserting the exponent given in equation (3.18) into equation (3.17), we obtain

$$\psi_{w.p.}(x,t) = \frac{1}{2\pi}\left(\frac{2a}{\pi}\right)^{1/4}\sqrt{\frac{\pi}{a}}\int_{-\infty}^{+\infty} e^{-k^2\left(\frac{1}{4a}+\frac{i\hbar t}{2m}\right)+k\left(ix+\frac{k_0}{2a}\right)-\frac{k_0^2}{4a}}dk$$

$$\text{or}\ \psi_{w.p.}(x,t) = \left(\frac{2a}{\pi}\right)^{1/4}\sqrt{\frac{1}{\left(1+\frac{2i\hbar ta}{m}\right)}}e^{-\frac{k_0^2}{4a}}e^{\frac{a\left(ix+\frac{k_0}{2a}\right)^2}{\left(1+\frac{2i\hbar ta}{m}\right)}} \tag{3.19}$$

Equation (3.19) represents the free-particle wavepacket at any later time t after the propagation event is initiated at $t = 0$.

Time evolution of probability density: As the probability density is the only experimentally realizable quantity, we make an attempt to obtain the same for the Gaussian wavepacket (representing the free particle). Assuming that $\frac{2\hbar ta}{m} = b$, we may rewrite equation (3.19) as

$$\psi_{w.p.}(x,t) = \left(\frac{2a}{\pi}\right)^{1/4} e^{-\frac{k_0^2}{4a}} \sqrt{\frac{1}{(1+ib)}}\, e^{\frac{a\left(ix+\frac{k_0}{2a}\right)^2}{(1+ib)}}$$

Consequently, the probability density for the particle at position x and at time t is given by

$$|\psi_{w.p.}(x,t)|^2 = \left(\frac{2a}{\pi}\right)^{1/2} e^{-\frac{k_0^2}{2a}} \sqrt{\frac{1}{(1+b^2)}}\, e^{\frac{a\left(ix+\frac{k_0}{2a}\right)^2}{(1+ib)} + \frac{a\left(-ix+\frac{k_0}{2a}\right)^2}{(1-ib)}} \tag{3.20}$$

The exponent can be conveniently rearranged as follows

$$\begin{aligned}
&\frac{a\left(ix+\frac{k_0}{2a}\right)^2}{(1+ib)} + \frac{a\left(-ix+\frac{k_0}{2a}\right)^2}{(1-ib)} \\
&= \frac{a}{(1+b^2)}\left[(1-ib)\left(ix+\frac{k_0}{2a}\right)^2 + (1+ib)\left(-ix+\frac{k_0}{2a}\right)^2\right] \\
&= \frac{a}{(1+b^2)}\left[-2x^2 + \frac{2bxk_0}{a} + \frac{2k_0^2}{4a^2}\right] \\
&= \frac{-2a}{(1+b^2)}\left[x^2 - \frac{bxk_0}{a} - \frac{k_0^2}{4a^2}\right] \\
&= \frac{-2a}{(1+b^2)}\left[x^2 - 2x\frac{k_0 b}{2a} + \left(\frac{k_0 b}{2a}\right)^2 - \left(\frac{k_0 b}{2a}\right)^2 - \frac{k_0^2}{4a^2}\right] \\
&= \frac{-2a}{(1+b^2)}\left[\left(x - \frac{k_0 b}{2a}\right)^2 - \frac{k_0^2}{4a^2}(1+b^2)\right] \\
&= -\frac{2a}{(1+b^2)}\left(x - \frac{k_0 b}{2a}\right)^2 + \frac{k_0^2}{2a}
\end{aligned}$$

Therefore, equation (3.20) can be further reduced using the above exponent:

$$|\psi_{w.p.}(x,t)|^2 = \left(\frac{2a}{\pi}\right)^{1/2} \sqrt{\frac{1}{(1+b^2)}} e^{-\frac{2a}{(1+b^2)}\left(x-\frac{k_0 b}{2a}\right)^2} \tag{3.21}$$

Equation (3.21) represents a Gaussian profile which is centered at

$$x_t = \frac{k_0 b}{2a} = \frac{k_0 2\hbar a t}{2ma} = \frac{k_0 \hbar t}{m} \quad \left(\text{because we assumed } \frac{2\hbar t a}{m} = b\right) \tag{3.22}$$

Thus, equation (3.21) shows that an initial Gaussian free-particle wavepacket remains Gaussian even after it is allowed to time-evolve freely (note that a free particle experiences zero potential). Equation (3.22) features the fact that the center of the Gaussian wavepacket is time-dependent (moves as a function of time). The velocity with which the center of this Gaussian wavepacket travels is called the **group velocity of the wavepacket** and is expressed as

$$V_g = \frac{k_0 \hbar}{m} \tag{3.23}$$

Furthermore, both the amplitude and width of the free-particle wavepacket also evolve as a function of time. In order to understand the time evolution of the width and amplitude of the Gaussian wavepacket representing a free particle, we first define the width of a Gaussian profile.

In general, the width of a Gaussian profile is defined by the **full width at half-maximum (FWHM)** of the Gaussian probability density distribution. Therefore, as shown in Figure 3.5(a), the initial (at $t = 0$) width (Δx_0) of the Gaussian wavepacket is defined by the FWHM of the initial probability density distribution function $|\psi_{w.p.}(x,0)|^2 = \left(\frac{2a}{\pi}\right)^{1/2} e^{-2ax^2}$ (from equation (3.15)). By definition of the FWHM, we may write (considering the fact that the amplitude

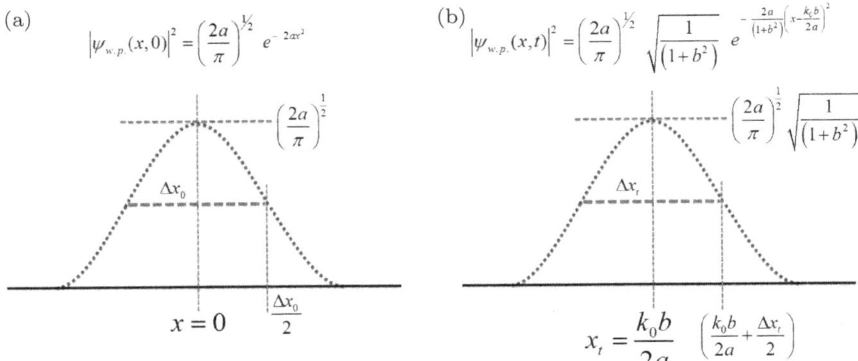

Figure 3.5. Full width at half-maximum (FWHM) of a Gaussian probability density distribution is defined here. In (a), we have considered the Gaussian probability density distribution at the initial time $t = 0$ (just before the onset of the propagation event), and in (b), we have considered the Gaussian probability density distribution at a later time t after the propagation event is initiated.

drops to half of its maximum at $x = \frac{\Delta x_0}{2}$):

$$\frac{1}{2}\left(\frac{2a}{\pi}\right)^{1/2} = \left(\frac{2a}{\pi}\right)^{1/2} e^{-2a\left(\frac{\Delta x_0}{2}\right)^2}$$

$$\text{or } -2a\left(\frac{\Delta x_0}{2}\right)^2 = \ln 1 - \ln 2$$

$$\text{or } a = \frac{2\ln 2}{\Delta x_0^2} \tag{3.24}$$

If we define the FWHM of the Gaussian profile $|\psi_{\text{w.p.}}(x,t)|^2$ as Δx_t, following a similar analysis as given above, we may write for the $|\psi_{w.p.}(x,t)|^2$ profile (as shown in Figure 3.5(b))

$$\frac{1}{2}\left(\frac{2a}{\pi}\right)^{1/2} \frac{1}{\sqrt{1+b^2}} = \left(\frac{2a}{\pi}\right)^{1/2} \frac{1}{\sqrt{1+b^2}} e^{-\frac{2a}{(1+b^2)}\left(\frac{\Delta x_t}{2}\right)^2}$$

$$\text{or } -\frac{a\Delta x_t^2}{2(1+b^2)} = \ln 1 - \ln 2$$

$$\text{or } \Delta x_t^2 = \frac{2\left(1+b^2\right)\ln 2}{a} = \frac{2\left(1+b^2\right)\ln 2\,\Delta x_0^2}{2\ln 2},$$

$$\text{using } a = \frac{2\ln 2}{\Delta x_0^2} \text{ (equation(3.24))}$$

$$\text{or } \Delta x_t^2 = \left(1+b^2\right)\Delta x_0^2$$

$$\text{or } \Delta x_t^2 = \left[1+\left(\frac{2\hbar t a}{m}\right)^2\right]\Delta x_0^2 = \left[\Delta x_0^2 + \frac{4\hbar^2 t^2 a^2 \Delta x_0^2}{m^2}\right]$$

$$\text{as } \frac{2\hbar t a}{m} = b$$

$$\text{or } \Delta x_t = \sqrt{\Delta x_0^2 + \frac{4\hbar^2 t^2 (2\ln 2)^2}{m^2 \Delta x_0^2}} \tag{3.25}$$

The above expression shows that **when a Gaussian free-particle wavepacket travels, it broadens its width as time progresses.** This is schematically depicted in Figure 3.4.

On the other hand, the amplitude of the $|\psi_{\text{w.p.}}(x,t)|^2$ profile is given by

$$\left(\frac{2a}{\pi}\right)^{1/2} \frac{1}{\sqrt{1+b^2}} = \frac{A_0^2}{\sqrt{1+\frac{4\hbar^2 t^2 (2\ln 2)^2}{m^2 \Delta x_0^4}}} \tag{3.26}$$

where $A_0^2 = \left(\frac{2a}{\pi}\right)^{1/2}$, which is the amplitude of the wavepacket at $t = 0$.

Clearly, equation (3.26) shows that **the probability amplitude of the Gaussian free-particle wavepacket decreases as time progresses**. This is also schematically depicted in Figure 3.4.

Representation in Position Space and Momentum Space

One can represent a quantum state at a particular time (t) either in position space using the wavefunction $\psi(x,t)$ or in momentum space using the wavefunction $\phi(k,t)$. As pointed out in Chapter 2, one representation can be

(Continued)

(Continued)

easily converted into the other using the Fourier transform or inverse Fourier transform:

$$\phi(k,t) = \frac{1}{\sqrt{2\pi}} \int_{-\infty}^{+\infty} \psi(x,t) e^{-ikx} dx$$

$$\psi(x,t) = \frac{1}{\sqrt{2\pi}} \int_{-\infty}^{+\infty} \phi(k,t) e^{ikx} dk$$

If the momentum-space variable is changed from k to p, one should use the following relationships:

$$\phi(p,t) = \frac{1}{\sqrt{2\pi\hbar}} \int_{-\infty}^{+\infty} \psi(x,t) e^{-i\frac{p}{\hbar}x} dx$$

$$\psi(x,t) = \frac{1}{\sqrt{2\pi\hbar}} \int_{-\infty}^{+\infty} \phi(p,t) e^{i\frac{p}{\hbar}x} dp$$

Proof:

$$\psi(x,t) = \frac{1}{\sqrt{2\pi\hbar}} \int_{-\infty}^{+\infty} \phi(p,t) e^{i\frac{p}{\hbar}x}\, dp$$

$$\text{or } \psi(x,t) = \frac{1}{\sqrt{2\pi\hbar}} \int_{-\infty}^{+\infty} \left[\frac{1}{\sqrt{2\pi\hbar}} \int_{-\infty}^{+\infty} \psi(x,t) e^{-i\frac{p}{\hbar}x}\, dx \right] e^{i\frac{p}{\hbar}x} dp$$

$$\text{or } \psi(x,t) = \frac{1}{\hbar\sqrt{2\pi}} \frac{1}{\sqrt{2\pi}} \int_{-\infty}^{+\infty} \left[\int_{-\infty}^{+\infty} \psi(x,t) e^{-i\frac{p}{\hbar}x}\, e^{i\frac{p}{\hbar}x} dx \right] dp$$

Note that the $e^{i\frac{p}{\hbar}x}$ factor can be placed anywhere because the variables of the two integrals are different.

$$\text{or } \psi(x,t) = \frac{1}{\hbar\sqrt{2\pi}} \frac{1}{\sqrt{2\pi}} \int_{-\infty}^{\infty} \left[\int_{-\infty}^{+\infty} \psi(x,t) dx \right] \hbar dk \text{ (as } p = \hbar k \text{ or } dp = \hbar dk)$$

$$\text{or } \psi(x,t) = \frac{1}{\sqrt{2\pi}} \int_{-\infty}^{+\infty} \left[\frac{1}{\sqrt{2\pi}} \int_{-\infty}^{+\infty} \psi(x,t)\, dx \right] dk$$

$$\text{or } \psi(x,t) = \frac{1}{\sqrt{2\pi}} \int_{-\infty}^{+\infty} \left[\frac{1}{\sqrt{2\pi}} \int_{-\infty}^{+\infty} \psi(x,t) e^{-ikx}\, e^{ikx}\, dx \right] dk$$

(Continued)

(Continued)

Similar to what was noted earlier, the e^{ikx} factor can be placed anywhere because the variables of the two integrals are different.

$$\text{or } \psi(x,t) = \frac{1}{\sqrt{2\pi}}\int_{-\infty}^{+\infty}\left[\frac{1}{\sqrt{2\pi}}\int_{-\infty}^{+\infty}\psi(x,t)e^{-ikx}\,dx\right]e^{ik}dk$$

$$\text{or } \psi(x,t) = \frac{1}{\sqrt{2\pi}}\int_{-\infty}^{+\infty}\phi(k,t)\,e^{ikx}\,dk$$

$$\text{where } \phi(k,t) = \frac{1}{\sqrt{2\pi}}\int_{-\infty}^{+\infty}\psi(x,t)e^{-ikx}dx$$

TDSE of a Free Particle in Momentum Space: An Example

In momentum space, the wavefunction $\phi(p,t)$ plays a role which is very similar to that played by the wavefunction $\psi(x,t)$ in position space. Therefore, conceptually, $|\phi(p,t)|^2dp$ refers to the probability of finding a particle with momentum in the interval p and $(p+dp)$ at time t. Since the position-space wavefunction $\psi(x,t)$ satisfies the TDSE, one can obtain an equivalent TDSE for the momentum-space wavefunction $\phi(p,t)$. A derivation of the TDSE for the momentum-space wavefunction of a free particle is given in the following. Inserting the form $\psi(x,t) = \frac{1}{\sqrt{2\pi\hbar}}\int_{-\infty}^{+\infty}\phi(p,t)e^{i\frac{p}{\hbar}x}\,dp$ into the TDSE $i\hbar\frac{\partial\psi(x,t)}{\partial t} = H\psi(x,t)$, we get

$$i\hbar\frac{\partial}{\partial t}\left[\frac{1}{\sqrt{2\pi\hbar}}\int_{-\infty}^{+\infty}\phi(p,t)e^{i\frac{p}{\hbar}x}dp\right] = -\frac{\hbar^2}{2m}\frac{\partial^2}{\partial x^2}\left[\frac{1}{\sqrt{2\pi\hbar}}\int_{-\infty}^{+\infty}\phi(p,t)e^{i\frac{p}{\hbar}x}dp\right]$$

(for a free particle, $V=0$)

$$\text{or } \int_{-\infty}^{+\infty}\frac{\partial\phi(p,t)}{\partial t}e^{i\frac{p}{\hbar}x}dp = -\frac{\hbar}{2im}\frac{\partial}{\partial x}\left[\int_{-\infty}^{+\infty}\phi(p,t)\left(i\frac{p}{\hbar}\right)e^{i\frac{p}{\hbar}x}dp\right]$$

$$= -\frac{\hbar}{2im}\left[\int_{-\infty}^{+\infty}\phi(p,t)\left(i\frac{p}{\hbar}\right)^2e^{i\frac{p}{\hbar}x}dp\right]$$

$$\text{or } \int_{-\infty}^{+\infty}i\hbar\frac{\partial\phi(p,t)}{\partial t}e^{i\frac{p}{\hbar}x}\,dp = \int_{-\infty}^{+\infty}\left(\frac{p^2}{2m}\right)\phi(p,t)e^{i\frac{p}{\hbar}x}dp$$

$$\text{or } i\hbar\frac{\partial\phi(p,t)}{\partial t} = \left(\frac{p^2}{2m}\right)\phi(p,t)$$

(Continued)

(Continued)

The above equation represents the TDSE of a free particle in momentum space. One can find a solution to the TDSE of a free article in momentum space as given in the following (check it yourself):

$$\phi(p,t) = \phi(p,0)e^{-i\frac{p^2t}{2m\hbar}}$$

Uncertainty Principle

The expectation value of all classical observables can be obtained if the wavefunction is known both in position space and momentum space. This is why both the position-space and momentum-space representations are equally important in quantum mechanics. However, with regard to the experimental measurement of position and momentum of a quantum particle, there is a fundamental limit which correlates with the respective probability density distributions. This fundamental limit is called Heisenberg's uncertainty principle:

$$\sigma_x \sigma_p \geq \frac{\hbar}{2}$$

where σ_x and σ_p represent the standard deviations of the probability density distribution in position space and momentum space, respectively. For a normalized wavefunction in position space, $\psi(x,t)$, the expectation value of position (following Postulate 3 in Chapter 2) is given by $\langle \hat{x} \rangle(t) = \int_{-\infty}^{+\infty} \psi^*(x,t)\hat{x}\psi(x,t)dx$. Furthermore, one can also write $\langle \hat{x}^2 \rangle (t) = \int_{-\infty}^{+\infty} \psi^*(x,t)\hat{x}^2\psi(x,t)dx$. Finally, the variance of the probability density distribution in position space is expressed by $\sigma_x^2(t) = \langle \hat{x}^2 \rangle (t) - \langle \hat{x} \rangle^2(t)$. Note that σ_x and σ_x^2 represent, respectively, the standard deviation and variance of the probability density distribution in position space. Similarly, for a normalized wavefunction in momentum space, $\phi(p,t)$, the expectation value of momentum is given by $\langle \hat{p} \rangle = \int_{-\infty}^{+\infty} \phi^*(p,t)\hat{p}\phi(p,t)dp$. Furthermore, one can also write $\langle \hat{p}^2 \rangle = \int^{+\infty} \phi^*(p,t)\hat{p}^2\phi(p,t)dp$. Finally, the variance of the probability density distribution in momentum space is expressed by $\sigma_p^2 = \langle \hat{p}^2 \rangle - \langle \hat{p} \rangle^2$.

Momentum-space representation: Thus far, we have analyzed the behavior of the Gaussian wavepacket for a free particle in position space. In this section, we inspect the same in momentum space.

We have already apprehended that the normalized initial wavepacket (at $t = 0$) for a free particle can be expressed in position space as (equation (3.15))

$$\psi_{w.p.}(x,0) = \left(\frac{2a}{\pi}\right)^{1/4} e^{-ax^2} e^{ik_0x}$$

where $a = \frac{2\,ln2}{\Delta x_0^2}$ (refer to equation (3.24)) and $k_0 = \frac{p_0}{\hbar}$. Here, as shown earlier, Δx_0 represents the FWHM of the initial probability density distribution of the free particle in position space.

The above initial wavepacket can also be represented in momentum space using the Fourier transform as

$$
\begin{aligned}
\phi_{w.p.}(p,0) &= \frac{1}{\sqrt{2\pi\hbar}} \int_{-\infty}^{+\infty} \psi_{w.p.}(x,0) e^{-i\frac{p}{\hbar}x} dx \\
&= \frac{1}{\sqrt{2\pi\hbar}} \left(\frac{2a}{\pi}\right)^{1/4} \int_{-\infty}^{+\infty} e^{-ax^2} e^{i\frac{p_0}{\hbar}x} e^{-i\frac{p}{\hbar}x} dx \\
\text{or } \phi_{w.p.}(p,0) &= \left(\frac{1}{2\pi a\hbar^2}\right)^{1/4} e^{-\frac{(p-p_0)^2}{4\hbar^2 a}}
\end{aligned}
$$

$$\text{(using standard Gaussian integral)} \tag{3.27}$$

As a result, the initial probability density distribution of the wavepacket in momentum space is given by

$$|\phi_{w.p.}(p,0)|^2 = \left(\frac{1}{2\pi a\hbar^2}\right)^{1/2} e^{-\frac{(p-p_0)^2}{2\hbar^2 a}} \tag{3.28}$$

It is clearly evident from equation (3.28) that the initial probability density distribution of the wavepacket for a free particle in momentum space is centered at $p = p_0$, as shown in Figure 3.6.

Using the definition of the FWHM (the amplitude drops to half of its maximum at $p = \frac{\Delta p_0}{2}$), one may also write

$$\frac{1}{2}\left(\frac{1}{2\pi a\hbar^2}\right)^{1/2} = \left(\frac{1}{2\pi a\hbar^2}\right)^{1/2} e^{-\frac{1}{2\hbar^2 a}\left(\frac{\Delta p_0}{2}\right)^2}$$

$$\text{or (verify it yourself) } \Delta p_0^2 = 8a\hbar^2 \ln 2$$

$$\text{or using } a = \frac{2\ln 2}{\Delta x_0^2}, \text{ we get}$$

$$\Delta x_0\, \Delta p_0 = 4\hbar \ln 2 \tag{3.29}$$

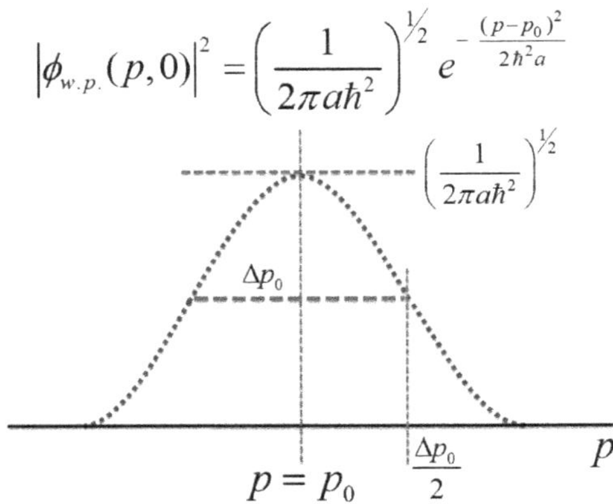

Figure 3.6. Full width at half-maximum (FWHM $= \Delta p_0$) of the initial probability density distribution of a Gaussian wavepacket in momentum space is illustrated. Note that the distribution function is centered at $p = p_0$.

Guiding Question

3.1. Thus far, we have used the following form to represent the wavefunction of an initial Gaussian wavepacket:

$$\psi_{w.p.}(x,0) = \left(\frac{2a}{\pi}\right)^{1/4} e^{-ax^2} e^{ik_0 x} \text{ where } a = \frac{2\ln 2}{\Delta x_0^2}$$

This form makes use of the FWHM (Δx_0) of its initial probability density distribution. With the help of this form, we have explored the spread of a Gaussian wavepacket in position space as a function of time. In quantum mechanics, often, another form based on the standard deviation (σ) of the probability distribution function is used to express a Gaussian function. When an initial Gaussian wavepacket is defined in terms of the standard deviation of its initial probability density distribution (σ_{x0}), the wavepacket adopts the following mathematical form in position space:

$$\psi_{w.p.}(x,0) = Ae^{-\frac{x^2}{4\sigma_{x0}^2}} e^{ik_0 x}$$

(a) Normalize the initial wavepacket.
(b) Prove that $\Delta x_0 = 2\sqrt{2\,ln\,2}\sigma_{x0}$, where Δx_0 represents the FWHM of the initial probability density distribution, while σ_{x0} represents the standard deviation of the initial probability density distribution.

(Continued)

(Continued)

(c) Prove that σ_{x0} corresponds to a half-width at approximately 60% of the maximum of the probability density distribution.
(d) Calculate $\langle \hat{x} \rangle$ and $\langle \hat{x}^2 \rangle$ at $t = 0$. Note that the variance of a probability density distribution is defined by $\sigma^2 = \langle \hat{x}^2 \rangle - \langle \hat{x} \rangle^2$. Find σ_{x0} using the definition of variance. Compare the result with the one obtained in Problem (b).
(e) Find expression for $\psi_{w.p.}(x, t)$ in terms of σ_{x0}.

Equation (3.29) represents one form of Heisenberg's uncertainty principle in terms of the FWHM of the corresponding probability density distribution. A more familiar form of Heisenberg's uncertainty principle, which is expressed in terms of the standard deviation of the corresponding probability density distribution, can be easily obtained by using the relation between the FWHM (Δ) and the standard deviation (σ) of the corresponding probability density distribution. For a Gaussian function, as $\Delta = 2\sqrt{2\ \ln\ 2}\,\sigma$, one can rewrite equation (3.29) as

$$2\sqrt{2\ln 2}\ \sigma_{x0}\ 2\sqrt{2\ln 2}\ \sigma_{p0} = 4\ln 2\,\hbar$$

$$\text{or } \sigma_{x0}\sigma_{p0} = \frac{\hbar}{2} \tag{3.30}$$

Equation (3.30) illustrates the fact that the initial Gaussian wavepacket has minimum uncertainty.

Finally, one can construct a free-particle wavepacket for any later time t as follows. A solution to the TDSE of a free particle in momentum space (as discussed in the "Representation in Position Space and in Momentum Space" box earlier) is given by

$$\phi_{w.p.}(p, t) = \phi_{w.p.}(p, 0)e^{-i\frac{p^2 t}{2m\hbar}}$$

Inserting equation (3.27) into the above expression, we get

$$\phi_{w.p.}(p, t) = \left(\frac{1}{2\pi a\hbar^2}\right)^{1/4} e^{-\frac{(p-p_0)^2}{4\hbar^2 a}}\ e^{-i\frac{p^2 t}{2m\hbar}}$$

As a result, the probability density distribution of the wavepacket in momentum space for any later time t is given by

$$|\phi_{w.p.}(p,t)|^2 = \left(\frac{1}{2\pi a\hbar^2}\right)^{1/2} e^{-\frac{(p-p_0)^2}{2\hbar^2 a}} \tag{3.31}$$

Equation (3.31) unravels the fact that in momentum space, the Gaussian wavepacket of a free particle for any later time t remains time-independent. On the contrary, we have already developed a clear consensus that both amplitude and width (or spread) of the same wavepacket in position space are time-dependent (refer to equation (3.21)). As a consequence, one can easily infer that **the Gaussian wavepacket for a free particle spreads out in position space while it remains unaltered in momentum space as time progresses**. This finding is closely related to Heisenberg's uncertainty principle, as shown in the following.

Consider the product of the FWHM of the probability density distribution of the Gaussian wavepacket in position space and in momentum space for any later time t:

$$\Delta x_t \Delta p_t = \Delta x_t \Delta p_0 = \sqrt{\Delta x_0^2 + \frac{4\hbar^2 t^2 (2\ln 2)^2}{m^2 \Delta x_0^2}} 4\ln 2 \frac{\hbar}{\Delta x_0}$$

We obtain the above expression using equations (3.25) and (3.29) and the fact that $\Delta p_t = \Delta p_0$

$$\text{or } \Delta x_t\, \Delta p_t = \frac{\hbar}{2}\sqrt{1 + \frac{4\hbar^2 t^2 (2\ln 2)^2}{m^2 \Delta x_0^4}} 8\ln 2$$

As we did before, to re-express the above product in terms of the standard deviation of the corresponding probability density distribution, we can use the relation $\Delta = 2\sqrt{2\ \ln\ 2}\sigma$ and rewrite the product as

$$2\sqrt{2\ln 2}\ \sigma_{x_t}\, 2\sqrt{2\ln 2}\ \sigma_{p_t} = \frac{\hbar}{2}\sqrt{1 + \frac{4\hbar^2 t^2 (2\ln 2)^2}{m^2 \Delta x_0^4}} 8\ln 2$$

$$\text{or } \sigma_{x_t}\sigma_{p_t} = \frac{\hbar}{2}\left(1 + \frac{4\hbar^2 t^2 (2\ln 2)^2}{m^2 \Delta x_0^4}\right)^{1/2}$$

$$= \frac{\hbar}{2}\left(1 + \frac{1}{2}\frac{4\hbar^2 t^2 (2\ln 2)^2}{m^2 \Delta x_0^4} + \cdots\cdots\right)$$

$$\text{Therefore, } \sigma_{x_t}\ \sigma_{p_t} > \frac{\hbar}{2} \tag{3.32}$$

Combining equations (3.30) and (3.32), we get our very familiar form of Heisenberg's uncertainty principle, $\sigma_x \sigma_p \geq \frac{\hbar}{2}$, for free-particle Gaussian wavepacket dynamics. As time progresses, the wavepacket spreads out in position space, while it remains unaltered in momentum space. As a result, the position measurements of the particle yield widely scattered values as time progresses. Note that equation (3.32) is an inequality, and consequently, there is no limit on how large σ_{x_t} can be.

3.6 General Form of a Gaussian Wavepacket

In this section, a general form of a Gaussian wavepacket, which is very frequently used in the semiclassical theory of chemical dynamics,[6] is deduced. A free-particle Gaussian wavepacket has the following analytical form (refer to equation (3.17)):

$$\psi_{w.p.}(x,t) = \frac{1}{2\pi}\left(\frac{2a}{\pi}\right)^{1/4}\sqrt{\frac{\pi}{a}}\int_{-\infty}^{+\infty} e^{-\frac{(k-k_0)^2}{4a} + ikx - i\frac{\hbar k^2 t}{2m}}\,dk \tag{3.33}$$

The exponent in the above expression can be rewritten as

$$-\frac{(k-k_0)^2}{4a} + ikx - i\frac{\hbar k^2 t}{2m}$$

$$= -\frac{(k-k_0)^2}{4a} + i(k-k_0)x - \frac{i\hbar(k^2-k_0^2)t}{2m} + ik_0x - \frac{i\hbar k_0^2 t}{2m}$$

Using the identity $(k^2 - k_0^2) = (k - k_0)^2 + 2k_0(k - k_0)$, we may further express the above argument as

$$-\frac{(k-k_0)^2}{4a} + i(k-k_0)x - \frac{i\hbar\left[(k-k_0)^2 + 2k_0(k-k_0)\right]t}{2m}$$
$$+ ik_0x - \frac{i\hbar k_0^2 t}{2m}$$

Assuming that $(k - k_0) = s$, the above argument can be rewritten as

$$\begin{aligned}&-\frac{s^2}{4a} + isx - \frac{i\hbar t}{2m}\left[s^2 + 2k_0 s\right] + ik_0x - \frac{i\hbar k_0^2 t}{2m}\\&= -s^2\left(\frac{1}{4a} + \frac{i\hbar t}{2m}\right) + is\left(x - \frac{\hbar t k_0}{m}\right) + ik_0x - \frac{i\hbar t k_0^2 t}{2m}\end{aligned}$$

Thus, inserting the above argument into the integral given in equation (3.33), we may write

$$\begin{aligned}\psi_{w.p.}(x,t) &= \frac{1}{2\pi}\left(\frac{2a}{\pi}\right)^{1/4}\sqrt{\frac{\pi}{a}}\int_{-\infty}^{+\infty} e^{-s^2\left(\frac{1}{4a}+\frac{i\hbar t}{2m}\right)+is\left(x-\frac{\hbar t k_0}{m}\right)+ik_0x-\frac{i\hbar k_0^2 t}{2m}}\,ds\\&= \frac{1}{2\pi}\left(\frac{2a}{\pi}\right)^{1/4}\sqrt{\frac{\pi}{a}}\sqrt{\frac{\pi}{\left(\frac{1}{4a}+\frac{i\hbar t}{2m}\right)}}\,e^{-\frac{\left(x-\frac{\hbar t k_0}{m}\right)^2}{4\left(\frac{1}{4a}+\frac{i\hbar t}{2m}\right)}}\,e^{\left(ik_0x-\frac{i\hbar k_0^2 t}{2m}\right)}\\&= \left(\frac{2a}{\pi}\right)^{1/4}\sqrt{\frac{1}{\left(1+\frac{2i\hbar t a}{m}\right)}}\,e^{-\frac{a}{\left(1+\frac{2i\hbar t a}{m}\right)}\left(x-\frac{\hbar t k_0}{m}\right)^2+ixk_0-\frac{i\hbar k_0^2 t}{2m}}\end{aligned} \tag{3.34}$$

Let us now define $\alpha_t = \frac{a}{\left(1+\frac{2i\hbar t a}{m}\right)}$. Furthermore, we have already realized that the center of the Gaussian wavepacket is given by $x_t = \frac{\hbar t k_0}{m}$ (refer to equation (3.22)). Inserting these two expressions into equation (3.34), we get

$$\psi_{w.p.}(x,t) = \left(\frac{2a}{\pi}\right)^{1/4}\left(1 + \frac{2i\hbar t a}{m}\right)^{-1/2} e^{-\alpha_t(x-x_t)^2 + ik_0(x-x_t) + ik_0x_t - \frac{i\hbar k_0^2 t}{2m}}$$

As $ik_0x_t - \frac{i\hbar k_0^2 t}{2m} = ik_0\frac{\hbar t k_0}{m} - \frac{i\hbar k_0^2 t}{2m} = \frac{i\hbar k_0^2 t}{m}\left(1-\frac{1}{2}\right) = \frac{i\hbar k_0^2 t}{2m}$, we may further reduce the above expression as

$$\psi_{w.p.}(x,t) = \left(\frac{2a}{\pi}\right)^{1/4}\left(1+\frac{2i\hbar ta}{m}\right)^{-1/2} e^{-\alpha_t(x-x_t)^2+ik_0(x-x_t)+\frac{i\hbar k_0^2 t}{2m}} \tag{3.35}$$

As the momentum of the particle is given by $p_0 = \hbar k_0$, we may write

$$\frac{i}{\hbar}\frac{p_0^2}{2m}t = \frac{i}{\hbar}\frac{k_0^2\hbar^2 t}{2m} = \frac{ik_0^2\hbar t}{2m}$$

Thus, inserting the above form into equation (3.35), we may write

$$\begin{aligned}
\psi_{w.p.}(x,t) &= \left(\frac{2a}{\pi}\right)^{1/4}\left(1+\frac{2i\hbar ta}{m}\right)^{-1/2} e^{-\alpha_t(x-x_t)^2+\frac{i}{\hbar}p_0(x-x_t)+\frac{i}{\hbar}\frac{p_0^2}{2m}t} \\
&= \left(\frac{2a}{\pi}\right)^{1/4} e^{-\frac{1}{2}\ln\left(1+\frac{2i\hbar ta}{m}\right)} e^{-\alpha_t(x-x_t)^2+\frac{i}{\hbar}p_0(x-x_t)+\frac{i}{\hbar}\frac{p_0^2}{2m}t} \\
&= \left(\frac{2a}{\pi}\right)^{1/4} e^{-a_t(x-x_t)^2+\frac{i}{\hbar}p_0(x-x_t)+\frac{i}{\hbar}\gamma_t}
\end{aligned}$$

where we have assumed $\gamma_t = \frac{p_0^2}{2m}t + \frac{i\hbar}{2}\ln\left(1+\frac{2i\hbar ta}{m}\right)$

General Mathematical Form for a Free-Particle Gaussian Wavepacket

$$\psi_{w.p.}(x,t) = \left(\frac{2a}{\pi}\right)^{1/4} \exp\left[-\alpha_t\left(x-x_t\right)^2 + \frac{i}{\hbar}p_0\left(x-x_t\right) + \frac{i}{\hbar}\gamma_t\right] \tag{3.36}$$

$$\text{where } \alpha_t = \frac{a}{\left(1+\dfrac{2i\hbar ta}{m}\right)} \tag{3.37}$$

$$p_0 = \hbar k_0 \tag{3.38}$$

$$x_t = \frac{\hbar t k_0}{m} \tag{3.39}$$

$$\text{and } \gamma_t = \frac{p_0^2}{2m}t + \frac{i\hbar}{2}\ln\left(1+\frac{2i\hbar ta}{m}\right) \tag{3.40}$$

3.7 Gaussian Wavepacket Dynamics under Linear Potential

Equation (3.36) represents the Gaussian wavepacket for a free particle which does not experience any potential. In many instances of significant chemical importance, a quantum particle moves under the influence of a potential. For example, Figure 3.7 shows the ground electronic and first excited electronic potential energy curves for sodium iodide (NaI) diatomic species. As briefly demonstrated in Chapter 1, taking the example of a simple harmonic oscillator, any potential energy curve for a diatomic species can conveniently be reduced to that of a one-body problem defined by the reduced mass $\mu = \frac{m_1 m_2}{m_1+m_2}$ and by the relative coordinate $x = (R - R_0)$. Here, m_1 and m_2 are the masses of two atoms, R_0 is the equilibrium bond distance, and R represents the instantaneous bond length during the vibration (manifesting nuclear motion). The following presentation has been prepared based on this reduced mathematical picture for the NaI diatomic species.

The photoexcitation of the ground state NaI to the first excited electronic state using a short laser pulse creates a (vibrational) wavepacket on the excited electronic potential energy curve (refer to red wavepacket (a) in Figure 3.7). This wavepacket represents (with its simplest interpretation based on the above-mentioned reduced one-body mathematical model of a diatomic species) a possible position of the particle with reduced mass immediately following the creation of the wavepacket (at $t = 0$ on the excited electronic potential energy curve. The wavepacket (a) experiences a non-zero-interaction potential in the excited state and, consequently, moves on the same curve, evolving to the wavepackets (b) and (c) at two different later times. This wavepacket propagation manifests the chemical dynamics of the electronically excited NaI species.

In general, it is not possible to explore analytically the wavepacket dynamics of a particle on any arbitrary potential energy curve (not even on the simplest diatomic NaI excited state potential energy curve given in Figure 3.7). For that, we have to rely on numerical methods. We will learn numerical methods and their Python implementations to solve one-dimensional wavepacket

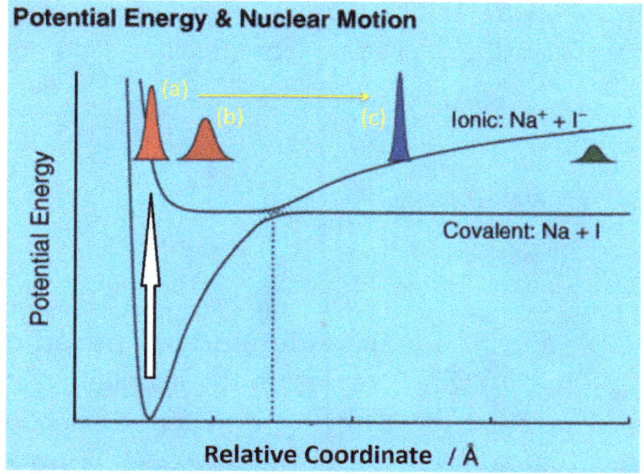

Figure 3.7. Ground electronic and first excited electronic potential energy curves of NaI are shown (with many other important details, such as the crossing point, which, for now, can be overlooked). The photoexcitation of the ground state species NaI using a short pulse to the first electronically excited state creates a wavepacket (refer to red packet (a)), which moves on the excited state potential energy curve. Packets (b) and (c) are two snapshots of the wavepacket as time progresses. The wavepacket, instead of moving only on the excited state potential energy curve, may also jump from the excited electronic potential energy curve to the ground electronic potential energy curve; however, for now, this detail can also be overlooked. The figure has been adopted (and modified to fulfill the demand of the present context) with permission from Ref. 7. Copyright (2001) American Chemical Society.

dynamics in Chapter 5 and in PythonChapter E. Fortunately, an analytical approach, very similar to what was used in the earlier section for a free particle, can be employed to explore the wavepacket dynamics of a particle experiencing **linear and quadratic potentials**. Our task in this section is to explore the wavepacket dynamics of a particle experiencing a linear potential. The problem of quadratic potential is left for the reader's self-study.

The general form of a wavepacket given in equation (3.36) represents a free-particle Gaussian wavepacket. If the particle experiences a potential (more specifically, **a potential gradient which exerts force on the particle**), clearly, its momentum does not remain constant (the influence of the potential on the momentum of the particle will be discussed soon). In that case, a general form of the Gaussian wavepacket representing a particle experiencing a potential is given by

$$\psi_{w.p.}(x,t) = A\exp\left[-\alpha_t\left(x-x_t\right)^2 + \frac{i}{\hbar}p_t\left(x-x_t\right) + \frac{i}{\hbar}\gamma_t\right] \tag{3.41}$$

Here, α_t, x_t, p_t, and γ_t are **time-dependent parameters** of the wavepacket. x_t and p_t, which represent the average position (expectation value of position $x_t = \langle\hat{x}\rangle(t)$) and momentum (expectation value of momentum $p_t = \langle\hat{P}\rangle(t)$) of the wavepacket, respectively, are real, time-dependent parameters of the wavepacket. The quantities α_t and γ_t, on the other hand, are time-dependent complex functions which determine the width and phase of the wavepacket, respectively. For a particular problem, these parameters are obtained by first inserting the general expression of the wavepacket given in equation (3.41) into the TDSE and then by comparing the coefficients of the terms with a similar power of $(x-x_t)$. This procedure is followed to understand the wavepacket dynamics in the presence of a linear potential.

For the present demonstration, a linear potential, expressed by $V(x) = -bx$ in which a positive constant b represents the slope of the potential, is considered. The corresponding TDSE can be written as

$$i\hbar\frac{\partial}{\partial t}\psi_{w.p.}(x,t) = \left[-\frac{\hbar^2}{2m}\left(\frac{\partial^2}{\partial x^2}\right) - bx\right]\psi_{w.p.}(x,t) \tag{3.42}$$

Next, as stated above, insert the expression given in equation (3.41) into equation (3.42) (note that, often, a trial wavefunction in quantum mechanics is called an *ansatz*; therefore, equation (3.41) represents an *ansatz* for the present problem):

$$i\hbar\frac{\partial}{\partial t}\left(\exp\left[-\alpha_t\left(x-x_t\right)^2 + \frac{i}{\hbar}p_t\left(x-x_t\right) + \frac{i}{\hbar}\gamma_t\right]\right)$$

$$= \left[-\frac{\hbar^2}{2m}\left(\frac{\partial^2}{\partial x^2}\right) - bx\right]$$
$$\times \exp\left[-\alpha_t (x - x_t)^2 + \frac{i}{\hbar}p_t (x - x_t) + \frac{i}{\hbar}\gamma_t\right] \quad (3.43)$$

To reduce equation (3.43), we need to calculate the respective derivatives:

$$i\hbar\frac{\partial}{\partial t}\psi_{w.p.}(x,t) = i\hbar A\frac{\partial}{\partial t}\left(\exp\left[-\alpha_t (x - x_t)^2 + \frac{i}{\hbar}p_t (x - x_t) + \frac{i}{\hbar}\gamma_t\right]\right)$$
$$= i\hbar\left(-\dot{\alpha}_t(x - x_t)^2 + 2\alpha_t(x - x_t)\dot{x}_t + \frac{i}{\hbar}\dot{p}_t(x - x_t)\right.$$
$$\left. - \frac{i}{\hbar}p_t\dot{x}_t + \frac{i}{\hbar}\dot{\gamma}_t\right)\psi_{w.p.}(x,t) \quad (3.44)$$

Here, $\dot{\alpha}_t = \frac{\partial \alpha_t}{\partial t}$ and so on.

On the other hand,

$$\left[-\frac{\hbar^2}{2m}\frac{\partial^2}{\partial x^2} - bx\right]\psi_{w.p.}(x,t)$$
$$= -\frac{\hbar^2}{2m}\frac{\partial}{\partial x}\left[\left(-2\alpha_t (x - x_t) + \frac{i}{\hbar}p_t\right)\psi_{w.p.}(x,t)\right]$$
$$- bx\psi_{w.p.}(x,t)$$
$$= -\frac{\hbar^2}{2m}\left[-2\alpha_t + \left(-2\alpha_t (x - x_t) + \frac{i}{\hbar}p_t\right)^2\right]$$
$$\psi_{w.p.}(x,t) - bx\psi_{w.p.}(x,t)$$
$$= -\frac{\hbar^2}{2m}\left[-2\alpha_t + 4\alpha_t^2 (x - x_t)^2 - 4\alpha_t (x - x_t)\frac{i}{\hbar}p_t - \frac{p_t^2}{\hbar^2}\right]$$
$$\psi_{w.p.}(x,t) - bx\psi_{w.p.}(x,t)$$
$$= -\frac{\hbar^2}{2m}\left[-2\alpha_t + 4\alpha_t^2 (x - x_t)^2 - 4\alpha_t (x - x_t)\frac{i}{\hbar}p_t - \frac{p_t^2}{\hbar^2}\right.$$
$$\left. - b(x - x_t) - bx_t\right]\psi_{w.p.}(x,t) \quad (3.45)$$

Now equating the similar powers of $(x - x_t)$ given in equation (3.44) and (3.45), we get following relationships:

(a) Equating coefficients of $(x - x_t)^2$, we get

$$-i\hbar\dot{\alpha}_t = -\frac{\hbar^2}{2m}4\alpha_t^2$$

$$\text{or,}\ \dot{\alpha}_t = -\frac{2i\hbar}{m}\alpha_t^2 \tag{3.46}$$

(b) Equating coefficients of $(x - x_t)$, we get

$$-\dot{p}_t = -b$$

$$\text{or,}\ \frac{dp_t}{dt} = b$$

$$\text{or}\ p_t = p_0 + bt \tag{3.47}$$

where, p_0 represents the initial (at $t = 0$) average momentum.

(c) Equating coefficients of $i(x - x_t)$, we get

$$2\hbar\alpha_t\dot{x}_t = \frac{2\hbar}{m}\alpha_t p_t$$

$$\text{or,}\ \dot{x}_t = \frac{p_t}{m} \tag{3.48}$$

Inserting equation (3.47) into equation (3.48), we get

$$\dot{x}_t = \frac{p_0}{m} + \frac{bt}{m} \tag{3.49}$$

$$\text{or,}\ \frac{dx_t}{dt} = \frac{p_0}{m} + \frac{bt}{m}$$

$$\text{or,}\ x_t - x_0 = \frac{p_0}{m}t + \frac{bt^2}{2m} \tag{3.50}$$

Here, x_0 represents the center (or the expectation value of position) of the Gaussian at $t = 0$. Furthermore, equation (3.49) shows that the velocity of the wavepacket increases as a function of time, resulting in the acceleration of the wavepacket as time progresses. This nicely teams up with the fact that the downward slope of the potential accelerates the wavepacket.

Key Points to Remember

- A particle in quantum mechanics is nothing but a wavepacket.

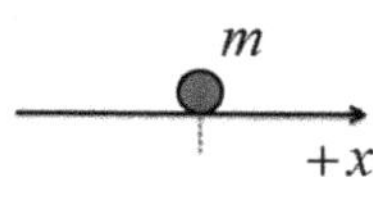

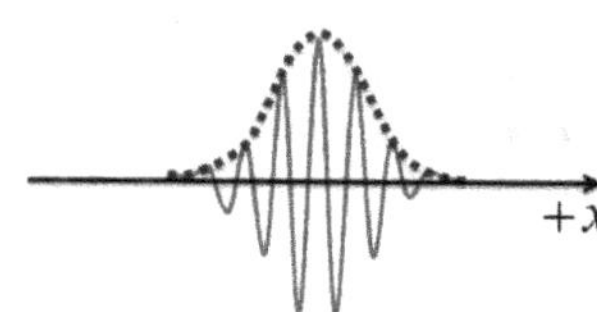

- A free particle wavepacket at any given time can be expressed as $\psi_{w.p.}(x,t) = \int_{-\infty}^{+\infty} A(k)e^{ikx}e^{-i\frac{E_k t}{\hbar}}dk$, where $A(k)$ can be derived from initial wavepacket, $A(k) = \frac{1}{2\pi}\int_{-\infty}^{+\infty}\psi_{w.p.}(x,0)e^{-ikx}dx$.
- A free particle Gaussian wavepacket spreads out in position space while it remains unaltered in momentum space as time progresses.
- The general expression of a one-dimensional Gaussian wavepacket representing a quantum particle (experiencing any potential) is given by $\psi_{w.p.}(x,t) = A \exp\left[-\alpha_t(x-x_t)^2 + \frac{i}{\hbar}p_t(x-x_t) + \frac{i}{\hbar}\gamma_t\right]$, where the time-dependent parameters α_t, x_t, p_t, and γ_t for a given problem can be obtained by inserting this *ansatz* into the TDSE.

Notes, References, and Further Reading

1. The real form of the solution can be given by $\psi(x) = C\cos(kx)$ or $\psi(x) = D\sin(kx)$. However, the complex form of the solution is used here (instead of the trigonometric one) for mathematical convenience and simplicity.
2. The definition of the phase velocity of a plane wave is already presented in the context of optical wave in Chapter 1.
3. Using the normalization condition one can write $\int_{-\infty}^{+\infty} A^* e^{-ikx}Ae^{ikx}dx = 1$, or $A^2\int_{-\infty}^{+\infty}dx = 1$, or $A = 0$. Therefore, $\psi(x,t) = Ae^{i\left(kx-\frac{Et}{\hbar}\right)}$ is not a normalizable wavefunction within the given limit. Consequently, it does not represent a physically realizable state of a free particle. In fact, it is not possible to have a free particle with a definite kinetic energy, E. The state expressed by the wavefunction $\psi(x,t) = Ae^{i\left(kx-\frac{Et}{\hbar}\right)}$ represents a stationary state (for which the probability density is time-independent), which cannot describe a state of a freely moving particle because the probability density of a freely moving particle must change as time progresses.
4. D. J. Tannor, Chapter 2: The free-particle wavepacket. *Introduction to Quantum Mechanics: A Time-Dependent Approach.* University Science Books, Sausalito (CA) (2007).
5. Various properties of Dirac delta function are discussed in adequate detail in B. H. Bransden and C. J. Joachain, Appendix A: Fourier integrals and the

Dirac delta function. *Quantum Mechanics,* 2nd edn. Pearson Education Ltd., Edinburgh (2000).
6. For more details, readers are referred to (a) E. J. Heller, Time-dependent approach to semiclassical dynamics. *J. Chem. Phys.* 62, 1544 (1975); (b) G. C. Schatz and M. A. Ratner, Chapter 7: Quantum scattering theory. *Quantum Mechanics in Chemistry.* Dover Publications, Inc. New York (2015).
7. J. S. Baskin and A. H. Zewail, Freezing atoms in motion: principles of Femtochemistry and demonstration of laser spectroscopy, *J. Chem. Educ.* 78, 737–751 (2001).

Exercises

3.1. A free particle has the initial wavefunction

$$\begin{aligned}\psi(x,0) &= A \text{ when } -a \leq x \leq +a \\ &= 0 \text{ when } x > |a|\end{aligned}$$

A and a are positive and real constants, and consider a to be very small. Find $\psi(x,t)$.

3.2. Assume that a free particle has the initial wavefunction (a Gaussian wavepacket created at $x = x_0$)

$$\psi(x,0) = Ae^{-a(x-x_0)^2}$$

A, a, and x_0 are positive and real constants. Find $\psi(x,t)$.

3.3. For a quadratic potential, explore the Gaussian wavepacket dynamics, considering the general form of the Gaussian wavepacket as

$$\psi_{w.p.}(x,t) = A\ \exp\left[-\alpha_t\left(x-x_t\right)^2 + \frac{i}{\hbar}p_t\left(x-x_t\right) + \frac{i}{\hbar}\gamma_t\right]$$

Here, α_t, x_t, p_t, and γ_t carry the usual meanings. *Hint: Obtain expressions for the parameters* α_t, x_t, p_t and γ_t *for the quadratic potential.*

3.4. Check whether the uncertainty principle remains valid for Gaussian wavepacket dynamics under quadratic and linear potentials.

PythonChapter C

Fourier Transform in Quantum Mechanics

Highlights: *Position-Space and Momentum-Space Representations of Wavefunction, Construction of Fourier Grid, Examples with Python Implementation, Fast Fourier Transform, scipy.fftpack, fftfreq, and fft, Fourier Transform of Cosine Function and Gaussian Function.*

C.1 Introduction

In quantum mechanics, Fourier transform is used to relate the position-space and momentum-space wavefunctions (see Chapter 3 for a detailed discussion).

Fourier transform of position-space wavefunction to momentum-space wavefunction:

$$\phi(k,t) = \frac{1}{\sqrt{2\pi}} \int_{-\infty}^{+\infty} \psi(x,t)\, e^{-ikx} dx$$

Inverse Fourier transform of momentum-space wavefunction to position-space wavefunction:

$$\psi(x,t) = \frac{1}{\sqrt{2\pi}} \int_{-\infty}^{+\infty} \phi(k,t)\, e^{ikx} dk$$

Here, the position space represents the x-space of the particle (one-dimensional position space is considered here). The momentum space represents the k-space of the particle, where

$$k\left(=\frac{2\pi}{\lambda}=2\pi\xi\right) \tag{C.1}$$

is called the **angular wavenumber** for the reason explained in the following.

A periodic spatial structure can be represented by the function $\cos(kx)$, as depicted in Figure C.1. Here, (kx) represents the phase of the spatial periodic structure, featuring an angle, and x represents the linear advancement. Thus, k represents the magnitude of the wavevector which characterizes the nature of the periodic spatial structure.

Note that a linear displacement of λ causes a phase advancement of 2π. Therefore, we can write

$$\begin{aligned} k(x+\lambda) &= kx+2\pi \\ \text{or } k &= \frac{2\pi}{\lambda} \\ \text{or } k &= 2\pi\left(\frac{1}{\lambda}\right) \\ \text{or } k &= 2\pi\xi \end{aligned}$$

where ξ is called **spatial frequency**, representing how quickly the periodic spatial structure repeats per unit length.[1]

In this PythonChapter, we introduce the efficient fast Fourier transform (FFT) algorithm and its implementation in Python to perform the Fourier transform of the position-space wavefunction to the momentum-space wavefunction and the inverse Fourier transform of

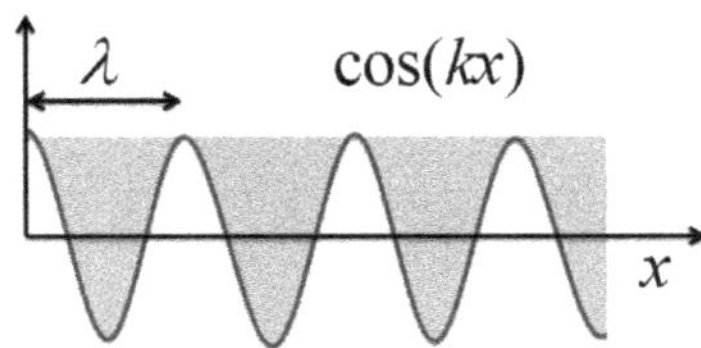

Figure C.1. Periodic spatial structure expressed by the function, $\cos(kx)$.

the momentum-space wavefunction to the position-space wavefunction. For details, readers are referred to a lovely textbook by Izaac and Wang.[2]

C.2 Construction of Fourier Grid

In order to represent the wavefunction either in x-domain or k-domain, the first step is to create the respective grid. We have understood in PythonChapter B that the **arange()** functionality of the **scipy** module can create the x-grid. How do we create the k-grid? The answer lies within the correspondence between x-domain and k-domain grids, which is discussed as follows:

(a) The same N number of grid points should be present in both x- and k-representations.
(b) However, Δk (the spacing between two adjacent grid points in k-space) cannot be randomly selected. The theory of **discrete Fourier transform (DFT)** mandates (proof is not given; the final result is directly used here) that Δk must be decided based on the following relation, which depends on Δx used to create the x-domain:

$$\Delta k \Delta x = \frac{2\pi}{N}$$

$$\text{or } \Delta k = \frac{2\pi}{N\,\Delta x} \tag{C.2}$$

Here, Δk and Δx represent the spacing between the adjacent grid points in the k-space and x-space, respectively, and N denotes the total number of grid points present in both spaces, as depicted in Figure C.2.

Figure C.2. x- and k-grids are depicted.

(c) Finally, **FFT** (which is a computing algorithm) implements the theory of DFT, allowing us to sample the spatial frequencies as

$$\xi = \frac{1}{N\,\Delta x} f \tag{C.3}$$

where f represents a list (more specifically an array) of spatial frequency components as given in the following:

$$f = \left[0, 1, 2, \ldots, \left(\frac{N}{2} - 1\right), \left(-\frac{N}{2}\right), \left(-\frac{N}{2} + 1\right), \ldots, -1\right] \quad \text{if } N \text{ is even} \tag{C.4}$$

$$\text{or } f = \left[0, 1, 2, \ldots, \left(\frac{(N-1)}{2}\right), \left(-\frac{(N-1)}{2}\right), \left(-\frac{(N-1)}{2} + 1\right), \ldots, -1\right] \quad \text{if } N \text{ is odd} \tag{C.5}$$

As a result, the final k-grid becomes $k = 2\pi\xi$. Note that the FFT algorithm, despite its very efficient implementation of DFT, converts a finite sequence of equally spaced samples of the x-space domain into a same length sequence of equally spaced samples of the k-space domain.

Based on the above-listed correspondence between x-domain and k-domain grids, one can easily find that the first element in the list of k-grid points corresponds to the zero spatial frequency component, the second element in the list of k-grid points corresponds to the spatial frequency component $\frac{2\pi}{N\Delta x}$, which features one oscillation in the entire spatial length ($N\Delta x$), and so on. This is schematically depicted in Figure C.3.

C.3 Illuminating Examples

Problem 1: Consider the x-grid, $[0, 1, 2, 3, 4, 5, 6, 7, 8, 9]$, which possesses total elements of (grid points) $N = 10$ (**even number**). The spacings in x-space are $\Delta x = 1$ (unit is the same as that of array elements). Now, based on the above discussion, one can write the

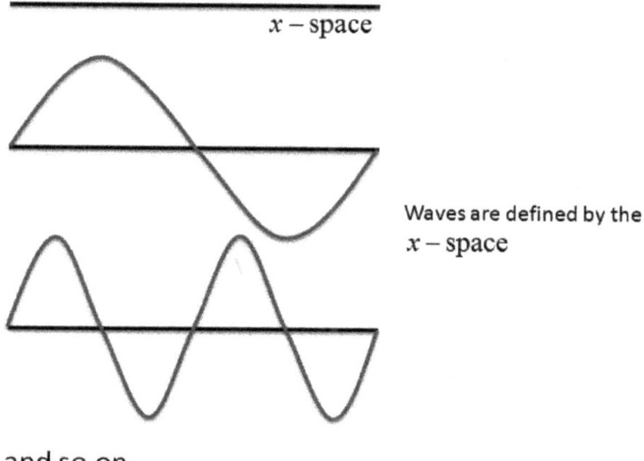

Figure C.3. A correspondence between the space domain and spatial frequency component is schematically shown.

spatial frequency components, which can be sampled through the FFT algorithm, as

$$\xi = \frac{1}{N\,\Delta x} f$$

or, for even N, one may explicitly write

$$\xi = \frac{1}{N\Delta x}\left[0, 1, 2, \ldots, \left(\frac{N}{2} - 1\right), \left(-\frac{N}{2}\right), \left(-\frac{N}{2} + 1\right), \ldots, -1\right]$$

As in the present problem, $N = 10$ and $\Delta x = 1$, we may write

$$\begin{aligned} \xi &= \frac{1}{10 \times 1}\left[0, 1, 2, \ldots, \left(\frac{10}{2} - 1\right), \left(-\frac{10}{2}\right), \left(-\frac{10}{2} + 1\right), \ldots, -1\right] \\ &= 0.1[0, 1, 2, 3, 4, -5, -4, -3, -2, -1] \\ &= [0, 0.1, 0.2, 0.3, 0.4, -0.5, -0.4, -0.3, -0.2, -0.1] \end{aligned} \tag{C.6}$$

As a result, note that 10 frequency components are sampled. Next, set up the corresponding k-grid points as

$$k = 2\pi\xi$$

Inserting the sampled frequency components (given in equation (C.6)) into the above expression, we get

$$k = 2\pi[0, 0.1, 0.2, 0.3, 0.4, -0.5, -0.4, -0.3, -0.2, -0.1]$$

or, approximately,

$$\begin{aligned} k &\approx 6.286[0, 0.1, 0.2, 0.3, 0.4, -0.5, -0.4, -0.3, -0.2, -0.1] \\ &\approx [0, 0.6286, 1.2571, 1.8858, 2.5144, -3.1430, -2.5144, \\ &\quad -1.8858, -1.2571, -0.6286] \end{aligned}$$

Note that in the present problem, the mandatory relation defined by the theory of DFT, $\Delta k \Delta x = \frac{2\pi}{N}$, is satisfied (check it yourself).

Python implementation

Python's module **scipy** offers the **scipy.fftpack** submodule,[3] which allows the user to perform FFT (which very efficiently implements the theory of DFT). In the present example, a k-grid is constructed for a given x-grid (which possesses $N = 10$ number of grid points with dx spacing) using the functionality **fftfreq(N,dx)** of the **scipy.fftpack** submodule. Note that the **fftfreq(N,dx)** functionality requires two inputs, the total number of grid points (N) present in the x-grid and the spacing (dx) of the x-grid points. The **fftfreq(N,dx)** functionality of the **scipy.fftpack** submodule returns a list (more specifically, an array) of spatial frequency components through the DFT of a sequence of position space domain samples of window length N and with spacing dx.

(a) Creating spatial frequency components

```
#Importing the required libraries
from scipy import arange
from scipy.fftpack import fftfreq
#Creating the x-grid
xmin=0
xmax=10
dx=1
x=arange(xmin,xmax,dx)
print(x)
#Creating k-grid
N=len(x)
xi=fftfreq(N,dx)
print(xi)
```

> ➢ Python's in-built functionality len() returns the length (or number of elements) of an array

```
[0 1 2 3 4 5 6 7 8 9]
[0.0 0.1 0.2 0.3 0.4 -0.5 -0.4 -0.3 -0.2 -0.1]
```

(a) Creating spatial frequency components

```
#Importing the required libraries
from scipy import arange,pi
from scipy.fftpack import fftfreq
#Creating the x-grid
xmin=0
xmax=10
dx=1
x=arange(xmin,xmax,dx)
print(x)
#Creating k-grid
N=len(x)
k=2*pi*fftfreq(N,dx)
print(k)
```

```
[0 1 2 3 4 5 6 7 8 9]
[0.0 0.62831853 1.25663706 1.88495559 2.51327412 -3.14159265
-2.51327412 -1.88495559 -1.25663706 -0.62831853]
```

Problem 2: Next, consider the x-grid to be [0, 1, 2, 3, 4, 5, 6, 7, 8, 9, 10]. The total number of elements (grid points) in this grid is $N = 11$ (odd number) and the spacing is $dx = 1$. As a result, the FFT algorithm provides the following list of spatial frequency components:

$$\xi = \tfrac{1}{N\Delta x}\, f(\text{for odd } N)$$

$$\text{or } \xi = \tfrac{1}{N\Delta x}\left[0, 1, 2, \dots, \left(\tfrac{(N-1)}{2}\right), \left(-\tfrac{(N-1)}{2}\right), \left(-\tfrac{(N-1)}{2} + 1\right), \dots, -1\right]$$

Inserting $N = 11$ and $dx = 1$, we get

$$\begin{aligned}
\xi &= \frac{1}{11 \times 1}\left[0, 1, 2, \dots, \left(\frac{(11-1)}{2}\right), \left(-\frac{(11-1)}{2}\right),\right. \\
&\quad \times \left.\left(-\frac{(11-1)}{2} + 1\right), \dots, -1\right] \\
&= 0.091[0, 1, 2, 3, 4, 5, -5, -4, -3, -2, -1] \\
&= [0, 0.091, 0.182, 0.273, 0.364, 0.455, -0.455, -0.364, -0.273, \\
&\quad -0.182, -0.091]
\end{aligned}$$

Note that, now, 11 frequency components are sampled. Next, we may set up the corresponding k-grid as

$$\begin{aligned}
k &= 2\pi[0, 0.091, 0.182, 0.273, 0.364, 0.455, -0.455, -0.364, -0.273, \\
&\quad -0.182, -0.091]
\end{aligned}$$

or, approximately,

$$\begin{aligned}
&\approx [0, 0.572, 1.144, 1.716, 2.288, 2.860, -2.860, -2.288, -1.716, \\
&\quad - 1.144, -0.572]
\end{aligned}$$

Note that, similar to the previous problem, in the present problem, the defining condition $\Delta k\, \Delta x = \frac{2\pi}{N}$ is satisfied (check it yourself).

Python implementation:

```
#Importing the required libraries
from scipy import arange,pi
from scipy.fftpack import fftfreq
#Creating the x-grid
xmin=0
xmax=11
dx=1
x=arange(xmin,xmax,dx)
print(x)
#Creating k-grid
N=len(x)
xi=fftfreq(N,dx)
print(xi)
k=2*pi*xi
print(k)
```

```
[0 1 2 3 4 5 6 7 8 9 10]
[0 0.09090909 0.18181818 0.27272727 0.36363636 0.45454545 -0.45454545 -0.36363636
-0.27272727 -0.18181818 -0.09090909]
[0 0.57119866 1.14239733 1.71359599 2.28479466 2.85599332 -2.85599332 -2.28479466
-1.71359599 -1.14239733 -0.57119866]
```

C.4 Fourier Transform of a Cosine Function

The Fourier transform of a discretized function is obtained by making use of the **fft(x-domain-function)** functionality of the **scipy.fftpack** submodule. Here, only input passed through the parenthesis is the discretized x-domain function. Before we present its Python implementation, we go over an important perspective of the Fourier transform, taking the example of a cosine function.

Let us assume that the x-domain function is given by

$$\psi(x) = \cos(k_0 x) = \frac{1}{2}\left(e^{ik_0 x} + e^{-ik_0 x}\right) \tag{C.7}$$

Here, the complex representation of the cosine function will be considered because it will make the subsequent mathematical derivation steps simple. The k-domain representation of the above function is obtained by the Fourier transform:

$$
\begin{aligned}
\phi(k) &= \frac{1}{\sqrt{2\pi}} \int_{-\infty}^{+\infty} \psi(x) e^{-ikx} dx \\
&= \frac{1}{\sqrt{2\pi}} \frac{1}{2} \int_{-\infty}^{+\infty} \left(e^{ik_0x} + e^{-ik_0x}\right) e^{-ikx} dx \\
&= \frac{1}{\sqrt{2\pi}} \frac{1}{2} \int_{-\infty}^{+\infty} e^{ik_0x} e^{-ikx} dx + \frac{1}{\sqrt{2\pi}} \frac{1}{2} \int_{-\infty}^{+\infty} e^{-ik_0x} e^{-ikx} dx
\end{aligned}
$$

$$
\text{or } \phi(k) = \frac{1}{\sqrt{2\pi}} \frac{1}{2} 2\pi\delta(k - k_0) + \frac{1}{\sqrt{2\pi}} \frac{1}{2} 2\pi\delta(k + k_0) \quad \text{(C.8)}
$$

Note that we have made use of an important representation of the Dirac delta function, which is discussed in Chapter 3: $\delta(k - k_0) = \frac{1}{2\pi} \int_{-\infty}^{+\infty} e^{ik_0x} e^{-ikx} dx$. Equation (C.8) shows that the Fourier transform of a cosine function is nothing but the Dirac delta function: function's value exists only when $k = k_0$ and $k = -k_0$. We prove this conclusion using the following Python implementation.

```
#Importing the required libraries
from scipy import arange,pi,cos
from scipy.fftpack import fftfreq,fft
from matplotlib.pyplot import plot,show,xlim
#Creating the x-grid
xmin=-100
xmax=100
dx=0.001
x=arange(xmin,xmax,dx)
#Defining a Cosine function in position space grid
y=cos(20*x)
#Creating k-grid
N=len(x)
k=2*pi*fftfreq(N,dx)
#Fourier transforming the Cosine function
y_k=fft(y)
plot(k,abs(y_k))
xlim(-30,30)
show()
```

> Note that we are plotting the absolute value of the *k*-domain function (obtained by using Python's built-in abs() functionality) on the *k*-grid.

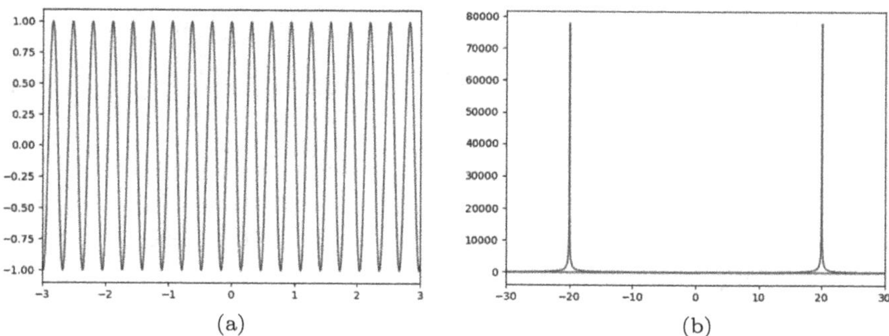

Figure C.4. (a) Cosine function has been plotted on the x-grid. (b) Absolute value of the function obtained after Fourier transforming the cosine function is plotted on the k-grid. Clearly both figures require further formatting (such as axis labeling, giving title, etc.). This task is left for readers.

As evident in Figure C.4, the $\cos(20x)$ function exhibits an angular wavenumber of 20 (unit). As a result, the delta function obtained after Fourier transforming the $\cos(20x)$ function is centered at $k = 20$ (or $k = -20$).

C.5 Fourier Transform of a Gaussian Wavepacket

Next assume that the x-domain function is expressed by a Gaussian wavepacket of the following form (refer to Chapter 3 for a discussion on such a wavepacket which possesses two parts: a carrier wave represented by $\cos(k_0 x)$ and an envelope represented by e^{-x^2})

$$\psi(x) = e^{-x^2}\cos(k_0 x) = \frac{1}{2}e^{-x^2}\left(e^{ik_0 x} + e^{-ik_0 x}\right) \tag{C.9}$$

Just like what we did before, here, consider the complex representation of the cosine part of the Gaussian for mathematical convenience. The Fourier transform of the Gaussian wavepacket can be easily obtained:

$$\begin{aligned}\phi(k) &= \frac{1}{\sqrt{2\pi}}\frac{1}{2}\int_{-\infty}^{+\infty} e^{-x^2}\left(e^{ik_0 x} + e^{-ik_0 x}\right)e^{-ikx}dx \\ &= \frac{1}{\sqrt{2\pi}}\frac{1}{2}\int_{-\infty}^{+\infty} e^{-x^2+i(k_0-k)x}dx + \frac{1}{\sqrt{2\pi}}\frac{1}{2}\int_{-\infty}^{+\infty} e^{-x^2}e^{-i(k_0+k)x}dx\end{aligned}$$

or, by making use of standard Gaussian integral, we get

$$\phi(k) = \frac{1}{\sqrt{2\pi}}\frac{1}{2}\sqrt{\pi}e^{-\frac{(k-k_0)^2}{4}} + \frac{1}{\sqrt{2\pi}}\frac{1}{2}\sqrt{\pi}\,e^{\frac{(k+k_0)^2}{4}} \tag{C.10}$$

Note that (as evident in equation (C.10)) the Fourier transform of a Gaussian renders a k-domain double Gaussian function, which is picked at $k = k_0$ and $k = -k_0$, respectively. This conclusion is numerically demonstrated using the following Python implementation.

```
#Importing the required libraries
from scipy import arange,pi,cos,exp
from scipy.fftpack import fftfreq,fft
from matplotlib.pyplot import plot,show,xlim
#Creating the x-grid
xmin=-100
xmax=100
dx=0.001
x=arange(xmin,xmax,dx)
#Defining a Gaussian wavepacket in position space grid
y=exp(-x**2)*cos(20*x)
#Creating k-grid
N=len(x)
k=2*pi*fftfreq(N,dx)
y_k=fft(y)
plot(k,abs(y_k))
xlim(-30,30)
show()
```

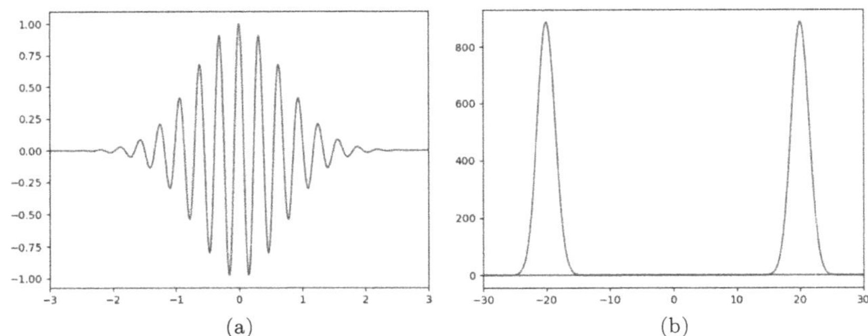

Figure C.5. (a) A Gaussian wavepacket is plotted on the x-grid. (b) Absolute value of the function obtained after Fourier transforming the Gaussian wavepacket is plotted on the k-grid. Similar to Figure C.4, this figure also needs better formatting – a task left for readers.

As evident in Figure C.5, the $e^{-x^2}\cos(20x)$ function exhibits a carrier wave angular wavenumber of 20 (unit). As a result, the Gaussian profile obtained after Fourier transforming the $e^{-x^2}\cos(20x)$ function is centered at $k = 20$ (or $k = -20$). Although analytical math or numerical method renders both positive and negative angular wavenumber components, one can simply analyze positive part of the Fourier transformed function to obtain practically useful information.

Key Points to Remember

- The **fftfreq(N,dx)** functionality of the **scipy.fftpack** submodule creates a list of spatial frequency components (ξ) which can further be used to create the k-grid points using the relation, $k = 2\pi\xi$.
- The **fft(Y)** functionality of the **scipy.fftpack** submodule Fourier transforms a discretized function **Y** from its x-space representation to k-space representation by implementing the Fast Fourier Transform algorithm which implements the theory of discrete Fourier transform (DFT).

Notes, References, and Further Reading

1. Note that ν is used to represent temporal frequency, and ξ is used to represent spatial frequency.
2. J. Izaac and J. Wang, Chapter 8: The Fourier transform. *Computational Quantum Mechanics*. Springer, Switzerland (2018).
3. Navigate the SciPy.fft documentation at `https://docs.scipy.org/doc/scipy/reference/fft.html#module-scipy.fft` for updated details of Python implementation of Fourier transformations.

Exercises

C.1. Create an x-grid ranging from -1 to $+1$. Find the corresponding k-grid. Print the spatial frequencies as a list. Confirm that the numbers of points on the both grids are equal. Define a function, $\sin(x)$. Find its values both on the x- and k-grids.

C.2. Plot a Gaussian function both on the x- and k-grids.

C.3. Plot the function $A\sin(10x) + B\sin(20x)$, both on the x- and k-grids. Vary A and B, and note your observation.

C.4. Write a Python program to confirm that the Fourier transform followed by the inverse Fourier transform of a Gaussian function reproduces the original function.

C.5. Write a Python program to Fourier transform a Lorentzian function.

C.6. Consider the position-space representation of the ground state wavefunction of a particle in a one-dimensional box of length L: $\psi(x) = \sqrt{\frac{2}{L}}\sin\left(\frac{\pi x}{L}\right)$. Numerically Fourier transform the wavefunction to obtain the momentum-space representation of the wavefunction.

C.7. Obtain the frequency components involved in the pulse created by $0.1\cos(x)+0.2\cos(1.1x)+0.3\cos(1.2x)+0.4\cos(1.3x)+0.3\cos(1.4x)+0.2\cos(1.1x)+0.1\cos(x)$.

C.8. Numerically Fourier transform the wavefunction of the $1s$ orbital of the hydrogen atom.

Chapter 4

Quantum Mechanics from a Linear Algebra Viewpoint

Highlights: *Hilbert Space, Norm, Inverse and Adjoint of Operator, Hermitian Operator, Unitary Operator, Commutator of Operators, Exponential Operator, Spectral Basis, Pseudo-Spectral Basis, Grid Representation, Matrix Algebra, Diagonalizing a Matrix, Eigenvalues and Eigenvectors, Matrix Representation of Hamiltonian Operator using Finite Difference Method.*

4.1 Introduction

In Chapter 3, we have analytically explored the dynamics of a free particle (which experiences zero potential) travelling in one-dimensional space: this renders the so-called "Gaussian wavepacket dynamics of free particle." A similar analytical approach can also be used to explore the Gaussian wavepacket dynamics of a particle experiencing a linear and a quadratic potential. For any arbitrary potential, however, an analytical approach fails, and we have to rely on numerical methods. In fact, the number of quantum dynamics problems that can be solved analytically is very limited. Therefore, it is quite instructive that we learn numerical methods, which will enable us to explore the quantum dynamics of a particle experiencing any potential.

Numerical methods to solve the TDSE is a gigantic subject which is fundamentally developed based on the matrix representation of quantum mechanical equations and on the realization that the mathematical language of quantum mechanics is linear algebra. In this chapter, we will develop a coherent sense of the wavefunction and the operator (two key constituents of quantum mechanics) from a linear algebraic viewpoint,[1] although, in earlier chapters, we have already made use of certain linear algebraic properties of wavefunctions and operators.

After reviewing the general properties of the quantum mechanically acceptable wavefunctions and the operators from a linear algebraic viewpoint, we will go over the basis set approach to quantum mechanics: this approach renders the matrix representation of wavefunction and operator. In the end, we will elucidate the methods to obtain eigenvalues and eigenfunctions of a quantum system by making use of the grid representation and finite difference method.

4.2 General Properties of the Wavefunction

(1) All well-behaved (physically acceptable) wavefunctions must be square normalizable (square integrable): The mathematical representation of this statement is

$$\int_{-\infty}^{+\infty} \psi^*(x)\psi(x)dx = \int_{-\infty}^{+\infty} |\psi(x)|^2 dx = 1, \text{ if } \psi(x) \text{ is a normalized wavefunction}$$

or if $\psi(x)$ is not a normalized wavefunction,

$$\int_{-\infty}^{+\infty} \psi^*(x)\psi(x)dx = \int_{-\infty}^{+\infty} |\psi(x)|^2 dx = \text{real positive constant (must be} < +\infty) \tag{4.1}$$

Here, $\psi(x)$ can be regarded as the wavefunction at $t = 0$, i.e., the initial wavefunction, $\psi(x, 0)$, just before the onset of the time-evolution process.

An illuminating Example

A Gaussian function (e^{-ax^2}) is a square-normalizable wavefunction in the limit $[-\infty, +\infty]$. However, a plane wave (e^{ikx}) form is not a square-normalizable wavefunction in the same limit. Prove it.

Answer: Consider the Gaussian wavefunction $\psi(x) = e^{-ax^2}$. To check whether it is square normalizable, one has to evaluate the integral, $\int_{-\infty}^{+\infty} |\psi(x)|^2 dx = \int_{-\infty}^{+\infty} e^{-2ax^2} dx = \sqrt{\frac{\pi}{2a}}$. As the value of the integral is a real positive constant (assuming that a is a real and positive constant), the Gaussian function is square normalizable in the given limit.

On the contrary, for the plane wave form, the integral becomes $\int_{-\infty}^{+\infty} |\psi(x)|^2 dx = \int_{-\infty}^{+\infty} e^{-ikx} e^{ikx} dx = [x]_{-\infty}^{+\infty} = \infty$. Therefore, $\psi(x) = Ae^{ikx}$ is not a square-normalizable wavefunction. As a result, it does not represent a physically realizable quantum state within the limit $[-\infty, +\infty]$. We have already developed this realization in Chapter 3.

One interesting mathematical fact about all square normalizable (integrable) wavefunctions is that they all follow properties of the **linear vector space** (also called the **Hilbert space**).[2] As exemplified in Figures 4.1(a) and 4.1(b), all wavefunctions related to a particle in a one-dimensional box form a Hilbert space, and similarly, all wavefunctions related to a quantum harmonic oscillator construct another Hilbert space. Operator acts on the wavefunction as linear transformation in a Hilbert space. This is why the mathematical language of quantum mechanics is linear algebra. We have already used, perhaps without noticing it, some of the interesting properties of the Hilbert space, such as the orthonormalization condition, in earlier chapters.

In linear algebraic language, the orthonormalization condition is expressed in terms of the inner product of two wavefunctions living in a specific Hilbert space. The inner product of two wavefunctions living in a specific Hilbert space represents a measure of their overlap in that Hilbert space. The complex conjugate of the inner product reverses the order:

$$\int_{-\infty}^{+\infty} \psi_1^*(x)\psi_2(x)dx = \left[\int_{-\infty}^{+\infty} \psi_2^*(x)\psi_1(x)dx\right]^* \tag{4.2}$$

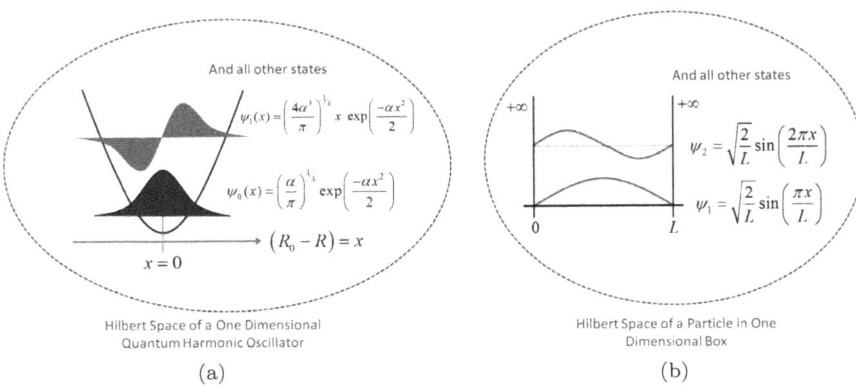

Figure 4.1. Examples of two different Hilbert spaces are given: one associated with (a) a one-dimensional quantum harmonic oscillator and another associated with (b) a particle in a one-dimensional box. Wavefunctions living in two different Hilbert spaces are not orthonormalized. Only wavefunctions living in a specific Hilbert space are orthonormalized.

The inner product always exists as long as both functions live in the same Hilbert space — a property of the Hilbert space. The "**norm**" or the normalization of a wavefunction (which has already been defined in PythonChapter B) is given by

$$\|\psi(x)\| = \left[\int_{-\infty}^{+\infty} |\psi(x)|^2 dx\right]^{\frac{1}{2}} \tag{4.3}$$

The above integral is nothing but the wavefunction's inner product with itself. Therefore, if $\|\psi(x)\| = 1$, the wavefunction $\psi(x)$ is called normalized. Two wavefunctions living in the same Hilbert space are called orthogonal when $\int_{-\infty}^{+\infty} \psi_1^*(x)\psi_2(x)dx = 0$. **A set of wavefunctions is called orthonormal when each wavefunction is normalized, and they are pair-wise orthogonal.**

(2) If $\psi_1, \psi_2, \ldots$ are square-integrable wavefunctions, any linear combination of these wavefunctions is also square integrable. The mathematical reason behind this statement is that any linear combination of wavefunctions belonging to the Hilbert space also lives in

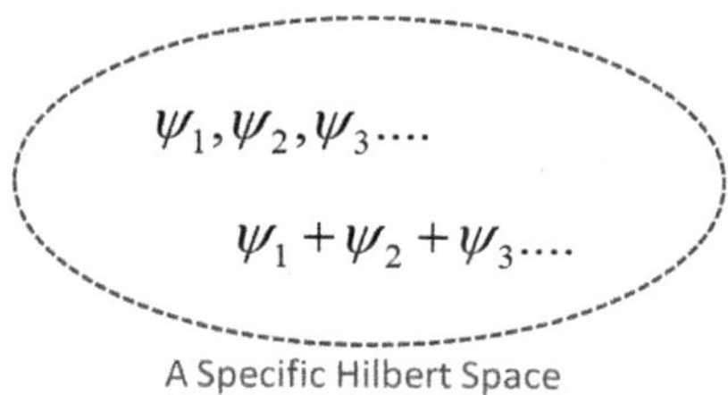

Figure 4.2. If certain wavefunctions live in a Hilbert space, their linear combinations also live in the same Hilbert space.

the same Hilbert space — a property of the Hilbert space. In fact, one defining condition of Hilbert space is that both wavefunctions and any of their linear combinations should be part of the same Hilbert space, as schematically depicted in Figure 4.2. This statement can be further rephrased: **Any linear combination of square-integrable wavefunctions is also square integrable.**

An illuminating discussion: If two time-independent wavefunctions ψ_m and ψ_n live in a Hilbert space, it means that they are solutions to the TISE of a specific quantum mechanical system related to that Hilbert space. In other words, the wavefunctions ψ_m and ψ_n are eigenstates of the Hamiltonian which defines the respective Hilbert space (and the quantum system). Furthermore, they must be square-integrable; otherwise, they would not live in a Hilbert space.

As noted in Chapter 1, the respective time-dependent wavefunctions can be represented by $\psi_m(x,t) = \psi_m(x)e^{-\frac{iE_m t}{\hbar}}$ and $\psi_n(x,t) = \psi_n(x)e^{-\frac{iE_n t}{\hbar}}$. One can easily prove that these time-dependent wavefunctions also live in the same Hilbert space (prove it yourself). Furthermore, the function obtained through a linear combination of these two wavefunctions, $\psi(x,t) = a_m\psi_m(x)e^{-\frac{iE_m t}{\hbar}} + a_n\psi_n(x)e^{-\frac{iE_n t}{\hbar}}$, also lives in the same Hilbert space. Prove it yourself. This means that $\psi(x,t)$ is a solution to the TDSE with the same Hamiltonian which has defined the Hilbert space. Prove it yourself.

An Illuminating Exercise

4.2. (a) Check whether the wavefunction $\psi(x) = x^{\frac{1}{2}}$ lives in a Hilbert space for the interval $[0, +1]$.

Answer: One can easily prove that this wavefunction is square-integrable for the interval $[0, +1]$:

$$\int_0^1 |\psi(x)|^2 dx = \int_0^1 x^{\frac{1}{2}} x^{\frac{1}{2}} dx = \int_0^1 x dx = \left[\frac{x^2}{2}\right]_0^1$$
$$= \frac{1}{2} = \text{a finite real positive constant}$$

Therefore, according to the condition given in equation (4.1), the wavefunction $\psi(x) = x^{\frac{1}{2}}$ lives in a Hilbert space for the given interval.

(b) Check if the $\frac{d}{dx}$ operator acts on the wavefunction $\psi(x) = x^{\frac{1}{2}}$, whether the resulting function lives in the Hilbert space for the interval $[0, +1]$.

Answer: The derivative operator converts the function $\psi(x) = x^{\frac{1}{2}}$ to another function $x^{\frac{-1}{2}}$. Interestingly, $x^{\frac{-1}{2}}$ does not live in the Hilbert space for the above-mentioned interval because

$$\int_0^1 x^{\frac{-1}{2}} x^{\frac{-1}{2}} dx = \int_0^1 \frac{1}{x} dx = [\ln(x)]_0^1 = -\infty,$$
$$\text{which is not a finite positive constant}$$

Therefore, sometimes a mathematical operator may take a function out of a Hilbert space. This causes a problem for the interpretation of quantum mechanics because if some operator converts a wavefunction living in a Hilbert space into a function outside the Hilbert space, the wavefunction is no longer useful in quantum mechanics. **A function living outside the Hilbert space does not carry statistical interpretation anymore, and this is why the wavefunction becomes useless in quantum mechanics**. Therefore, any mathematical operator cannot be used in quantum mechanics. Only a restricted set of mathematical operators should be used in quantum mechanics. Any operator which will be used in quantum mechanics must not take a wavefunction out of its Hilbert space. This restriction of operator space is the origin of the Hermitian operator in quantum mechanics.

4.3 General Properties of the Operator

(1) Inverse of an operator: Inverse of an operator $\hat{A}$ is defined by $\hat{A}^{-1}$ such that if $\hat{A}\psi = \phi$, then $\hat{A}^{-1}\phi = \psi$, or $\hat{A}(\hat{A}^{-1}\phi) = \hat{A}\psi = \phi$, or $\hat{A}\hat{A}^{-1} = \hat{1}$. Here, $\hat{1}$ is an operator which infers to "multiply by 1."

(2) Adjoint of an operator: The adjoint (or the Hermitian conjugate or the Hermitian adjoint) of an operator $\hat{A}$ is defined by $\hat{A}^{\dagger}$ such that

$$\int_{-\infty}^{+\infty} \phi^* \hat{A}\psi dx = \int_{-\infty}^{+\infty} (\hat{A}^{\dagger}\phi)^* \psi dx \tag{4.4}$$

Here, we have assumed that both ϕ and ψ live in the same Hilbert space.

Illuminating Examples

4.3. Find the adjoint of the position operator $\hat{x}$.

Answer: $\int_{-\infty}^{+\infty} \phi^* \hat{x}\psi dx = \int_{-\infty}^{+\infty} \phi^* x\psi dx = \int_{-\infty}^{+\infty} x\phi^* \psi dx$ (as x is simply a multiplication operator, it can be placed anywhere in the integrand.)

$$\text{or } \int_{-\infty}^{+\infty} x\phi^*\psi dx = \int_{-\infty}^{+\infty} (x\phi)^*\psi dx; \text{ as } x \text{ is real, } x = x^*.$$

$$\text{or } \int_{-\infty}^{+\infty} (x\phi)^*\psi dx = \int_{-\infty}^{+\infty} (\hat{x}^{\dagger}\phi)^*\psi dx$$

$$\text{Thus, } \hat{x} = \hat{x}^{\dagger}$$

4.4. Find the adjoint of the operator $\hat{A} = e^{-it}$.

Answer: $\int_{-\infty}^{+\infty} \phi^* \hat{A}\psi dx = \int_{-\infty}^{+\infty} \phi^* e^{-it}\psi dx = \int_{-\infty}^{+\infty} e^{-it}\phi^*\psi dx$. As e^{-it} does not depend on x (consequently, it features a multiplication operator), its position in the integrand can be changed without any constraint.

$$\text{or, } \int_{-\infty}^{+\infty} e^{-it}\phi^*\psi dx = \int_{-\infty}^{+\infty} (e^{it}\phi)^*\psi dx = \int_{-\infty}^{+\infty} (\hat{A}^{\dagger}\phi)^*\psi dx$$

$$\text{Thus, } \hat{A}^{\dagger} = e^{it}$$

4.5. Find the adjoint of the operator $\hat{A} = \frac{d}{dx}$.

Answer: $\int_{-\infty}^{+\infty} \phi^* \hat{A}\psi dx = \int_{-\infty}^{+\infty} \phi^* \frac{d}{dx}\psi dx = [\phi^*\psi]_{-\infty}^{+\infty} - \int_{-\infty}^{+\infty} \frac{d}{dx}\phi^*\psi dx$ (integrating by parts)[3]

$$\text{or } \int_{-\infty}^{+\infty} \phi^* \hat{A}\psi dx = 0 + \int_{-\infty}^{+\infty} \left[-\frac{d}{dx}\phi\right]^* \psi dx = \int_{-\infty}^{+\infty} [\hat{A}^{\dagger}\phi]^*\psi dx$$

(Continued)

(*Continued*)

The first term vanishes because both ϕ^* and ψ live in a Hilbert space, so they are square-integrable and become zero at both infinite limits.

$$\text{Thus, } \hat{A}^\dagger = -\frac{d}{dx}$$

4.6. Find the adjoint of the operator $\hat{A} = i\frac{d}{dx}$.

Answer: $\int_{-\infty}^{+\infty} \phi^* \hat{A}\psi dx = \int_{-\infty}^{+\infty} \phi^* i\frac{d}{dx}\psi dx = [\phi^* i\psi]_{-\infty}^{+\infty} - \int_{-\infty}^{+\infty} i\frac{d}{dx}\phi^*\psi dx = 0 + \int_{-\infty}^{+\infty} \left[i\frac{d}{dx}\phi\right]^* \psi dx = \int_{-\infty}^{+\infty} \left[\hat{A}^\dagger\phi\right]^* \psi dx$

$$\text{Thus, } \hat{A}^\dagger = i\frac{d}{dx}$$

From the above examples, note that for $\hat{A} = x$ and $\hat{A} = i\frac{d}{dx}$, $\hat{A} = \hat{A}^\dagger$ (i.e., **the operator itself is its own adjoint**).

(3) Hermitian operator: An operator that is its own adjoint is called a self-adjoint or Hermitian operator. The above illuminating examples show that both x and $i\frac{d}{dx}$ are Hermitian operators. An important property of a Hermitian operator is that the **eigenvalue of a Hermitian operator is real**. Prove it yourself. As a (time-independent) Hamiltonian operator is a Hermitian operator (prove it yourself), its eigenvalue represents the real (total) energy of the system for a particular quantum state. Furthermore, a closer look at a Hermitian operator, $\hat{A} = \left(i\frac{d}{dx}\right)$, reveals that

$$\begin{aligned}\int_{-\infty}^{+\infty} \phi^* \hat{A}\psi dx &= \int_{-\infty}^{+\infty} \phi^* \left(i\frac{d}{dx}\right) \psi dx = \int_{-\infty}^{+\infty} \left(i\frac{d\phi}{dx}\right)^* \psi dx \\ &= \int_{-\infty}^{+\infty} \psi \left(i\frac{d\phi}{dx}\right)^* dx\end{aligned}$$

The above equality can also be represented in the following general form:

$$\int_{-\infty}^{+\infty} f^* \hat{A} g dx = \int_{-\infty}^{+\infty} g(\hat{A}f)^* dx$$

Here, $\hat{A}$ is a Hermitian operator and both f^* and g live in the same Hilbert space. Note that we have made use of this equality in Chapter 2 for the derivation of the Ehrenfest theorem.

(4) Unitary operator: An operator $\hat{U}$ is a unitary operator if $\hat{U}^{-1} = \hat{U}^\dagger$ or, in other words, if the inverse of an operator is equal to the adjoint of the same operator. An important property of a unitary operator is that if $\hat{A}$ is a Hermitian operator, $e^{i\hat{A}}$ is a unitary operator. Prove it yourself. To prove it, one needs to know that the exponential function of an operator can be expressed in terms of the Taylor series expansion, as discussed in the following.

Self Study

4.7. $\hat{U} = e^{-i\frac{\hat{H}t}{\hbar}}$ is called the time-evolution operator. Prove that $\hat{U} = e^{-i\frac{\hat{H}t}{\hbar}}$ is a unitary operator. Here, $\hat{H}$ is a Hamiltonian operator.

(5) The exponential function of an operator (or exponential operator) is defined in terms of its Taylor series expansion. For example,

$$e^{\hat{A}} = \hat{1} + \hat{A} + \frac{\hat{A}^2}{2!} + \frac{\hat{A}^3}{3!} + \cdots \infty \tag{4.5}$$

Often, in time-dependent quantum mechanics, an exponential operator is used to move the initial wavefunction in time and space. For example, $e^{-i\frac{\hat{H}t}{\hbar}}$ represents the time propagator, which propagates the wavefunction in time. The operator $e^{-i\frac{\hat{H}t}{\hbar}}$ is called the time-evolution operator because it can be used to find the time evolution of $\psi(x,t)$ if $\psi(x,0)$ is known:

$$\psi(x,t) = e^{-i\frac{\hat{H}t}{\hbar}}\psi(x,0) \tag{4.6}$$

Equation (4.6) implies that if the initial wavefunction $\psi(x,0)$ is known, with the help of the time-evolution operator, one can find the wavefunction at any later time $\psi(x,t)$: Thus, the time evolution of a quantum system can be theoretically predicted. One can easily derive equation (4.6).

For the left-hand side (LHS) of equation (4.6): We assume that $\psi(x,t)$ is unknown but $\psi(x,0)$ is known. Then, using the Taylor series expansion, one can write

$$\psi(x,t) = \psi(x,0) + \left[\frac{\partial\psi(x,t)}{\partial t}\right]_{t=0} t + \frac{1}{2!}\left[\frac{\partial^2\psi(x,t)}{\partial t^2}\right]_{t=0} t^2 + \cdots\infty \tag{4.7}$$

For the right-hand side (RHS) of equation (4.6): Let us make use of the Taylor series expansion of the time-evolution operator:

$$\begin{aligned} e^{-i\frac{\hat{H}t}{\hbar}}\psi(x,0) &= e^{\frac{\hat{H}t}{i\hbar}}\psi(x,0) \\ &= \left[\hat{1} + \left(\frac{\hat{H}t}{i\hbar}\right) + \frac{1}{2}\left(\frac{\hat{H}t}{i\hbar}\right)^2 + \cdots\infty\right]\psi(x,0) \end{aligned} \tag{4.8}$$

From the time-dependent Schrödinger equation (TDSE), we may obtain the following equivalent relationship (this is true because for the entire presentation of this book, the Hamiltonian operator is considered time-independent):

$$\begin{aligned} i\hbar\tfrac{\partial}{\partial t}\psi(x,t) &= \hat{H}\psi(x,t) \\ \text{or } \tfrac{\hat{H}}{i\hbar} &\equiv \tfrac{\partial}{\partial t} \end{aligned} \tag{4.9}$$

Inserting the above equivalent relation into equation (4.8), we get

$$e^{-i\frac{\hat{H}t}{\hbar}}\psi(x,0) = \psi(x,0) + \left[\tfrac{\partial\psi(x,t)}{\partial t}\right]_{t=0} t + \tfrac{1}{2!}\left[\tfrac{\partial^2\psi(x,t)}{\partial t^2}\right]_{t=0} t^2 + \cdots\infty$$

$$\text{Therefore, LHS} = \text{RHS}$$

From the above derivation, it is also obvious that $\psi(x,t) = e^{-i\frac{\hat{H}t}{\hbar}}\psi(x,0)$ is a solution to the TDSE. We will make use of this solution to explore quantum dynamics in this book, as discussed in Chapter 5.

(6) Commutator of two operators:

The commutator of two operators is defined by

$$[\hat{A}, \hat{B}] = \hat{A}\hat{B} - \hat{B}\hat{A} \tag{4.10}$$

If $\hat{A}$ and $\hat{B}$ commute, then $[\hat{A}, \hat{B}] = \hat{0}$, or $\hat{A}\hat{B} = \hat{B}\hat{A}$.

Therefore, the order of the operation does not matter if two operators commute.

(7) All quantum mechanical operators are linear operators in the sense that

$$\hat{A}(\psi_1 + \psi_2) = \hat{A}\psi_1 + \hat{A}\psi_2, \text{ where } \hat{A} \text{ is a linear operator.}$$

(8) The product of operators is defined as operating from right to left: $\hat{B}\hat{A}\psi = \hat{B}(\hat{A}\psi)$. Similarly, the nth power of an operator is represented by

$$\hat{A}^n = \hat{A}\hat{A}\hat{A}\ldots$$

4.4 Basis Set Approach to Quantum Mechanics

In the above two sections, we have reviewed several general properties of quantum mechanically acceptable wavefunctions and operators and have defined the necessary linear algebra terminologies using simple analytical mathematics, such as integration, differentiation, exponentiation, and Taylor series expansion. While an analytical approach helps us develop the basic theoretical framework needed to understand the quantum mechanically acceptable wavefunction space and operator space, as pointed out earlier, our ultimate target in this chapter, however, is to prepare the platform for the numerical implementation of a quantum mechanical equation. To achieve the target, in this section, we will apprehend that a numerical implementation of quantum mechanical equations requires matrix representations of wavefunctions and operators. Their matrix representations originate from the concept of a **basis set approach to quantum mechanics**. This concept is first introduced here.

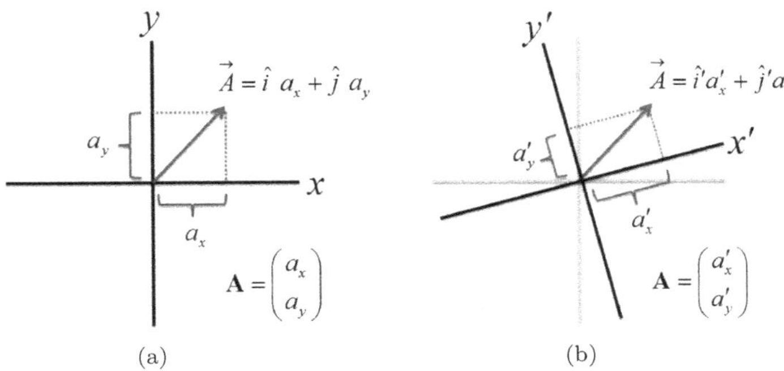

Figure 4.3. The same vector in a two-dimensional space can be represented by different bases. Basis $(\hat{\imath}, \hat{\jmath})$ is used in (a) and basis $(\hat{\imath}', \hat{\jmath}')$ is used in (b).

In the three-dimensional space, a vector is often represented by its components with respect to the xyz-axes (Cartesian coordinates). For simplicity, consider a vector $\vec{A} = \hat{\imath}a_x + \hat{\jmath}a_y$ in the xy-two-dimensional space. Its components with respect to the xy-axes are shown in Figure 4.3(a). Here, $\hat{\imath}$ and $\hat{\jmath}$ are unit vectors along the x - and y-axes, respectively. This vector can also be represented using a column matrix **A**, as shown in Figure 4.1(a). The elements of the matrix are nothing but the components with respect to the basis $(\hat{\imath}, \hat{\jmath})$, respectively.

One may select another two-dimensional coordinate system $x'y'$ (as shown in Figure 4.3(b)) and can represent the same vector as $\vec{A} = \hat{\imath}'a'_x + \hat{\jmath}'a'_y$. Thus, one can represent the same vector $\vec{A}$ using different bases, as long as the bases and corresponding components are known (or can be evaluated).

As a physical state in quantum mechanics is represented by a wavefunction which can be expanded with respect to certain orthonormal bases (using the property of the Hilbert space), the above vectorial argument also applies to a quantum mechanical state of a system. This is called the **basis set expansion of the wavefunction** (or the **basis set approach to quantum mechanics**). As a result, a wavefunction behaves like a vector in the basis set representation.

Let us first consider the initial wavefunction $\psi(x, 0)$, which is (in short) represented by $\psi(x)$ — the wavefunction before the onset of the time-evolution process. Assume that the wavefunction $\psi(x)$ can be expanded using the orthonormal bases of ϕ_i:

$$\psi(x) = \sum_{i=0}^{\infty} c_i\phi_i(x) = c_0\phi_0(x) + c_1\phi_1(x) + \cdots\infty \tag{4.11}$$

where $\phi_i(x)$ represent the orthonormal bases for which $\int_{-\infty}^{+\infty} \phi_j^*(x)$ $\phi_i(x)dx = \delta_{jt}$, representing the Kronecker delta (= 1 if $i = j$ and = 0 if $i \neq j$), and c_i are the respective components (expansion coefficients). Furthermore, as apprehended already, both ψ and ϕ_i must live in the same Hilbert space. Equation (4.11) formally represents an infinite sum. Therefore, a formal mathematical framework of quantum mechanics in the basis set approach represents an infinite-dimensional Hilbert space.

However, no practical computation can deal with infinite-dimensional space. It can only deal with a finite dimension of the bases. Therefore, the sum in equation (4.11) must be truncated to some finite number N (which may well be very large but must still be finite). This requirement can be fulfilled by making use of a reduced Hilbert space (or one can call it a computational Hilbert space). In that case, one would assume that all the general properties of the wavefunctions and the operators (discussed already in earlier sections of this chapter) also remain valid in the reduced Hilbert space. In fact, there are mathematical formulations which rigorously support this assumption.[4] Thus, we have the following matrix representation of a wavefunction in reduced (N-dimensional) Hilbert space with respect to the orthonormal bases of ϕ_i:

$$\psi(x) = \begin{pmatrix} c_0 \\ c_1 \\ c_2 \\ \cdot \\ \cdot \\ \cdot \\ c_{N-1} \end{pmatrix}$$

An Illuminating Example

4.8. For two orthonormal bases, $\phi_1(x)$ and $\phi_2(x)$, the wavefunction can be represented as $\psi(x) = c_1\phi_1(x)+c_2\phi_2(x)$ (using the basis set approach). Express the wavefunction in matrix form.

Answer: The above equation can be simply represented as a column matrix (which is a vector in linear algebra) representing a state in quantum mechanics with respect to the (ϕ_1, ϕ_2) basis:

$$\psi(x) = \begin{pmatrix} c_1 \\ c_2 \end{pmatrix}$$

Clearly, as mentioned above, if the bases are changed, the elements of the column matrix will also change; however, the wavefunction remains the same.

4.4.1 *If the bases are known, how do we evaluate the components (expansion coefficients)?*

The wavefunction $\psi(x)$ can be expanded using the bases of $\phi_i(x)$ (the basis set expansion of the wavefunction):

$$\psi(x) = \sum_{i=0}^{N} c_i\phi_i(x) \tag{4.12}$$

where, as usual, $\phi_i(x)$ represent the orthonormal bases for which $\int_{-\infty}^{+\infty} \phi_j^*(x)\phi_i(x)dx = \delta_{ji}$ and c_i are the respective expansion coefficients. One can find the expansion coefficients c_i by first multiplying the equation (4.12) by $\phi_j^*(x)$ from left and then by integrating over the entire x-space:

$$\int_{-\infty}^{+\infty} \phi_j^*(x)\psi(x)dx = \int_{-\infty}^{+\infty} \phi_j^*(x)\left[\sum_{i=0}^{N} c_i\phi_i(x)\right] dx$$

$$= \sum_{i=0}^{N} c_i \int_{-\infty}^{+\infty} \phi_j^*(x)\phi_i(x)dx = \sum_{i=0}^{N} c_i\delta_{ji} = c_j$$

Therefore, the jth expansion coefficient is given by

$$c_j = \int_{-\infty}^{+\infty} \phi_j^*(x)\psi(x)dx \tag{4.13}$$

The above exercise shows that if the bases are known, one can compute the respective expansion coefficients. **But what bases do we select?** There are two answers to this question.

(a) Spectral basis: Each basis function ϕ_i may correspond to the orthonormal (stationary) eigenstate of the Hamiltonian of the system (obtained from the solution to the time-independent Schrödinger equation (TISE), $\hat{H}\phi_i = E_i\phi_i$, before the onset of the time evolution). By nature, as discussed in Chapter 1, the TISE gives a set of solutions. These solutions represent the spectrum of a quantum system. This is why these bases are called spectral bases. By nature, they are all delocalized (global) in the entire space, as schematically shown in Figure 4.4(a): Each ϕ is a function extending from $-\infty$ to $+\infty$.

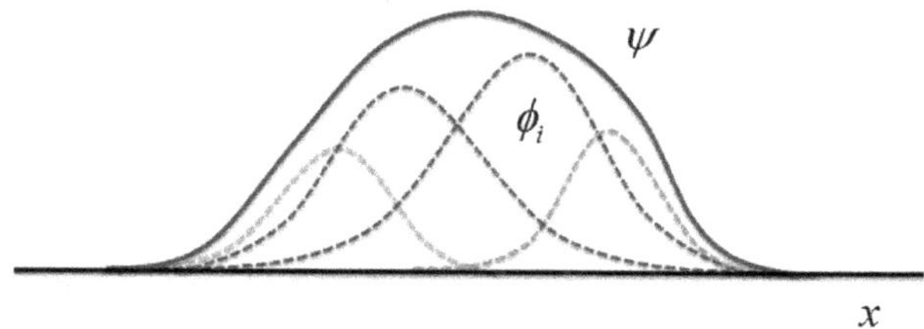

(b) Psudo-spectral Bases to Represent the Wavefunction

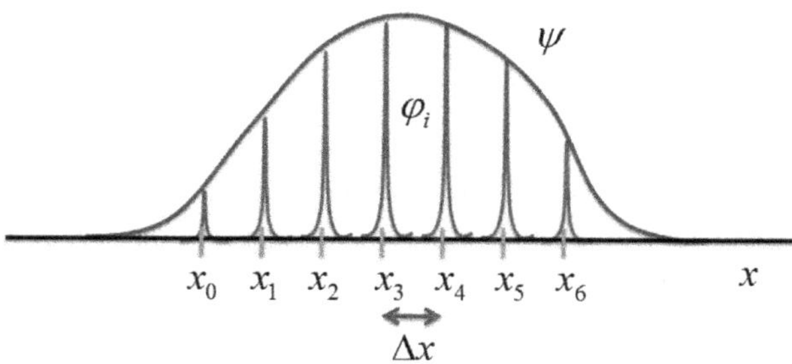

Figure 4.4. Schematic diagrams depicting (a) the spectral basis representation of the wavefunction and (b) the pseudo-spectral basis representation of the wavefunction. Note that in the latter, Δx is selected to be very small; however, in (b), the spacing Δx is intentionally drawn large to improve the visual clarity of the figure. See text for more details.

(b) Pseudo-spectral basis: The direct numerical implementation of the spectral basis is not always straightforward. This is why almost all currently available numerical methods for solving the TDSE do not make use of the spectral basis. Rather, a pseudo-spectral basis approach is used under the grid representation. As we will make use of the grid representation of a wavefunction to find numerical solutions to the TDSE in this textbook, we will go over the pseudo-spectral basis approach in detail.

A wavefunction, by its nature, is continuous (as stated by a postulate of quantum mechanics) in position space. Under the grid representation, however, a continuous wavefunction is expressed on a set of position grid points. This is already discussed in Python-Chapter B. The discretization of the wavefunction on a grid is done by dividing the entire position space into a uniform grid (as schematically shown in Figure 4.4(b)). The amplitude at a grid point represents the respective coefficient of the localized basis function. These localized basis functions are called pseudo-spectral bases. They are in contrast to the conventional spectral bases, which are delocalized in nature (as schematically evident in Figures 4.4(a) and 4.4(b)).

One simple example of pseudo-spectra bases is the Kronecker delta ($\delta_{ji} = 1$ if $i = j$ and $= 0$ if $i \neq j$) functions, which create an orthonormal set. A wavefunction $\psi(x)$ can be expanded in the basis of δ_{ji} such that

$$\psi(x_j) = \sum_{i=0} c_i \delta_{ij}$$

where c_i represents the amplitude at the ith grid point. These coefficients are nothing but the discretized values of the wavefunction on the x-grid:

$$\psi(x_0) = c_0$$
$$\psi(x_1) = c_1$$

$$\psi(x_2) = c_2$$

$$\vdots$$

$$\psi(x_{N-1}) = c_{N-1}$$

The above discretized wavefunction is conveniently represented by a column matrix as follows:

$$\psi(x) = \begin{pmatrix} c_0 \\ c_1 \\ c_2 \\ \cdot \\ \cdot \\ \cdot \\ c_{N-1} \end{pmatrix}$$

Note that, as stated earlier, $\psi(x)$ represents the initial ($t = 0$) wavefunction, $\psi(x, 0)$, which is the wavefunction just before the onset of the time-evolution process. Clearly, during the time-evolution process, the elements of the column matrix (which represent the wavefunction at any later time $\psi(x, t)$) will change. As a result, in quantum dynamics, the expansion coefficient, c_i, becomes time-dependent. This will be clearly evident in PythonChapter E.

Once a wavefunction is represented by a column matrix using the pseudo-spectra basis approach under the grid representation, our next important question arises: **What is the matrix representation of the quantum mechanical operators under the grid representation?** Before we address this question, however, we will quickly review the salient features of matrix algebra[1] because we will have to make use of matrix algebra to answer this question.

4.5 Brief Review of Matrix Algebra

In linear algebra, a matrix represents an array of numbers or mathematical expressions, arranged in rows and columns, as shown in the following:

$$\begin{array}{c} \\ 1 \\ 2 \\ 3 \\ \vdots \\ \text{Rows} \rightarrow m \end{array} \begin{array}{c} \begin{array}{ccccc} 1 & 2 & 3 & \cdots & n \leftarrow \text{Columns} \end{array} \\ \begin{pmatrix} a_{11} & a_{12} & a_{13} & \cdot & \cdot \\ a_{21} & a_{22} & \cdot & \cdot & \cdot \\ a_{31} & \cdot & \cdot & \cdot & \cdot \\ \cdot & \cdot & \cdot & \cdot & \cdot \\ \cdot & \cdot & \cdot & \cdot & \cdot \end{pmatrix}_{m\times n} \end{array}$$

When $m = n$, the matrix is called a square matrix. Certain rules are followed to carry out algebraic operations (such as addition and multiplication) with matrices.

(a) Matrix addition: Two matrices can be added if they are of same dimensions. Their corresponding elements are added as follows:

$$\mathbf{A} = \begin{pmatrix} a_{11} & a_{12} \\ a_{21} & a_{22} \end{pmatrix} \quad \mathbf{B} = \begin{pmatrix} b_{11} & b_{12} \\ b_{21} & b_{22} \end{pmatrix}$$

$$\mathbf{A} + \mathbf{B} = \begin{pmatrix} a_{11} + b_{11} & a_{12} + b_{12} \\ a_{21} + b_{21} & a_{22} + b_{22} \end{pmatrix}$$

The final matrix elements are represented by $(\mathbf{A}+\mathbf{B})_{ij} = \mathbf{A}_{ij} + \mathbf{B}_{ij}$.

(b) Scalar multiplication: Each element is multiplied by the scalar λ:

$$\mathbf{A} = \begin{pmatrix} a_{11} & a_{12} \\ a_{21} & a_{22} \end{pmatrix}$$

$$\lambda\mathbf{A} = \begin{pmatrix} \lambda a_{11} & \lambda a_{12} \\ \lambda a_{21} & \lambda a_{22} \end{pmatrix}$$

(c) Matrix multiplication:

$$\mathbf{A} = \begin{pmatrix} a_{11} & a_{12} \\ a_{21} & a_{22} \end{pmatrix} \quad \mathbf{B} = \begin{pmatrix} b_{11} & b_{12} \\ b_{21} & b_{22} \end{pmatrix}$$

$$\mathbf{AB} = \begin{pmatrix} a_{11}b_{11} + a_{12}b_{21} & a_{11}b_{12} + a_{12}b_{22} \\ a_{21}b_{11} + a_{22}b_{21} & a_{21}b_{12} + a_{22}b_{22} \end{pmatrix}$$

The final matrix elements are represented by $(\mathbf{AB})_{ik} = \sum_{j=1}^{N} \mathbf{A}_{ij}\mathbf{B}_{jk}$.

(d) Transpose of a matrix: The transpose of a matrix is obtained by interchanging its rows and columns, $(\mathbf{A}_{ij})^T = \mathbf{A}_{ji}$:

$$\mathbf{A} = \begin{pmatrix} a_{11} & a_{12} \\ a_{21} & a_{22} \end{pmatrix}$$

$$\mathbf{A}^T = \begin{pmatrix} a_{11} & a_{21} \\ a_{12} & a_{22} \end{pmatrix}$$

Thus, the transpose of a column matrix is a row matrix:

$$\mathbf{A} = \begin{pmatrix} a_1 \\ a_2 \\ a_3 \end{pmatrix}$$

$$\mathbf{A}^T = (a_1 \; a_2 \; a_3)$$

A column matrix is called a vector in linear algebra. This is why, from a linear algebraic viewpoint and under grid representation, a wavefunction is a vector. A square matrix is called symmetric if $\mathbf{A} = \mathbf{A}^T$ and antisymmetric if $\mathbf{A} = -\mathbf{A}^T$.

(e) Complex conjugate of a matrix: The complex conjugate of a matrix consists of the complex conjugate of every element:

$$\text{If } \mathbf{A} \begin{pmatrix} a_1 \\ a_2 \\ a_3 \end{pmatrix}, \text{ then } \mathbf{A}^* = \begin{pmatrix} a_1^* \\ a_2^* \\ a_3^* \end{pmatrix}$$

(f) Adjoint (or Hermitian adjoint) of a matrix: It is similar to the transpose, but all elements are replaced by their respective complex conjugates, $(\mathbf{A}_{ij})^\dagger = (\mathbf{A}_{ji})^*$:

$$\text{if } \mathbf{A} = \begin{pmatrix} 1 \\ i \\ 2i \end{pmatrix}, \text{ then } \mathbf{A}^\dagger = (1 \ \ -i \ \ -2i)$$

When $\mathbf{A}^\dagger = \mathbf{A}$, the matrix is said to be **Hermitian**. It is obvious from the above example that a column matrix cannot be a Hermitian matrix. Only a square matrix can be Hermitian if the condition $\mathbf{A}^\dagger = \mathbf{A}$ is fulfilled.

(g) Vector dot product: Two vectors $\vec{A}$ and $\vec{B}$ can be expressed by their components with respect to a certain basis $(\hat{\imath}, \hat{\jmath}, \hat{k})$:

$$\vec{A} = \hat{\imath}a_x + \hat{\jmath}a_y + \hat{k}a_z$$
$$\vec{B} = \hat{\imath}b_x + \hat{\jmath}b_y + \hat{k}b_z$$

The algebraic definition of dot product is $\vec{A} \cdot \vec{B} = a_x b_x + a_y b_y + a_z b_z$,
In matrix form,

$$\mathbf{A} = \begin{pmatrix} a_x \\ a_y \\ a_z \end{pmatrix} \quad \mathbf{B} = \begin{pmatrix} b_x \\ b_y \\ b_z \end{pmatrix}$$

$$\mathbf{A}^\dagger\mathbf{B} = (a_x^* \ a_y^* \ a_z^*) \begin{pmatrix} b_x \\ b_y \\ b_z \end{pmatrix}$$
$$= a_x^* b_x + a_y^* b_y + a_z^* b_z$$

The length of a vector is called the norm of the vector: $|\vec{A}| = (\vec{A} \cdot \vec{A})^{1/2} = \sqrt{|a_x|^2 + |a_y|^2 + |a_z|^2}$

(h) Inner and outer products: A column matrix represents a vector (in linear algebra) and a state in quantum mechanics. According to Dirac's bra–ket notation,

$$|A\rangle = \text{ket} = \text{column matrix} = \mathbf{A} = \begin{pmatrix} a_1 \\ a_2 \\ a_3 \end{pmatrix}$$

The adjoint of this vector is called **bra**: $\langle A| = bra = \text{row matrix} = \mathbf{A}^\dagger = (a_1^* \ a_2^* \ a_3^*)$.

The inner product is called **braket** (in Dirac notation):

$$\langle A|A\rangle = (a_1^* \ a_2^* \ a_3^*) \begin{pmatrix} a_1 \\ a_2 \\ a_3 \end{pmatrix} = |a_1|^2 + |a_2|^2 + |a_3|^2$$

The norm of the vector $\sqrt{\langle A|A\rangle} = \sqrt{|a_1|^2 + |a_2|^2 + |a_3|^2}$.

The outer product is called **ketbra**:

$$|A\rangle\langle A| = \begin{pmatrix} a_1 \\ a_2 \\ a_3 \end{pmatrix} (a_1^* \ a_2^* \ a_3^*) = \begin{pmatrix} a_1a_1^* & a_1a_2^* & a_1a_3^* \\ a_2a_1^* & a_2a_2^* & a_2a_3^* \\ a_3a_1^* & a_3a_2^* & a_3a_3^* \end{pmatrix}$$

$$= \begin{pmatrix} |a_1|^2 & a_1a_2^* & a_1a_3^* \\ a_2a_1^* & |a_2|^2 & a_2a_3^* \\ a_3a_1^* & a_3a_2^* & |a_3|^2 \end{pmatrix}$$

An Illuminating Example

Consider an eigenvector (a ket-vector), $|A\rangle = \mathbf{A} = \begin{pmatrix} 1 \\ i \\ 2i \end{pmatrix}$. To normalize it, first let us find the norm : Norm $= \sqrt{|1|^2 + |i|^2 + |2i|^2} = \sqrt{6}$. Therefore, the normalized eigenvector is $|A\rangle = \mathbf{A} = \frac{1}{\sqrt{6}}\begin{pmatrix} 1 \\ i \\ 2i \end{pmatrix} = \begin{pmatrix} \frac{1}{\sqrt{6}} \\ \frac{i}{\sqrt{6}} \\ \frac{2i}{\sqrt{6}} \end{pmatrix}$. The corresponding normalized bra-vector (which is the adjoint of **A**) is

(*Continued*)

(*Continued*)

$$\langle A| = \mathbf{A}^{\dagger} = \begin{pmatrix} \frac{1}{\sqrt{6}} & \frac{-i}{\sqrt{6}} & \frac{-2i}{\sqrt{6}} \end{pmatrix}$$

Then, the inner product is given by

$$\langle A|A\rangle = \mathbf{A}^{\dagger}\mathbf{A} = \begin{pmatrix} \frac{1}{\sqrt{6}} & \frac{-i}{\sqrt{6}} & \frac{-2i}{\sqrt{6}} \end{pmatrix} \begin{pmatrix} \frac{1}{\sqrt{6}} \\ \frac{i}{\sqrt{6}} \\ \frac{2i}{\sqrt{6}} \end{pmatrix} = 1$$

(i) Trace: A trace of a square matrix is the sum of all its diagonal elements:

$$\mathrm{tr}(A) = \sum_i A_{ii}$$

(j) Determinant of a square matrix:

$$\begin{vmatrix} a_{11} & a_{12} \\ a_{21} & a_{22} \end{vmatrix} = (a_{11}a_{22} - a_{21}a_{12})$$

(k) Inverse of a square matrix: If $\mathbf{A} = \begin{pmatrix} a_{11} & a_{12} \\ a_{21} & a_{22} \end{pmatrix}$, then $\mathbf{A}^{-1}$, or the inverse of the matrix, is defined by $\mathbf{A}\mathbf{A}^{-1} = \mathbf{I}$, where $\mathbf{I} = \begin{pmatrix} 1 & 0 \\ 0 & 1 \end{pmatrix}$ is a unit matrix or identity matrix (in which all diagonal elements are 1 and all off-diagonal elements are 0).

(l) Eigenvalues and eigenvectors of a square matrix: The eigenvalues of a matrix can be computed using the characteristic equation of the matrix:

$$|\mathbf{A} - \lambda\mathbf{I}| = \mathbf{0}$$

Here, $\mathbf{0}$ is the zero matrix (a null matrix in which all elements are zero). For each eigenvalue, there exists an eigenvector.

An Illuminating Example

4.9. Consider a matrix, $\mathbf{A} = \begin{pmatrix} 2 & 1 \\ 1 & 2 \end{pmatrix}$. Find the eigenvalues and eigenvectors of this matrix using its characteristic equation.

Answer: The characteristic equation of this matrix is given by

$$\begin{vmatrix} 2-\lambda & 1 \\ 1 & 2-\lambda \end{vmatrix} = 0$$

$$\text{or } (2-\lambda)^2 - 1 = 0$$

$$\text{or } \lambda^2 - 4\lambda + 3 = 0$$

$$\text{or, } \lambda_\pm = \tfrac{4\pm\sqrt{16-12}}{2} = \tfrac{4\pm 2}{2} = 3, 1$$

Thus, two eigenvalues of the matrix $\mathbf{A}$ are 3 and 1. Now, the eigenvector corresponding to each eigenvalue can be evaluated.

For $\lambda = 1$, the eigenvalue equation becomes

$$\begin{pmatrix} 2 & 1 \\ 1 & 2 \end{pmatrix}\begin{pmatrix} a_1 \\ a_2 \end{pmatrix} = 1\begin{pmatrix} a_1 \\ a_2 \end{pmatrix}$$

$$\text{or } \begin{pmatrix} 2a_1 + a_2 \\ a_1 + 2a_2 \end{pmatrix} = \begin{pmatrix} a_1 \\ a_2 \end{pmatrix}$$

Equating the components of the vector on the left- and right-hand sides, we get

$$2a_1 + a_2 = a_1$$

$$\text{or } a_1 = -a_2$$

If $a_1 = 1$, then $a_2 = -1$. In that case, the eigenvector corresponding to the eigenvalue 1 is $\begin{pmatrix} 1 \\ -1 \end{pmatrix}$. On the other hand, for $\lambda = 3$, the eigenvalue equation becomes

$$\begin{pmatrix} 2 & 1 \\ 1 & 2 \end{pmatrix}\begin{pmatrix} b_1 \\ b_2 \end{pmatrix} = 3\begin{pmatrix} b_1 \\ b_2 \end{pmatrix}$$

$$\text{or } \begin{pmatrix} 2b_1 + b_2 \\ b_1 + 2b_2 \end{pmatrix} = \begin{pmatrix} 3b_1 \\ 3b_2 \end{pmatrix}$$

(*Continued*)

(Continued)

Equating the components of the vector on the left- and right-hand sides, we get

$$2b_1 + b_2 = 3b_1$$

$$\text{or } b_1 = b_2$$

As we did before, if $b_1 = 1$, then $b_2 = 1$. In that case, the eigenvector corresponding to the eigenvalue 3 is $\begin{pmatrix} 1 \\ 1 \end{pmatrix}$.

The norm of the eigenvector $\begin{pmatrix} 1 \\ -1 \end{pmatrix}$ is

$$\sqrt{(1 \;\; -1)\begin{pmatrix} 1 \\ -1 \end{pmatrix}} = \sqrt{1+1} = \sqrt{2}$$

Thus, the normalization constant is $\frac{1}{\sqrt{2}}$, and the normalized eigenvector is given by

$$\frac{1}{\sqrt{2}}\begin{pmatrix} 1 \\ -1 \end{pmatrix} = \begin{pmatrix} \frac{1}{\sqrt{2}} \\ -\frac{1}{\sqrt{2}} \end{pmatrix}$$

Similarly, the norm of the eigenvector $\begin{pmatrix} 1 \\ 1 \end{pmatrix}$ is $\sqrt{2}$, and thus, the normalized eigenvector is given by

$$\frac{1}{\sqrt{2}}\begin{pmatrix} 1 \\ 1 \end{pmatrix} = \begin{pmatrix} \frac{1}{\sqrt{2}} \\ \frac{1}{\sqrt{2}} \end{pmatrix}$$

4.10. Consider a matrix, $\begin{pmatrix} a & 0 \\ 0 & b \end{pmatrix}$. Find the eigenvalues and eigenvectors of this matrix.

(m) Diagonalization of a normalized matrix: In the above illuminating example, we have analytically evaluated the eigenvalues and eigenvectors of a simple matrix, $\mathbf{A} = \begin{pmatrix} 2 & 1 \\ 1 & 2 \end{pmatrix}$, using its characteristic equation. For eigenvalue $\lambda = 1$, the normalized eigenvector is $\begin{pmatrix} \frac{1}{\sqrt{2}} \\ -\frac{1}{\sqrt{2}} \end{pmatrix}$, and for eigenvalue $\lambda = 3$, the normalized eigenvector is $\begin{pmatrix} \frac{1}{\sqrt{2}} \\ \frac{1}{\sqrt{2}} \end{pmatrix}$. For the reasons to be revealed later, we may now create a new (2×2)

matrix using these two eigenvectors as

$$\mathbf{U} = \begin{pmatrix} \frac{1}{\sqrt{2}} & \frac{1}{\sqrt{2}} \\ -\frac{1}{\sqrt{2}} & \frac{1}{\sqrt{2}} \end{pmatrix}$$

We may now find $\mathbf{U}^{-1}$:

$$\mathbf{U}\mathbf{U}^{-1} = \begin{pmatrix} 1 & 0 \\ 0 & 1 \end{pmatrix}$$

$$\text{or} \quad \begin{pmatrix} \frac{1}{\sqrt{2}} & \frac{1}{\sqrt{2}} \\ -\frac{1}{\sqrt{2}} & \frac{1}{\sqrt{2}} \end{pmatrix} \begin{pmatrix} a & b \\ c & d \end{pmatrix} = \begin{pmatrix} 1 & 0 \\ 0 & 1 \end{pmatrix}$$

$$\text{or} \quad \begin{pmatrix} \frac{a}{\sqrt{2}} + \frac{c}{\sqrt{2}} & \frac{b}{\sqrt{2}} + \frac{d}{\sqrt{2}} \\ \frac{-a}{\sqrt{2}} + \frac{c}{\sqrt{2}} & \frac{-b}{\sqrt{2}} + \frac{d}{\sqrt{2}} \end{pmatrix} = \begin{pmatrix} 1 & 0 \\ 0 & 1 \end{pmatrix}$$

Equating the elements, we get

$a+c = \sqrt{2}$; $b+d = 0$; $-a+c = 0$; $-b+d = \sqrt{2}$, and eventually we get $a = \frac{1}{\sqrt{2}}, b = -\frac{1}{\sqrt{2}}, c = \frac{1}{\sqrt{2}}, d = \frac{1}{\sqrt{2}}$.

$$\text{As a result,} \quad \mathbf{U}^{-1} = \begin{pmatrix} \frac{1}{\sqrt{2}} & -\frac{1}{\sqrt{2}} \\ \frac{1}{\sqrt{2}} & \frac{1}{\sqrt{2}} \end{pmatrix}$$

$$\text{and } \mathbf{U}^{\dagger} = \begin{pmatrix} \frac{1}{\sqrt{2}} & -\frac{1}{\sqrt{2}} \\ \frac{1}{\sqrt{2}} & \frac{1}{\sqrt{2}} \end{pmatrix}$$

Note that $\mathbf{U}^{\dagger} = \mathbf{U}^{-1}$; therefore, $\mathbf{U}$ is a unitary matrix. One may now find

$$\mathbf{A}\mathbf{U} = \begin{pmatrix} 2 & 1 \\ 1 & 2 \end{pmatrix} \begin{pmatrix} \frac{1}{\sqrt{2}} & \frac{1}{\sqrt{2}} \\ -\frac{1}{\sqrt{2}} & \frac{1}{\sqrt{2}} \end{pmatrix} = \begin{pmatrix} \frac{1}{\sqrt{2}} & \frac{3}{\sqrt{2}} \\ -\frac{1}{\sqrt{2}} & \frac{3}{\sqrt{2}} \end{pmatrix}$$

Furthermore, we may continue to find

$$\mathbf{U}^{-1}\mathbf{A}\mathbf{U} = \begin{pmatrix} \frac{1}{\sqrt{2}} & -\frac{1}{\sqrt{2}} \\ \frac{1}{\sqrt{2}} & \frac{1}{\sqrt{2}} \end{pmatrix} \begin{pmatrix} \frac{1}{\sqrt{2}} & \frac{3}{\sqrt{2}} \\ -\frac{1}{\sqrt{2}} & \frac{3}{\sqrt{2}} \end{pmatrix}$$
$$= \begin{pmatrix} 1 & 0 \\ 0 & 3 \end{pmatrix} = \begin{pmatrix} \lambda_1 & 0 \\ 0 & \lambda_2 \end{pmatrix} = \boldsymbol{\lambda}$$

Thus, quite intriguingly, the above exercise gives a useful perspective on diagonalizing a matrix. The diagonalization can be performed using the following linear transformation:

$$\mathbf{U}^{-1}\mathbf{A}\mathbf{U} = \boldsymbol{\lambda} \quad \text{or} \quad \mathbf{U}^{\dagger}\mathbf{A}\mathbf{U} = \boldsymbol{\lambda} \tag{4.14}$$

Here, the role of the unitary matrix $\mathbf{U}$ is that it transforms the matrix $\mathbf{A}$ into a new diagonal matrix $\boldsymbol{\lambda}$ in which the diagonal elements represent the eigenvalues of the matrix $\mathbf{A}$ and the columns of the matrix $\mathbf{U}$ represents the eigenvectors of the matrix $\mathbf{A}$.

Thereby, one may get the eigenvalues and eigenvectors of a quantum system simply by diagonalizing the Hamiltonian matrix. If the matrix is of lower order (say (2×2) or (3×3)), an analytical solution to the characteristic equation of the matrix can be easily obtained; however, a more convenient approach to obtaining eigenvalues and eigenvectors of a Hamiltonian is the numerical linear transformation method (which will be discussed in PythonChapter D). Note that the procedure to represent a Hamiltonian operator in matrix form under the grid representation is yet to be discussed.

4.6 Matrix Representation of the Hamiltonian Operator

The Hamiltonian operator contains a kinetic energy term which is a differential operator in position space. Therefore, we first present the matrix representation of the derivative operators $\frac{d}{dx}$ and $\frac{d^2}{dx^2}$ under

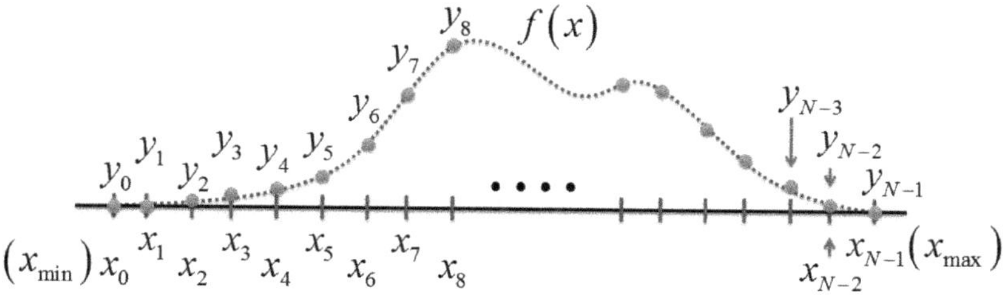

Figure 4.5. Discretized function $f(x)$ on the x-grid.

the grid representation. The finite difference (FD) method is a popular numerical technique which can be used to represent a derivative operator in matrix form under the grid representation.[5] This is done based on the discretization of a continuous function on a grid. In the discretization process, the x-coordinate (say, $x_{\min} \leq x \leq x_{\max}$) is divided by a suitably small step size Δx to produce a uniform discrete x-grid, as shown in Figure 4.5, rendering N discrete values of x within the $x_{\min} \leq x \leq x_{\max}$ range. Here, as shown in PhythonChapter B,

$$N = 1 + \frac{x_{\max} - x_{\min}}{\Delta x} \tag{4.15}$$

When a continuous function $y = f(x)$ is represented on the x-grid, we get the discretized function as follows:

$$\begin{aligned}
y_0 &= f(x_0) \\
y_1 &= f(x_1) \\
y_2 &= f(x_2) \\
y_3 &= f(x_3) \\
&\cdot \\
&\cdot \\
&\cdot \\
y_{N-1} &= f(x_{N-1})
\end{aligned}$$

As mentioned earlier, the above discretized function can be represented by a column matrix:

$$\mathbf{y} = \begin{pmatrix} y_0 \\ y_1 \\ y_2 \\ \cdot \\ \cdot \\ \cdot \\ y_{N-1} \end{pmatrix}_{N\times 1}$$

Once a function is discretized on the x-grid, two important questions are obvious: How do we obtain the first and second derivatives of the discretized function on the x-grid? Or, in other words, what are the matrix representations of the derivative operators $\frac{d}{dx}$ and $\frac{d^2}{dx^2}$ on the x-grid?

$$\frac{dy}{dx} = \begin{pmatrix} & & \\ & ? & \\ & & \end{pmatrix} \begin{pmatrix} y_0 \\ y_1 \\ y_2 \\ \cdot \\ \cdot \\ \cdot \\ y_{N-1} \end{pmatrix}_{N\times 1} \quad \text{and}$$

$$\frac{d^2y}{dx^2} = \begin{pmatrix} & & \\ & ? & \\ & & \end{pmatrix} \begin{pmatrix} y_0 \\ y_1 \\ y_2 \\ \cdot \\ \cdot \\ \cdot \\ y_{N-1} \end{pmatrix}_{N\times 1}$$

Matrix representation of the first derivative: Let us assume that the Taylor series expansion can be used to express the value of the function around a point, say x_0:

$$f(x_0 + \Delta x) = f(x_0) + \frac{\left[\frac{df(x)}{dx}\right]_{x_0}}{1!}\Delta x + \frac{\left[\frac{d^2 f(x)}{dx^2}\right]_{x_0}}{2!}\Delta x^2 + \frac{\left[\frac{d^3 f(x)}{dx^3}\right]_{x_0}}{3!}\Delta x^3 + \ldots\infty$$

$$\text{or } \left[\frac{df(x)}{dx}\right]_{x_0} = \frac{f(x_0 + \Delta x) - f(x_0)}{\Delta x} - \frac{\left[\frac{d^2 f(x)}{dx^2}\right]_{x_0}}{2!}\Delta x - \frac{\left[\frac{d^3 f(x)}{dx^3}\right]_{x_0}}{3!}\Delta x^2 + \cdots\infty$$

As Δx is very small, to the first-order approximation, one can neglect higher-order terms and can write

$$\left[\frac{df(x)}{dx}\right]_{x_0} = \frac{f(x_0 + \Delta x) - f(x_0)}{\Delta x} + O(\Delta x) \tag{4.16}$$

where $O(\Delta x)$ is called the truncation error, which is introduced due to the fact that an infinite sum has been truncated to a finite sum. If Δx term is very small, then the $O(\Delta x)$ term will be very small and the big O-term can be negligible.

Equation (4.16) represents the expression for the first derivative at a point (x_0) under the first-order FD approximation, in which the error scales linearly with Δx. Equation (4.16) is called the forward difference expression for the first derivative of a function because, for this calculation, we need the function value at the point where we are interested in calculating the derivative and the function value at a small step forward.

The analogous backward difference expression for the same can be obtained:

$$\left[\frac{df(x)}{dx}\right]_{x_0} = \frac{f(x_0) - f(x_0 - \Delta x)}{\Delta x} + O(\Delta x)$$

Note that the backward difference expression depends on the function value at the point where we are interested in calculating the derivative and the function value at some small step backward.

Next, rewrite the forward and backward difference Taylor series expansions together:

$$f(x_0 + \Delta x) = f(x_0) + \frac{\left[\frac{df(x)}{dx}\right]_{x_0}}{1!}\Delta x + \frac{\left[\frac{d^2 f(x)}{dx^2}\right]_{x_0}}{2!}\Delta x^2 + \frac{\left[\frac{d^3 f(x)}{dx^3}\right]_{x_0}}{3!}\Delta x^3 + \cdots \infty$$

$$f(x_0 - \Delta x) = f(x_0) - \frac{\left[\frac{df(x)}{dx}\right]_{x_0}}{1!}\Delta x + \frac{\left[\frac{d^2 f(x)}{dx^2}\right]_{x_0}}{2!}\Delta x^2 - \frac{\left[\frac{d^3 f(x)}{dx^3}\right]_{x_0}}{3!}\Delta x^3 + \cdots \infty$$

Subtracting the above two expressions from each other, we obtain another variant (called the central difference expression) for the first derivative of the function:

$$f(x_0 + \Delta x) - f(x_0 - \Delta x) = 2\frac{\left[\frac{df(x)}{dx}\right]_{x_0}}{1!}\Delta x + 2\frac{\left[\frac{d^3 f(x)}{dx^3}\right]_{x_0}}{3!}\Delta x^3 + \cdots \infty$$

$$\text{or} \quad \left[\frac{df(x)}{dx}\right]_{x_0} = \frac{f(x_0 + \Delta x) - f(x_0 - \Delta x)}{2\Delta x} + \frac{\left[\frac{d^3 f(x)}{dx^3}\right]_{x_0}}{3!}\Delta x^2 + \cdots \infty$$

$$\text{or} \quad \left[\frac{df(x)}{dx}\right]_{x_0} = \frac{f(x_0 + \Delta x) - f(x_0 - \Delta x)}{2\Delta x} + O(\Delta x^2) \tag{4.17}$$

The error in the central difference method scales quadratically with the step size in the expression given in equation (4.17), making it a better approximation than the forward or backward methods for the same step size. Using the central difference method, one may write a set of coupled linear equations (obtaining the derivative at each grid point):

$$
\begin{aligned}
y_1' &= \frac{1}{2\Delta x}[y_2 - y_0]\\
y_2' &= \frac{1}{2\Delta x}[y_3 - y_1]\\
y_3' &= \frac{1}{2\Delta x}[y_4 - y_2]\\
&\;\;\vdots\\
y_{N-2}' &= \frac{1}{2\Delta x}[y_{N-1} - y_{N-3}]
\end{aligned}
$$

Here, $y' \equiv \frac{dy}{dx}$.

The above set of coupled linear equations can be easily expressed in matrix form:

$$
\begin{pmatrix} y_1' \\ y_2' \\ y_3' \\ \cdot \\ \cdot \\ \cdot \\ y_{N-2}' \end{pmatrix}_{(N-2)\times 1}
= \frac{1}{2\Delta x}
\begin{pmatrix}
-1 & 0 & 1 & 0 & 0 & \cdot & \cdot & \cdot & \cdot & \cdot \\
0 & -1 & 0 & 1 & 0 & \cdot & \cdot & \cdot & \cdot & \cdot \\
0 & 0 & -1 & 0 & 1 & \cdot & \cdot & \cdot & \cdot & \cdot \\
\cdot & \cdot & \cdot & \cdot & \cdot & & & & & \\
\cdot & \cdot & \cdot & \cdot & \cdot & & & & & \\
\cdot & \cdot & \cdot & \cdot & \cdot & & & & & \\
0 & 0 & 0 & 0 & 0 & \cdot & \cdot & -1 & 0 & 1
\end{pmatrix}_{(N-2)\times N}
\times
\begin{pmatrix} y_0 \\ y_1 \\ y_2 \\ y_3 \\ \cdot \\ \cdot \\ \cdot \\ y_{N-2} \\ y_{N-1} \end{pmatrix}_{N\times 1}
$$

Thus, using the central difference method, we obtain the matrix form of the $\frac{d}{dx}$ operator as

$$\frac{d}{dx} \equiv \frac{1}{2\Delta x}\begin{pmatrix} -1 & 0 & 1 & 0 & 0 & \cdot & \cdot & \cdot & \cdot & \cdot \\ 0 & -1 & 0 & 1 & 0 & \cdot & \cdot & \cdot & \cdot & \cdot \\ 0 & 0 & -1 & 0 & 1 & \cdot & \cdot & \cdot & \cdot & \cdot \\ \cdot & \cdot & \cdot & \cdot & \cdot & & & & & \\ \cdot & \cdot & \cdot & \cdot & \cdot & & & & & \\ \cdot & \cdot & \cdot & \cdot & \cdot & & & & & \\ \cdot & \cdot & \cdot & \cdot & \cdot & \cdot & \cdot & -1 & 0 & 1 \end{pmatrix}_{(N-2)\times N}$$

Matrix representation of the second derivative: Our next task is to find a matrix form of the second derivative using the FD method. For this, we may now add the forward and backward expressions of Taylor series expansions:

$$f(x_0+\Delta x)+f(x_0-\Delta x) = 2f(x_0) + 2\frac{\left[\frac{d^2 f(x)}{dx^2}\right]_{x_0}}{2!}\Delta x^2 + 2\frac{\left[\frac{d^4 f(x)}{dx^4}\right]_{x_0}}{4!}\Delta x^4 + \cdots\infty$$

or, after rearranging, we get

$$\left[\frac{d^2 f(x)}{dx^2}\right]_{x_0} = \frac{f(x_0+\Delta x) - 2f(x_0) + f(x_0-\Delta x)}{\Delta x^2} + O(\Delta x^2) \tag{4.18}$$

The above expression, which represents the central difference expression for the second derivative at point x_0, exhibits an error which scales quadratically with Δx. Based on the above expression, one may write a set of coupled linear equations:

$$y_1'' = \frac{1}{\Delta x^2}[y_2 - 2y_1 + y_0]$$
$$y_2'' = \frac{1}{\Delta x^2}[y_3 - 2y_2 + y_1]$$
$$y_3'' = \frac{1}{\Delta x^2}[y_4 - 2y_3 + y_2]$$
$$\cdot$$
$$\cdot$$
$$\cdot$$
$$y_{N-2}'' = \frac{1}{\Delta x^2}[y_{N-1} - 2y_{N-2} + y_{N-3}]$$

Here, $y'' \equiv \frac{d^2y}{dx^2}$.

Alternatively, the above set of linear equations can be placed in matrix form:

$$\begin{pmatrix} y_1'' \\ y_2'' \\ y_3'' \\ \cdot \\ \cdot \\ \cdot \\ y_{N-2}'' \end{pmatrix}_{(N-2)\times 1} = \frac{1}{\Delta x^2}\begin{pmatrix} 1 & -2 & 1 & 0 & 0 & \cdot & \cdot & \cdot & \cdot & \cdot \\ 0 & 1 & -2 & 1 & 0 & \cdot & \cdot & \cdot & \cdot & \cdot \\ 0 & 0 & 1 & -2 & 1 & \cdot & \cdot & \cdot & \cdot & \cdot \\ \cdot & \cdot & \cdot & \cdot & \cdot & & & & & \\ \cdot & \cdot & \cdot & \cdot & \cdot & & & & & \\ \cdot & \cdot & \cdot & \cdot & \cdot & & & & & \\ 0 & 0 & 0 & 0 & 0 & \cdot & \cdot & 1 & -2 & 1 \end{pmatrix}_{(N-2)\times N}$$
$$\times \begin{pmatrix} y_0 \\ y_1 \\ y_2 \\ y_3 \\ \cdot \\ \cdot \\ \cdot \\ y_{N-2} \\ y_{N-1} \end{pmatrix}_{N\times 1}$$

Thus, using the central difference method, we obtain the matrix form of the $\frac{d^2}{dx^2}$ as:

$$\frac{d^2}{dx^2} \equiv \frac{1}{\Delta x^2}\begin{pmatrix} 1 & -2 & 1 & 0 & 0 & \cdot & \cdot & \cdot & \cdot & \cdot \\ 0 & 1 & -2 & 1 & 0 & \cdot & \cdot & \cdot & \cdot & \cdot \\ 0 & 0 & 1 & -2 & 1 & \cdot & \cdot & \cdot & \cdot & \cdot \\ \cdot & \cdot & \cdot & \cdot & \cdot & & & & & \\ \cdot & \cdot & \cdot & \cdot & \cdot & & & & & \\ \cdot & \cdot & \cdot & \cdot & \cdot & & & & & \\ 0 & 0 & 0 & 0 & 0 & \cdot & \cdot & 1 & -2 & 1 \end{pmatrix}_{(N-2)\times N}$$

An Illuminating Example

Δx

x_0 x_1 x_2 x_3 x_4

Consider the above x-grid to find the matrix form of the central difference expression for the second derivative:

$$\begin{pmatrix} y_1'' \\ y_2'' \\ y_3'' \end{pmatrix}_{3\times 1} = \frac{1}{\Delta x^2}\begin{pmatrix} 1 & -2 & 1 & 0 & 0 \\ 0 & 1 & -2 & 1 & 0 \\ 0 & 0 & 1 & -2 & 1 \end{pmatrix}_{3\times 5}\begin{pmatrix} y_0 \\ y_1 \\ y_2 \\ y_3 \\ y_4 \end{pmatrix}_{5\times 1}$$

If $y_0 = 0$ and $y_4 = 0$ (defining condition for the function that it becomes zero at the boundary), then one may write

$$\begin{pmatrix} y_1'' \\ y_2'' \\ y_3'' \end{pmatrix}_{3\times 1} = \underbrace{\frac{1}{\Delta x^2}\begin{pmatrix} -2 & 1 & 0 \\ 1 & -2 & 1 \\ 0 & 1 & -2 \end{pmatrix}_{3\times 3}}_{\frac{d^2}{dx^2}}\begin{pmatrix} y_1 \\ y_2 \\ y_3 \end{pmatrix}_{3\times 1}$$

(*Continued*)

(Continued)

Thus, note that the second-derivative matrix becomes a square matrix under the above defining condition of the function (the boundary condition). Furthermore, it is a Hermitian matrix (prove it yourself).

With the analysis of the finite difference method given above, one may now represent the Hamiltonian operator in matrix form under the grid representation. We begin with the one-dimensional time-independent Hamiltonian operator:

$$\hat{H} = \left[-\frac{\hbar^2}{2m}\frac{d^2}{dx^2} + V(x)\right]$$

Employing the grid representation, one may express a discretized wavefunction as a column matrix:

$$\psi(x) = \begin{pmatrix} \psi_0 \\ \psi_1 \\ \psi_2 \\ \cdot \\ \cdot \\ \cdot \\ \psi_{N-1} \end{pmatrix}_{(N-1)\times 1}$$

where $\psi_0 = \psi(x_0)$, $\psi_1 = \psi(x_1)$, and so on. Next, one may define the boundary condition for the wavefunction, ψ_0 and $\psi_{N-1} = 0$. This boundary condition originates from the fact that when the wavefunction lives in the Hilbert space, it must be square normalizable, and as a result, the function value must be zero at the boundary. Therefore, within the boundary condition, the discretized wavefunction has the following form:

$$\psi(x) = \begin{pmatrix} \psi_1 \\ \psi_2 \\ \psi_3 \\ \cdot \\ \cdot \\ \cdot \\ \psi_{N-2} \end{pmatrix}_{(N-2)\times 1}$$

We may now recall the central difference formula for the second derivative and represent the kinetic energy operator as

$$-\frac{\hbar^2}{2m}\frac{d^2}{dx^2} = -\frac{\hbar^2}{2m\Delta x^2}\begin{pmatrix} 1 & -2 & 1 & 0 & 0 & \cdot & \cdot & \cdot & \cdot & \cdot \\ 0 & 1 & -2 & 1 & 0 & \cdot & \cdot & \cdot & \cdot & \cdot \\ 0 & 0 & 1 & -2 & 1 & \cdot & \cdot & \cdot & \cdot & \cdot \\ \cdot & \cdot & \cdot & \cdot & \cdot & & & & & \\ \cdot & \cdot & \cdot & \cdot & \cdot & & & & & \\ \cdot & \cdot & \cdot & \cdot & \cdot & & & & & \\ 0 & 0 & 0 & 0 & 0 & \cdot & \cdot & 1 & -2 & 1 \end{pmatrix}_{(N-2)\times N}$$

The above matrix is neither a square matrix nor a Hermitian matrix (prove it yourself). However, within the boundary condition for the wavefunction used above, the kinetic energy operator reduces to a Hermitian matrix (prove it yourself):

$$\frac{\hbar^2}{2m}\frac{d^2}{dx^2} = -\frac{\hbar^2}{2m\Delta x^2}\begin{pmatrix} -2 & 1 & 0 & 0 & 0 & \cdot & \cdot & \cdot & \cdot & \cdot \\ 1 & -2 & 1 & 0 & 0 & \cdot & \cdot & \cdot & \cdot & \cdot \\ 0 & 1 & -2 & 1 & 0 & \cdot & \cdot & \cdot & \cdot & \cdot \\ \cdot & \cdot & \cdot & \cdot & \cdot & & & & & \\ \cdot & \cdot & \cdot & \cdot & \cdot & & & & & \\ \cdot & \cdot & \cdot & \cdot & \cdot & & & & & \\ 0 & 0 & 0 & 0 & 0 & \cdot & \cdot & 0 & 1 & -2 \end{pmatrix}_{(N-2)\times(N-2)}$$

On the other hand, the potential operator $V(x)$ is nothing but a multiplicative operator. As $V(x)\psi(x)$ represents the product of two discretized x-dependent functions, one can express the product in the following matrix product form:

$$\begin{pmatrix} V_0\psi_0 \\ V_1\psi_1 \\ V_2\psi_2 \\ \cdot \\ \cdot \\ \cdot \\ V_{N-1}\psi_{N-1} \end{pmatrix} = \begin{pmatrix} V_0 & 0 & 0 & \cdot & \cdot & \cdot & 0 \\ 0 & V_1 & 0 & \cdot & \cdot & \cdot & 0 \\ 0 & 0 & V_2 & \cdot & \cdot & \cdot & 0 \\ \cdot & \cdot & \cdot & \cdot & \cdot & \cdot & \cdot \\ \cdot & \cdot & \cdot & \cdot & \cdot & \cdot & \cdot \\ \cdot & \cdot & \cdot & \cdot & \cdot & \cdot & \cdot \\ 0 & 0 & 0 & \cdot & \cdot & \cdot & V_{N-1} \end{pmatrix} \begin{pmatrix} \psi_0 \\ \psi_1 \\ \psi_2 \\ \cdot \\ \cdot \\ \cdot \\ \psi_{N-1} \end{pmatrix}$$

Consequently, the potential operator under the grid representation adopts a diagonal matrix form:

$$V(x) \equiv \begin{pmatrix} V_0 & 0 & 0 & \cdot & \cdot & & \\ 0 & V_1 & 0 & \cdot & \cdot & & \\ 0 & 0 & V_2 & \cdot & \cdot & & \\ \cdot & \cdot & \cdot & \cdot & \cdot & & \\ \cdot & \cdot & \cdot & \cdot & \cdot & & \\ & & & & & \cdot & \\ & & & & & & V_{N-1} \end{pmatrix}_{N\times N}$$

where $V_0 = V(x_0)$, $V_1 = V(x_1)$, and so on (representing the local values of the potential). Within the boundary condition for the wavefunction, this potential energy matrix further reduces to

$$V(x) \equiv \begin{pmatrix} V_1 & 0 & 0 & \cdot & \cdot & & \\ 0 & V_2 & 0 & \cdot & \cdot & & \\ 0 & 0 & V_3 & \cdot & \cdot & & \\ \cdot & \cdot & \cdot & \cdot & \cdot & & \\ \cdot & \cdot & \cdot & \cdot & \cdot & & \\ & & & & & \cdot & \\ & & & & & & V_{N-2} \end{pmatrix}_{(N-2)\times(N-2)}$$

Therefore, the matrix form of the Hamiltonian operator under the grid representation and using the boundary condition is given by

$$
\mathbf{H} = -\frac{\hbar^2}{2m\Delta x^2}\begin{pmatrix}
-2 & 1 & 0 & 0 & 0 & \cdot & \cdot & \cdot & \cdot & \cdot \\
1 & -2 & 1 & 0 & 0 & \cdot & \cdot & \cdot & \cdot & \cdot \\
0 & 1 & -2 & 1 & 0 & \cdot & \cdot & \cdot & \cdot & \cdot \\
\cdot & \cdot & \cdot & \cdot & \cdot & & & & & \\
\cdot & \cdot & \cdot & \cdot & \cdot & & & & & \\
\cdot & \cdot & \cdot & \cdot & \cdot & & & & & \\
0 & 0 & 0 & 0 & 0 & \cdot & \cdot & 0 & 1 & -2
\end{pmatrix}_{(N-2)\times(N-2)}
$$

$$
+\begin{pmatrix}
V_1 & 0 & 0 & \cdot & \cdot & \\
0 & V_2 & 0 & \cdot & \cdot & \\
0 & 0 & V_3 & \cdot & \cdot & \\
\cdot & \cdot & \cdot & \cdot & \cdot & \\
\cdot & \cdot & \cdot & \cdot & \cdot & \\
 & & & & \cdot & \\
 & & & & & V_{N-2}
\end{pmatrix}_{(N-2)\times(N-2)}
$$

The eigenvalues and eigenvectors of this Hamiltonian matrix can be numerically computed using the diagonalization method — a subject which is discussed in PythonChapter D.

Key Points to Remember

- Only a square-integrable wavefunction is acceptable in quantum mechanics because it lives in the Hilbert space. A set of wavefunctions living in Hilbert space is called orthonormal when each wavefunction is normalized, and the wavefunctions are pair-wise orthogonal.
- If $\hat{A}$ is a Hermitian operator, $e^{i\hat{A}}$ becomes a unitary operator.
- $e^{-i\frac{\hat{H}t}{\hbar}}$ is called the time-evolution operator because it can find $\psi(x,t)$ if $\psi(x,0)$ is known:

$$\psi(x,t) = e^{-i\frac{\hat{H}t}{\hbar}}\psi(x,0)$$

- Under the grid representation and within a boundary condition, a wavefunction is expressed as a column matrix:

$$\psi(x) = \begin{pmatrix} \psi_1 \\ \psi_2 \\ \psi_3 \\ \cdot \\ \cdot \\ \cdot \\ \psi_{N-2} \end{pmatrix}_{(N-2)\times 1}$$

- Under the grid representation, using the central difference formula of the finite difference method and within the boundary condition, the Hamiltonian matrix obtains the following matrix form:

$$\mathbf{H} = -\frac{\hbar^2}{2m\Delta x^2}\begin{pmatrix} -2 & 1 & 0 & 0 & 0 & \cdot & \cdot & \cdot & \cdot & \cdot \\ 1 & -2 & 1 & 0 & 0 & \cdot & \cdot & \cdot & \cdot & \cdot \\ 0 & 1 & -2 & 1 & 0 & \cdot & \cdot & \cdot & \cdot & \cdot \\ \cdot & \cdot & \cdot & \cdot & \cdot & & & & & \\ \cdot & \cdot & \cdot & \cdot & \cdot & & & & & \\ \cdot & \cdot & \cdot & \cdot & \cdot & & & & & \\ 0 & 0 & 0 & 0 & 0 & \cdot & \cdot & 0 & 1 & -2 \end{pmatrix}_{(N-2)\times(N-2)}$$

$$+ \begin{pmatrix} V_1 & 0 & 0 & \cdot & \cdot & & \\ 0 & V_2 & 0 & \cdot & \cdot & & \\ 0 & 0 & V_3 & \cdot & \cdot & & \\ \cdot & \cdot & \cdot & \cdot & \cdot & & \\ \cdot & \cdot & \cdot & \cdot & \cdot & & \\ & & & & & \cdot & \\ & & & & & & V_{N-2} \end{pmatrix}_{(N-2)\times(N-2)}$$

Notes, References, and Further Reading

1. For further discussions, refer to (a) D. J. Tannor, Chapter 8: Linear algebra and quantum mechanics. *Introduction to Quantum Mechanics: A Time-Dependent Approach.* University Science Books, Sausalito (CA) (2007); (b) D. J. Griffiths, Appendix: Linear algebra and Chapter 3: Formalism. *Introduction to Quantum Mechanics*, 2nd edn. Pearson Education, Inc., New Jersey (2005); (c) D. A. McQuarrie, MathChapter F: Determinants and MathChapter G: Matrices. *Quantum Chemistry*, 2nd edn. University Science Books, California (2008).
2. In several quantum mechanics books, we very frequently come across a statement that all square-normalizable wavefunctions (which are also physically acceptable wavefunctions) live in the Hilbert space. Let us understand what exactly this means. We take an analogy of our solar system. If our solar system is viewed as a space where all living beings live, then all living beings should follow the properties of the solar system, or, in other words, they should have the same origin of life (the process by which life has originated from simple organic compounds). Similarly, all physically acceptable wavefunctions live in a mathematical space called the Hilbert space or the linear vector space. So, a quantum mechanical equation is defined in a specific Hilbert space. A quantum mechanical operator should not take a wavefunction out of the Hilbert space. Hilbert space is a mathematical construct. Its physical meaning is that as long as a wavefunction lives in a specific Hilbert space, it carries statistical interpretation, and as a result, it is acceptable in quantum mechanics.
3. The general form of integration by parts is $\int_a^b uv'dx = [uv]_a^b - \int_a^b u'vdx$.
4. See the discussion in D. J. Tannor, Chapter 8: Linear algebra and quantum mechanics. *Introduction to Quantum Mechanics: A Time-Dependent Approach.* University Science Books, Sausalito (CA) (2007).
5. For details, see (a) R. J. LeVeque, *Finite Difference Methods for Ordinary and Partial Differential Equations: Steady State and Time-Dependent Problems.* Society for Industrial and Applied Mathematics (Philadelphia) (2007); (b) J. Izaac and J. Wang, Chapter 5: Differentiation and initial value problems. *Computational Quantum Mechanics.* Springer, Switzerland (2018).

Exercises

4.1 Prove that the eigenvalue of a Hermitian operator is real.

4.2 Prove that both the Hamiltonian and momentum operators are Hermitian operators.

4.3 Prove that if $\hat{A}$ is a Hermitian operator, then $e^{i\hat{A}}$ is a unitary operator.

4.4 Express (a) the time-independent Schrödinger equation (TISE) and (b) the time-dependent Schrödinger equation (TDSE) in matrix form.

4.5 Prove that the following inequality holds: $e^{\frac{-i\hat{H}t}{\hbar}} \neq e^{\frac{-i\hat{T}t}{\hbar}} e^{\frac{-i\hat{V}t}{\hbar}}$, where $\hat{H}$ is the Hamiltonian operator, $\hat{T}$ is the kinetic energy operator, and $\hat{V}$ is the potential energy operator.

4.6 Find the commutator $[\hat{p}_x, \hat{x}]$

4.7 Find the eigenstates of the Hamiltonian $\mathbf{H}_0 = \begin{pmatrix} E_1 & 0 \\ 0 & E_2 \end{pmatrix}$.

PythonChapter D

Constructing a Matrix and Finding Its Eigenvalues and Eigenvectors

Highlights: *Creating a Square Matrix Data Structure using Python Program, Finding Eigenvalues and Eigenvectors using SciPy.LinAlg Submodule, Band Storage Form of Hermitian Tridiagonal Matrix, Finding Eigenvalues and Eigenvectors of Hermitian Tridiagonal Matrix using Band Storage Form.*

D.1 Introduction

We have already understood that under the (one-dimensional) grid representation of the wavefunction, employing the finite difference method and within the boundary conditions, the kinetic energy part of the Hamiltonian matrix adopts a Hermitian tridiagonal form:

$$\hat{T} \equiv -\frac{\hbar^2}{2m}\frac{d^2}{dx^2} = -\frac{\hbar^2}{2m\Delta x^2}$$

$$\times \begin{pmatrix} -2 & 1 & 0 & 0 & 0 & \cdot & \cdot & \cdot & \cdot & \cdot \\ 1 & -2 & 1 & 0 & 0 & \cdot & \cdot & \cdot & \cdot & \cdot \\ 0 & 1 & -2 & 1 & 0 & \cdot & \cdot & \cdot & \cdot & \cdot \\ \cdot & \cdot & \cdot & \cdot & \cdot & & & & & \\ \cdot & \cdot & \cdot & \cdot & \cdot & & & & & \\ \cdot & \cdot & \cdot & \cdot & \cdot & & & & & \\ 0 & 0 & 0 & 0 & 0 & \cdot & \cdot & 0 & 1 & -2 \end{pmatrix}_{(N-2)\times(N-2)}$$

On the other hand, the corresponding (time-independent) potential part of the Hamiltonian matrix adopts a diagonal form:

$$\hat{V} \equiv \begin{pmatrix} V_1 & 0 & 0 & \cdot & \cdot & & \\ 0 & V_2 & 0 & \cdot & \cdot & & \\ 0 & 0 & V_3 & \cdot & \cdot & & \\ \cdot & \cdot & \cdot & \cdot & \cdot & & \\ \cdot & \cdot & \cdot & \cdot & \cdot & & \\ & & & & & \cdot & \\ & & & & & & V_{N-2} \end{pmatrix}_{(N-2)\times(N-2)}$$

Once one constructs the Hamiltonian matrix of a quantum system, the eigenvalues and eigenvectors of that quantum system can be easily obtained by diagonalizing the Hamiltonian matrix. In this PythonChapter, we will learn how to construct the Hamiltonian matrix of a quantum system and, subsequently, how to use Python programming to obtain the eigenvalues and eigenvectors of the quantum system.

Chapter 4 demonstrates that the diagonalization of a matrix $\mathbf{A}$ can be performed using the linear transformation $\mathbf{U}^{-1}\mathbf{A}\mathbf{U} = \boldsymbol{\lambda}$. The action of the unitary matrix $\mathbf{U}$ transforms the matrix $\mathbf{A}$ into a new diagonal matrix $\boldsymbol{\lambda}$ in which the diagonal elements represent the eigenvalues of the matrix $\mathbf{A}$ and the columns of the matrix $\mathbf{U}$ represent the eigenvectors of the matrix $\mathbf{A}$.

Considering a simple example, the unitary matrix $\mathbf{U} = \begin{pmatrix} \frac{1}{\sqrt{2}} & \frac{1}{\sqrt{2}} \\ -\frac{1}{\sqrt{2}} & \frac{1}{\sqrt{2}} \end{pmatrix}$ can diagonalize the (2×2) matrix $\mathbf{A} = \begin{pmatrix} 2 & 1 \\ 1 & 2 \end{pmatrix}$ to unravel its eigenvalues and eigenvectors. In this PythonChapter, our first Python exercise is to find the eigenvalues and eigenvectors of the same $\mathbf{A} = \begin{pmatrix} 2 & 1 \\ 1 & 2 \end{pmatrix}$ matrix numerically. In order to perform the task, however, we need to learn how to represent a matrix in Python programming.

In general, the data structure which is used to represent a mathematical matrix in computer programming is called an **array**.

An array can be of N dimensions, but a two-dimensional array is called a matrix, and a one-dimensional array is called a vector. Python does not have intrinsic (built-in) functionalities to deal with arrays (and different array operations or broadly linear algebra numerical techniques). For that, one needs to **import** the required functionalities from **scipy**,[1] which provides all linear algebra routines (numerical techniques) in its submodule **scipy.linalg**.[2] In this PythonChapter, we introduce the **scipy.linalg** submodule and go over a few examples of the eigenvalue problem.

D.2 Finding Eigenvalues and Eigenvectors of a Simple Square Matrix

There are several approaches which can be used in conjunction with the **scipy.linalg** submodule to find the eigenvalues and eigenvectors of a square matrix; however, we make use of the following approach:

> **First**, create a square ($N \times N$) null matrix, **then** replace (reassign) the elements with their desired values to create the final matrix which one would like to construct. **In the end**, one can easily find the eigenvalues and eigenvectors of the constructed matrix using the **eig** functionality of the **scipy.linalg** submodule.

Example 1. Find the eigenvalues and eigenvectors of $\begin{pmatrix} 2 & 1 \\ 1 & 2 \end{pmatrix}$.

(a) Step 1: Construct the null matrix of the same dimension.

```
from scipy import zeros      #zeros functionality is imported from scipy module
A=zeros((2,2))               #a (2x2) null matrix is constructed
print(A)                     #print the matrix to check Python's way of representing a matrix

[[0. 0.]
 [0. 0.]]
```

Python's module **scipy** has several functionalities to define an array of any dimension. For example, **zeros((N,N),dtype)** creates an $(N \times N)$ array with all elements as zero (**dtype** feature is optional, not used in the above example).

(b) Step 2: Replace elements to construct the target matrix.

```
from scipy import zeros
A=zeros((2,2))
A[0,0]=2                    #reassign the element in the first row and first column
A[0,1]=1                    #reassign the element in the first row and second column
A[1,0]=1                    #reassign the element in the second row and first column
A[1,1]=2                    #reassign the element in the second row and second column
print(A)

[[2. 1.]
 [1. 2.]]
```

If A represents an $(N \times N)$ array, $A[i, j]$ represents the element of the ith row and jth columns of that array. Here, as usually encountered in Python's built-in **list** functionality, the indices i and j start with zero. For example, $A[0, 0]$ returns the element in the first row and first column, and $A[1, 2]$ returns the element in the second row and third column. The array indexing procedure of the **scipy** module is quite similar to Python's built-in **list** indexing procedure: The only difference is that one has to specify the indices of all array dimensions at the same time (and they are separated by commas).

In the above context, note that $A[:, j]$ and $A[i, :]$ return the entire jth column and the entire ith row, respectively (a colon indicates everything of that particular dimension). One can also change the element of an array even after its construction: Similar to Python's built-in element assignment feature of list, one can set (or reset) the value of the $A[i, j]$ element using **A[i,j]=value** construct of **scipy** . This is exactly what we have done in the above-mentioned example. We have reassigned the elements to construct our desired matrix from a null matrix.

(c) Step 3: Finding eigenvalues and eigenvectors.

```
from scipy import zeros
from scipy.linalg import eig
A=zeros((2,2))
A[0,0]=2
A[0,1]=1
A[1,0]=1
A[1,1]=2
E,V=eig(A)
print(E)
print(V)
```

> Here, note that an array data structure (such as zeros) is created using the built-in functionality of the scipy module. But the scipy.linalg submodule is required to employ any linear algebra routine (such as eig) for further array operations.

```
[3.+0.j 1.+0.j]

[[ 0.70710678 -0.70710678]
 [ 0.70710678 0.70710678]]
```

Note that the above program returns two eigenvalues, 1 and 3, for the matrix $\begin{pmatrix} 2 & 1 \\ 1 & 2 \end{pmatrix}$. In the **scipy.linalg** submodule, the construct **E,V=eig()** computes the eigenvalues (E) and normalized right eigenvectors (V) of the matrix given in the round bracket. A more precise construct to compute the eigenvalues and eigenvectors is **E,V=eig(A,left=false,right=true)**. All features other than A are optional, and this is why they are not used in the above example. If the **left=true** feature is used, then the **normalized left eigenvectors** are returned, and if the **right=true** (which is the default) is used, the **normalized right eigenvectors** are returned.

In many common applications of quantum mechanics, only the right eigenvector (*not* the left eigenvector) is considered. Hence, in general, the term "eigenvector" is used to refer to a right eigenvector. For the left eigenvector, clear mentioning of the term "left eigenvector" is needed. Here, note that in the eigenvalue equation $\mathbf{A}|\psi\rangle = \lambda|\psi\rangle$, $|\psi\rangle$ represents a right eigenvector, which is nothing but a column matrix (or a ket vector in Dirac's bra–ket notation). On the other hand, in the eigenvalue equation $\langle\psi|\mathbf{A} = \langle\psi|\lambda$, $\langle\psi|$ is a left eigenvector, which is nothing but a row matrix (or bra vector in

Dirac's bra–ket notation). Here, $\langle\psi|$ is the adjoint of $|\psi\rangle$. For further details, see Chapter 4.

In the **E,V=eig()** construct, the E and V are returned in specific array forms. E returns a one-dimensional array with elements representing the eigenvalues. Thus, one can use $E(i)$ to access the ith eigenvalue. On the other hand, V provides the normalized right eigenvectors in matrix form, in which the jth column $V[:,j]$ corresponds to the normalized right eigenvector associated with the jth eigenvalue $E(j)$. Thus, the eigenvectors associated, respectively, with the eigenvalues 3 and 1 can be separately returned using the following Python program.

```
from scipy import zeros
from scipy.linalg import eig
A=zeros((2,2))
A[0,0]=2
A[0,1]=1
A[1,0]=1
A[1,1]=2
E,V=eig(A)
print(E[0])
print(V[:,0])
print(E[1])
print(V[:,1])
```

```
(3+0j)
[0.70710678 0.70710678]
(1+0j)
[-0.70710678  0.70710678]
```

Note that a normalized right eigenvector (which is a ket vector) should be a column matrix, but the program prints it like a row matrix. This is just the default visualization (Python prints like this, but it should be regarded as a column matrix).

Now, the above numerical results can be compared with the analytically obtained eigenvalues and eigenvectors of the matrix $\mathbf{A} = \begin{pmatrix} 2 & 1 \\ 1 & 2 \end{pmatrix}$ (refer to the exercise done in Chapter 4). This task is left for readers.

(d) Testing properties of the eigenvectors:

One can easily prove that the above-obtained eigenvectors are orthonormalized:

```
from scipy import zeros
from scipy.linalg import eig
A=zeros((2,2))
A[0,0]=2
A[0,1]=1
A[1,0]=1
A[1,1]=2
E,V=eig(A)
print(V[:,0]@ V[:,0])        #inner product between eigenvectors V[:,0] and V[:,0]
print(V[:,1]@ V[:,1])        #inner product between eigenvectors V[:,1] and V[:,1]
print(V[:,0]@ V[:,1])        #inner product between eigenvectors V[:,0] and V[:,1]

0.9999999999999998
0.9999999999999998
0.0
```

The inner product of the eigenvectors $V1$ and $V2$ can be easily obtained by the construct **V1@V2**. The above example demonstrates that $V[:,0]$ and $V[:,1]$ are orthonormalized.

Self-Study: Find the eigenvalues and eigenvectors of the following matrices:

$$\begin{pmatrix} 5 & 2 \\ 9 & 2 \end{pmatrix}, \begin{pmatrix} 2 & 1 & 0 \\ 1 & 4 & 0 \\ 2 & 5 & 2 \end{pmatrix}, \begin{pmatrix} 2 & 0 & 1 \\ 1 & -1 & 0 \\ 3 & 0 & 4 \end{pmatrix}, \quad \text{and} \quad \begin{pmatrix} 0 & 1 \\ 1 & 0 \end{pmatrix}$$

Important Recapitulation

In quantum mechanics, often, we work with a Hermitian matrix in the tridiagonal form. A generic structure of a Hermitian tridiagonal matrix is given as follows.

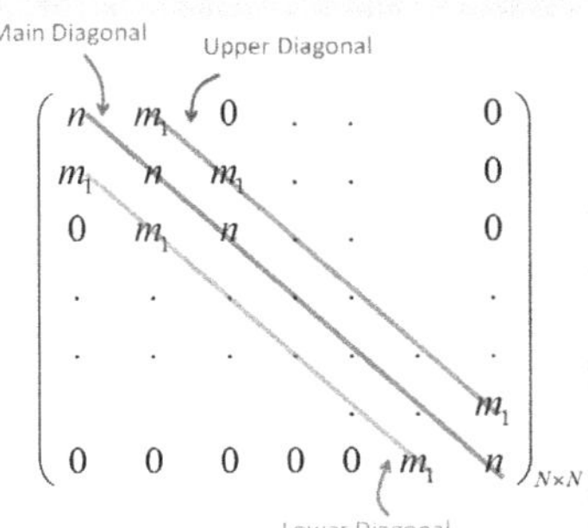

(*Continued*)

(Continued)

A square matrix possessing non-zero elements only in the main diagonal is called a diagonal matrix. The main diagonal of a matrix consists of elements that lie on the diagonal that runs from its top left to its bottom right. A tridiagonal matrix has non-zero elements on its main diagonal, upper diagonal, and lower diagonal. A tridiagonal matrix is also called a band-diagonal matrix because its only non-zero elements are confined to a diagonal band.

Example 2. Automate the matrix construction, and find the eigenvalues and eigenvectors of the tridiagonal square matrix $\begin{pmatrix} 2 & -1 & 0 \\ -1 & 2 & -1 \\ 0 & -1 & 2 \end{pmatrix}$.

(a) Automating the construction of a matrix.

```
from scipy import zeros,arange
A=zeros((3,3))
for i in arange(3):
        A[i,i]=2
for i in arange(2):
        A[i,i+1]=-1
        A[i+1,i]=-1
print(A)

[[ 2. -1.  0.]
 [-1.  2. -1.]
 [ 0. -1.  2.]]
```

> ➢ The arange(N) functionality of scipy carries a single argument,*N*, which defines the stop of the sequence. As usual, stop is excluded from the output sequence, which starts at zero and has an increment of one.

In the above example, we have used Python's built-in **for-loop** with **scipy's** functionality **arange()** to automate the matrix construction. In PythonChapter A, we have used the **arange()** functionality of **scipy** with three arguments **(start,stop,step)** to create a sequence of numbers (technically, **arange()** produces an array). Furthermore, we have noted that the sequence of numbers produced by the **arange()** functionality excludes **stop** but includes **start**. Another simplified version of the **arange()** functionality of **scipy** is to provide only a single argument. For example, **arange(N)** creates a sequence of numbers starting at zero and stopping at $(N-1)$ with a step size of 1. This is a well-practiced way of creating a sequence of numbers which **starts at zero and has an increment of one**.

The **for loop** heading, **for i in arange()**, creates a loop over all elements in the list created by the **arange()** functionality. In each pass of the loop, the **variable i refers** to an element in the list, starting with zeroth element (0), followed by the first element (1), and so on. In each loop, the respective element values of the matrix (presented by $A[i,i]$, $A[i,i+1]$, and $A[i+1,i]$) are reassigned.

(b) Directly finding the eigenvalues and eigenvectors of the tridiagonal square matrix.

```
from scipy.linalg import eig
from scipy import zeros,arange
A=zeros((3,3))
for i in arange(3):
        A[i,i]=2
for i in arange(2):
        A[i,i+1]=-1
        A[i+1,i]=-1
E,V=eig(A)
print(E)
print(V)
```

```
[3.41421356+0.j 2.       +0.j 0.58578644+0.j]

[[-5.00000000e-01 -7.07106781e-01  5.00000000e-01]
 [ 7.07106781e-01  4.05925293e-16  7.07106781e-01]
 [-5.00000000e-01  7.07106781e-01  5.00000000e-01]]
```

Self-Study: Prove that $\begin{pmatrix} 2 & -1 & 0 \\ -1 & 2 & -1 \\ 0 & -1 & 2 \end{pmatrix}$ is a Hermitian matrix.

D.3 Finding Eigenvalues and Eigenvectors of a Very Large Hermitian Tridiagonal Matrix

In Chapter 4, we have realized that the defining condition for a Hermitian matrix is that it must be self-adjoint, i.e., $A_{ij} = A^*_{ji}$. Due to this symmetry, a tridiagonal Hermitian matrix $(N \times N)$ can be stored as a $(2 \times N)$ band storage form. There are two conventions which can be used to store an $(N \times N)$ Hermitian tridiagonal matrix in the $(2 \times N)$ band storage form.[3]

With upper diagonal elements:

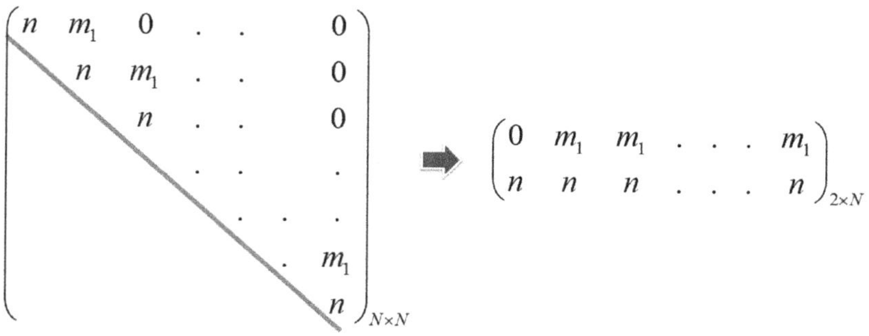

$$\begin{pmatrix} n & m_1 & 0 & . & . & 0 \\ & n & m_1 & . & . & 0 \\ & & n & . & . & 0 \\ & & & . & . & . \\ & & & & . & . \\ & & & & . & m_1 \\ & & & & & n \end{pmatrix}_{N\times N} \Rightarrow \begin{pmatrix} 0 & m_1 & m_1 & . & . & . & m_1 \\ n & n & n & . & . & . & n \end{pmatrix}_{2\times N}$$

With lower diagonal elements:

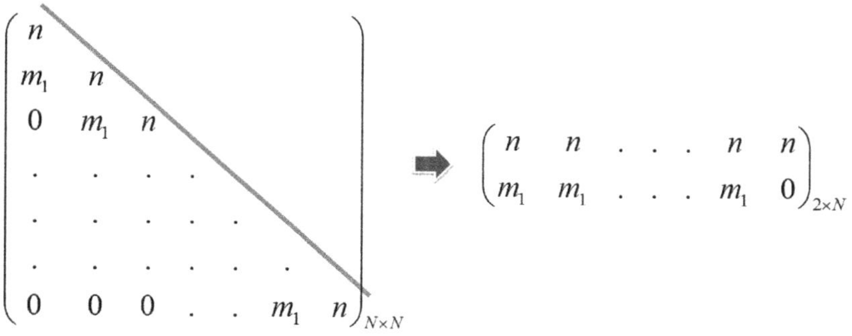

$$\begin{pmatrix} n & & & & & \\ m_1 & n & & & & \\ 0 & m_1 & n & & & \\ . & . & . & . & & \\ . & . & . & . & . & \\ . & . & . & . & . & . \\ 0 & 0 & 0 & . & . & m_1 & n \end{pmatrix}_{N\times N} \Rightarrow \begin{pmatrix} n & n & . & . & . & n & n \\ m_1 & m_1 & . & . & . & m_1 & 0 \end{pmatrix}_{2\times N}$$

The advantage of above band storage forms is that they save a significant amount of computer memory while working with a large Hermitian tridiagonal matrix. Once the Hermitian tridiagonal matrix is stored in one of the above band storage forms, the **eig_banded()** functionality of the **scipy.linalg** submodule can easily find the eigenvalues and eigenvectors of the $(N \times N)$ tridiagonal Hermitian matrix.

Example 1. Find the eigenvalues and eigenvectors of a tridiagonal Hermitian matrix $\begin{pmatrix} 2 & -1 & 0 \\ -1 & 2 & -1 \\ 0 & -1 & 2 \end{pmatrix}$ using the band storage form.

(a) Step 1: Construct the band storage form.

```
from scipy import zeros,arange
A=zeros((3,3))
#Constructing (3x3) tridiagonal matrix
for i in arange(3):
        A[i,i]=2
for i in arange(2):
        A[i,i+1]=-1
        A[i+1,i]=-1
#Printing the (3x3) matrix
print(A)
#Constructing (2x3) band storage form
A_band_up=zeros((2,3))
for i in arange(2):
        A_band_up[0,i+1]=-1.0
for i in arange(3):
        A_band_up[1,i]=2.0
#Print (2x3) band storage form
print(A_band_up)
```

> Note that in this example, the band storage form is constructed using the upper diagonal elements.

```
[[ 2. -1.  0.]
 [-1.  2. -1.]
 [ 0. -1.  2.]]

[[ 0. -1. -1.]
 [ 2.  2.  2.]]
```

(b) Step 2: Finding eigenvalues and eigenvectors.

```
from scipy.linalg import eig,eig_banded
from scipy import zeros,arange
A=zeros((3,3))
#Constructing (2x3) band storage form using upper diagonal elements
A_band_up=zeros((2,3))
for i in arange(2):
        A_band_up[0,i+1]=-1.0
for i in arange(3):
        A_band_up[1,i]=2.0
E,V=eig_banded(A_band_up)
print(E)
print(V)
```

```
[0.58578644 2.        3.41421356]
[[-5.00000000e-01 -7.07106781e-01  5.00000000e-01]
 [-7.07106781e-01  3.12250226e-16 -7.07106781e-01]
 [-5.00000000e-01  7.07106781e-01  5.00000000e-01]]
```

Assuming that the matrix is stored in **A_band_up** using the upper diagonal form, the **eig_banded(A_band_up,lower=False)** functionality of the **scipy.linalg** submodule determines the eigenvalues and eigenvectors of a Hermitian tridiagonal matrix. If no argument for **lower** is specified, the **default upper diagonal** form is used. One can, however, construct a (2×3) band storage form using lower diagonal elements and make use of the **lower=True** argument in the **eig_banded()** functionality of the **scipy.linalg** submodule to obtain the eigenvalues and eigenvectors of a Hermitian tridiagonal matrix. This is demonstrated in the following.

```
from scipy.linalg import eig,eig_banded
from scipy import zeros,arange
#Constructing (2x3) band storage form using lower diagonal elements
A_band_lower=zeros((2,3))
for i in range(3):
        A_band_lower[0,i]=2.0
for i in range(2):
        A_band_lower[1,i]=-1.0
print(A_band_lower)
E,V=eig_banded(A_band_lower,lower=True)
print(E)
print(V)
```

> Note that the lower=True argument has the small letter l in lower and the capital letter T in True.

```
[[ 2.  2.  2.]
 [-1. -1.  0.]]

[0.58578644 2.         3.41421356]

[[-5.00000000e-01 -7.07106781e-01  5.00000000e-01]
 [-7.07106781e-01  3.12250226e-16 -7.07106781e-01]
 [-5.00000000e-01  7.07106781e-01  5.00000000e-01]]
```

Self-Study: To realize the usefulness of a band storage form, construct a (3000×3000) Hermitian tridiagonal matrix and find its eigenvalues and eigenvectors using the **eig()** functionality of the **scipy.linalg** submodule. Note the approximate time the computer takes to perform the calculation. Next, construct a (2×3000) band storage form of the same Hermitian tridiagonal matrix and find its eigenvalues and eigenvectors using the **eig_banded()** functionality of the **scipy.linalg** submodule. Note the approximate time the computer takes to perform the calculation using the band storage

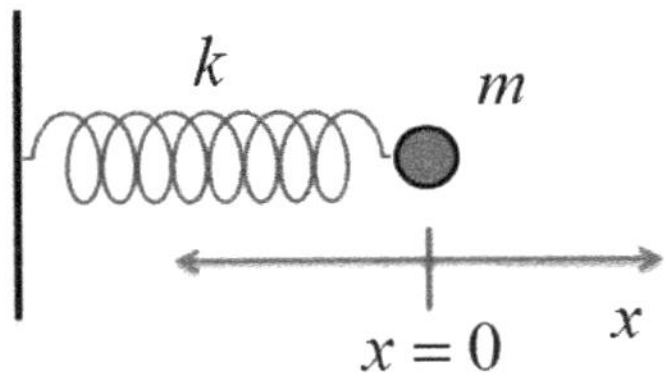

Figure D.1. A quantum particle with mass m is undergoing oscillation.

form. Compare the two computational times (for a surprising outcome!).

Example 2. Numerically find the three lowest-lying energy states of a one-dimensional quantum harmonic oscillator. Compare the results with the ones obtained through analytical derivation.

One-dimensional quantum harmonic oscillators are discussed in adequate detail in most of the standard quantum chemistry textbooks. See, for example, McQuarrie's *Quantum Chemistry.*[4] The Hamiltonian of a one-dimensional quantum harmonic oscillator is written as

$$\hat{H} = \left[-\frac{\hbar^2}{2m}\frac{d^2}{dx^2} + \frac{1}{2}kx^2\right]$$

where m is the mass of the particle which is undergoing the oscillation, k is the force constant representing the stiffness of the vibrating spring, as shown in Figure D.1, and the angular frequency of the oscillation is ω, which is given by $\omega = \sqrt{\frac{k}{m}}$.

For convenience, often, the atomic and molecular quantum mechanical calculations are performed using a unique unit system — called the atomic unit (a.u.). A detailed discussion on the atomic unit is given in PythonChapter E. Here, we directly make use of this unit system to re-express the Hamiltonian of a one-dimensional harmonic oscillator. In atomic unit, for reason to be revealed in PythonChapter E, $\hbar$ assumes a value of 1 a.u. For the present problem, we further assume that $m = k = 1$ a.u. Then, in atomic unit and using the defining conditions of the problem, the Hamiltonian of the one-dimensional harmonic oscillator becomes

$$\hat{H} = \left[-0.5\frac{d^2}{dx^2} + 0.5x^2\right]$$

Note that the purpose of using atomic units and considering $m = k = 1$ a.u. for the present problem is that these greatly simplify the Hamiltonian so that one can easily grasp the essence of this exercise. Thus, under the grid representation and using the finite difference method, one can now express the Hamiltonian in matrix form as (refer to the related discussion given in Chapter 4)

$$\mathbf{H} = \mathbf{T} + \mathbf{V} = -0.5\frac{1}{\Delta x^2}\begin{pmatrix} -2 & 1 & 0 & \cdot & \cdot & & 0 \\ 1 & -2 & 1 & \cdot & \cdot & & 0 \\ 0 & 1 & -2 & \cdot & \cdot & & 0 \\ \cdot & \cdot & \cdot & \cdot & \cdot & & \cdot \\ \cdot & \cdot & \cdot & \cdot & \cdot & \cdot & \cdot \\ & & & & \cdot & \cdot & 1 \\ 0 & 0 & 0 & 0 & 0 & 1 & -2 \end{pmatrix}_{N\times N}$$

$$+ \begin{pmatrix} V_0 & 0 & 0 & \cdot & \cdot & & 0 \\ 0 & V_1 & 0 & \cdot & \cdot & & 0 \\ 0 & 0 & V_2 & \cdot & \cdot & & 0 \\ \cdot & \cdot & \cdot & \cdot & \cdot & & \cdot \\ \cdot & \cdot & \cdot & \cdot & \cdot & \cdot & \cdot \\ & & & & \cdot & \cdot & 0 \\ 0 & 0 & 0 & 0 & 0 & 0 & V_{N-1} \end{pmatrix}_{N\times N}$$

Here, $V_i = 0.5x_i^2$, where x_i represents the x-grid points. Our next task is to represent this Hamiltonian matrix in Python code.

```
#Importing the Required Libraries
from scipy import arange,zeros
#Creating the x-grid in Atomic Unit
xmin=-0.5
xmax=0.7
dx=0.2
x=arange(xmin,xmax,dx)
#Defining the Harmonic Potential Function
k=1
Vdiag=0.5*k*(x**2)
#Constructing the Hamiltonian Matrix
m=1
h_cut=1
N=len(x)
T=zeros((N,N))#Kinetic Part
for i in arange(N):
        T[i,i]=(-0.5)*((h_cut**2)/m)*(1/dx**2)*(-2)
for i in arange(N-1):
        T[i,i+1]=(-0.5)*((h_cut**2)/m)*(1/dx**2)*(1)
        T[i+1,i]=(-0.5)*((h_cut**2)/m)*(1/dx**2)*(1)
V=zeros((N,N)) #Potential Part
for i in arange(N):
        V[i,i]=Vdiag[i]
H=T+V#Total Hamiltonian
print(T)
print(V)
```

```
[[ 25.  -12.5  0.    0.    0.    0.  ]
 [-12.5 25.  -12.5  0.    0.    0.  ]
 [  0.  -12.5 25.  -12.5  0.    0.  ]
 [  0.    0.  -12.5 25.  -12.5  0.  ]
 [  0.    0.    0.  -12.5 25.  -12.5]
 [  0.    0.    0.    0.  -12.5 25. ]]
```

> ➢ Note that, as expected, the kinetic part of the Hamiltonian adopts a tridiagonal Hermitian matrix.

```
[[0.125 0.    0.    0.    0.    0.   ]

 [0.    0.045 0.    0.    0.    0.   ]
 [0.    0.    0.005 0.    0.    0.   ]
 [0.    0.    0.    0.005 0.    0.   ]
 [0.    0.    0.    0.    0.045 0.   ]
 [0.    0.    0.    0.    0.    0.125]]
```

> ➢ Note that, as expected, the potential part of the Hamiltonian adopts a diagonal Hermitian matrix.

The above Python program is for demonstration purposes only so that one can identify the nature of the kinetic and potential parts of the Hamiltonian. We need more x-grid points to perform a realistic calculation. Furthermore, as noted earlier, it is more convenient to use a band storage form of the large Hamiltonian matrix to save on the computational costs of computing its eigenvalues and eigenvectors. This is why we make use of the $(2 \times N)$ band storage form with the upper diagonal elements of the Hamiltonian matrix. In this form, the diagonal elements of the Hamiltonian matrix can be called using $H[i, i]$, and the upper diagonal elements can be called using $H[i, i+1]$. The corresponding Python program is given as follows.

```
#Importing the Required Libraries
from scipy import arange,zeros
from scipy.linalg import eig_banded
from matplotlib.pyplot import plot,show,xlim,ylim
#Creating the x-grid in Atomic Unit
xmin=-500
xmax=500.2
dx=0.2
x=arange(xmin,xmax,dx)
#Defining the Harmonic Potential Function
k=1
Vdiag=0.5*k*(x**2)
#Constructing the Hamiltonian Matrix
m=1
h_cut=1
N=len(x)
T=zeros((N,N)) #Kinetic Part
for i in arange(N):
        T[i,i]=(-0.5)*((h_cut**2)/m)*(1/dx**2)*(-2)
for i in arange(N-1):
        T[i,i+1]=(-0.5)*((h_cut**2)/m)*(1/dx**2)*(1)
        T[i+1,i]=(-0.5)*((h_cut**2)/m)*(1/dx**2)*(1)
V=zeros((N,N))#Potential Part
for i in arange(N):
        V[i,i]=Vdiag[i]
H=T+V#Total Hamiltonian
#Representing the Hamiltonian Matrix in BandStorage Form
H_band_up=zeros((2,N))
for i in arange(N-1):
        H_band_up[0,i+1] =H[i,i+1]
for i in arange(N):
        H_band_up[1,i] =H[i,i]
#Calculating the Energy and Normalized Wavefunction of the Eigenstates
en,vec=eig_banded(H_band_up)
#Printing Energy of Three Lowest-Lying Eigenstates
print(en[0])
print(en[1])
print(en[2])
#Plotting the Potential and Normalized Wavefunctions
plot(x,Vdiag)
plot(x,(en[0]+vec[:,0]))
plot(x,(en[1]+vec[:,1]))
plot(x,(en[2]+vec[:,2]))
xlim(-4,4)
ylim(0,3.1)
show()
```

> Note that each normalized wavefunction is displaced on the energy axis (Y-axis) by its respective energy for clarity in presentation.

```
0.49874685129999463
1.4937215171042428
2.48363863050061
```

Note that the three lowest-lying eigenstate energies are printed upon execution of the above program. An analytical solution to the TISE of the one-dimensional quantum harmonic oscillator gives the eigenstate energies of the oscillator as

$$\mathrm{E}_v = \left(v + \frac{1}{2}\right)\hbar\omega = \left(v + \frac{1}{2}\right)\hbar\sqrt{\frac{k}{m}},$$

where v represents the quantum number.

In atomic units ($\hbar = 1$) and using the defining conditions of the present problem ($k = m = 1$), the above energy expression becomes

$$\mathrm{E}_v = \left(v + \frac{1}{2}\right) \quad \text{a.u.}$$

As a result, we may obtain the energy of the three lowest-lying eigenstates of the quantum harmonic oscillator as 0.5, 1.5, and 2.5 a.u. Note that the numerical solution given above renders (approximately) the same result.

The above program also plots the normalized wavefunctions for the three lowest-lying eigenstates of the quantum harmonic oscillator along with the potential energy curve, as shown in the following figure.

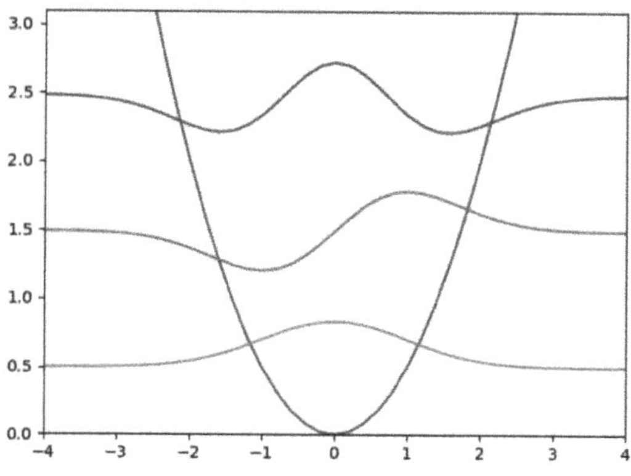

➢ Clearly, this figure requires further formatting for better presentation with axis labeling, giving a title, etc. This task is left for readers.

Guiding Question

D.1. Numerically compute the zero-point vibrational energy (ground vibrational state energy) of the H_2 molecule assuming the quantum harmonic oscillator approximation. Compare the result with the analytical value which can be obtained from $E_0 = \frac{1}{2}h\nu$. The following information is given. The vibrational frequency $\bar{\omega}$ of the H_2 molecule is $4403.56\,\text{cm}^{-1}$. One atomic unit of frequency $= 4.13 \times 10^{16}\,\text{s}^{-1}$. Planck constant $h = 4.13566 \times 10^{-15}$ eV.s.

Hints: As the frequency is given in cm^{-1}, first, convert the frequency to $s^{-1}(\nu(\text{s}^{-1}) = c\bar{\omega}(\text{cm}^{-1}))$ and then, convert the frequency to atomic unit using the given quantity. On the other hand, using the expression, $k = (2\pi\nu)^2\mu$, one can calculate the force constant of the H_2 oscillator in atomic units. Here, $\mu = \frac{m_1 m_2}{m_1+m_2}$ a.u.

Key Points to Remember

- The **E,V=eig(A)** functionality of the **scipy.linalg** finds eigenvalue **E[i]** and corresponding eigenvector **V[:,i]** of a square matrix **A**.
- For a very large tridiagonal Hermitian matrix (e.g., Hamiltonian matrix for a realistic quantum system), use of its band storage form significantly reduces computational time for finding its eigenvalue and corresponding eigenvector.

Notes, References, and Further Reading

1. For a brief illuminating discussion on matrix algebra using Python programming, see C. Hill, Chapter 6: NumPy – 6.5 linear algebra. *Learning Scientific Programming with Python,* 2nd edn. Cambridge University Press, UK (2020). Although, in this book, the discussion is developed based on NumPy, the same is applicable for SciPy.
2. For all features of the scipy.linalg submodule, see the linear algebra documentation of scipy (https://docs.scipy.org/doc/scipy/reference/linalg.html). Here, note that many of the linear algebra routines that are available in the scipy linear algebra submodule are also available in the numpy linear algebra submodule (numpy.linalg); however, in this PythonChapter, only the scipy.linalg submodule is used. The linear algebra submodule of scipy directly interfaces with the Fortran LAPAC library (which is one of the most efficient, fastest, and most optimized linear algebra packages developed in Fortran over several decades).
3. For further illuminating discussion, see J. Izaac and J. Wang, Chapter 7: The eigenvalue problem. *Computational Quantum Mechanics.* Springer (Switzerland) (2018).
4. D. A. McQuarrie, Chapter 5: The harmonic oscillator and vibrational spectroscopy. *Quantum Chemistry,* 2nd edn. University Science Books, California (2008).

Exercises

D.1. Numerically find the ground state normalized wavefunction and the energy of a particle in a one-dimensional box using the direct matrix method and grid representation. Plot the wavefunction and compare the numerical result with the analytical one. Refer to McQuarrie's *Quantum Chemistry* for the analytical results.

D.2. Numerically find the ground state normalized wavefunction and the energy of an electron experiencing an atomic potential, $V(x) = -\frac{1}{\sqrt{1+x^2}}$. Plot the wavefunction.

D.3. Numerically find the zero-point vibrational energy for the following diatomic molecules using the quantum harmonic oscillator approximation. Compare the results with the analytically obtained zero-point vibrational energy $E_0 = \frac{1}{2}h\nu$, where h and ν are the Planck's constant and the frequency of oscillation (in s^{-1}), respectively. Given: 1 atomic unit of frequency = $4.13 \times 10^{16}\,s^{-1}$, Planck's constant $h = 4.13566 \times 10^{-15}$ eV.s, $\nu(s^{-1}) = c\bar{\omega}(cm^{-1})$, and force constant $k = (2\pi\nu)^2\mu$.

Molecule	**Frequency (cm^{-1})**
$H^{19}F$	4138.32
$^{14}N^{16}O$	1904.20
$^{35}Cl^{35}Cl$	559.72

D.4. Numerically compare the three lowest-lying eigenstate energies and wavefunctions of a diatomic species under the quantum harmonic oscillator and the quantum anharmonic oscillator approximations. Assume that $\mu = k = 1$ a.u. Consider the Morse potential $V(x) = D_e(1 - e^{-ax})^2$, where $D_e = 1.5$ eV and $a = 1\,\text{Å}^{-1}$ for the quantum anharmonic oscillator.

D.5. Find the vibrational spectrum of the $H^{35}Cl$ molecule, including the fundamental transition (v = 0 → v = 1) and the first overtone transition (v = 0 → v = 2). You do not need to predict the intensity pattern; determine only the expected position of the spectral line. Given: The vibrational frequency $\bar{\omega}$ of the $H^{35}Cl$ molecule is 2990.94 cm^{-1}.

Chapter 5

Numerical Solution to the TDSE

Highlights: *General Properties of Time-Evolution Operator and its Numerical Implementation under Grid Representation, Split Operator Method.*

5.1 Introduction

We have already understood from Chapter 1 that if a potential does not depend on time, the one-dimensional time-dependent Schrödinger equation (TDSE) $i\hbar\frac{\partial\psi(x,t)}{\partial t} = \left[-\frac{\hbar^2}{2m}\frac{\partial^2}{\partial x^2} + V(x)\right]\psi(x,t)$ can be solved using the variable separation method. In this method, the eigenvectors and corresponding eigenvalues (energies) of the quantum system before the onset of time evolution are first calculated using the time-independent Schrödinger equation, $\hat{H}\psi_n(x,t) = E_n\psi_n(x,t)$, to obtain stationary state wavefunctions, $\psi_n(x,t) = \psi_n(x)e^{\frac{-iE_nt}{\hbar}}$, where n denotes the quantum number (quantum state). Thereafter, the non-stationary state wavefunction is expressed as $\psi(x,t) = \sum_n c_n(t)\ \psi_n(x)\ e^{\frac{-iE_nt}{\hbar}}$, where $c_n(t)$ represents the time-dependent expansion coefficient (referring to the contribution of each stationary state to the total wavefunction of the non-stationary state). Finally this non-stationary state wavefunction is inserted into the TDSE to obtain an approximate expression for $c_n(t)$.

The above-mentioned procedure to explore quantum dynamics is very straightforward, and in fact, it is used in developing the time-dependent perturbation theory.[1] However, there are many situations in which the exploration of quantum dynamics via eigenvalues and eigenvectors may not be practical either because a very large number of eigenstates are needed or because the calculation of these states is computationally too expensive. In that case, it is convenient to compute the time evolution of a quantum system for a given initial state directly without making use of the eigenstates of the quantum system. This can be achieved using the time-evolution operator, which is expressed as

$$\hat{U}(t) = e^{\frac{-i\hat{H}t}{\hbar}} \tag{5.1}$$

In the direct time-evolution approach, we start from an initial state of the quantum system represented by $\psi(x, 0)$. Then, using the time-evolution operator, one directly monitors the time evolution of the wavefunction as

$$\psi(x, t) = \hat{U}(t)\psi(x, 0) = e^{\frac{-i\hat{H}t}{\hbar}}\psi(x, 0) \tag{5.2}$$

In this chapter, a simple numerical implementation of the time-evolution operator is discussed to obtain a numerical solution to the TDSE. Before we present the numerical implementation of the time-evolution operator, we go over some important general properties of this operator.

5.2 General Properties of the Time-Evolution Operator

(1) $\hat{U}(t) = e^{\frac{-i\hat{H}t}{\hbar}}$ is called the time propagator or the time-evolution operator because it propagates the wavefunction in time. If the initial wavefunction (at $t = 0$) and the Hamiltonian of the system are known, one may find the wavefunction at any later time t using equation (5.2), unraveling the dynamics of a quantum system. We have already proved the equality expressed in equation (5.2) by making use of the TDSE (see Chapter 4 for derivation). Therefore, $\psi(x, t) = e^{\frac{-i\hat{H}t}{\hbar}}\psi(x, 0)$ is a solution to the TDSE.

(2) As noted already, the time-evolution operator is an "exponential of an operator," which is defined using a Taylor series expansion:

$$e^{\frac{-i\hat{H}t}{\hbar}} = \hat{1} + \left(\frac{-i\hat{H}t}{\hbar}\right) + \frac{1}{2}\left(\frac{-i\hat{H}t}{\hbar}\right)^2 + \cdots \infty \tag{5.3}$$

(3) The time-evolution operator is reversible or symmetric in time:

$$\hat{U}(\Delta t) = e^{\frac{-i\hat{H}\Delta t}{\hbar}} \text{ and } \hat{U}(-\Delta t) = e^{\frac{i\hat{H}\Delta t}{\hbar}}$$

Thus,

$$\hat{U}(-\Delta t)\hat{U}(\Delta t)\psi(x,0) = e^{\frac{i\hat{H}\Delta t}{\hbar}} e^{\frac{-i\hat{H}\Delta t}{\hbar}} \psi(x,0) = \psi(x,0)$$

Therefore, by reversing the time, we can recover the initial state of the system.

(4) The time-evolution operator is a unitary operator, i.e., $(e^{\frac{-i\hat{H}t}{\hbar}})^{-1} = (e^{\frac{-i\hat{H}t}{\hbar}})^{\dagger}$: its inverse is equal to its adjoint. The unitarity of the time-evolution operator originates from the fact that the Hamiltonian ($\hat{H}$) is a Hermitian operator. Proof of unitarity is given as follows:
The left-hand side (LHS) can be expressed as

$$\left(e^{\frac{-i\hat{H}t}{\hbar}}\right)^{-1} = e^{\frac{i\hat{H}t}{\hbar}} = \hat{1} + \frac{i\hat{H}t}{\hbar} + \frac{1}{2}\left(\frac{i\hat{H}t}{\hbar}\right)^2 + \cdots \infty$$

The right-hand side (RHS) can be expressed as

$$\left(e^{\frac{-i\hat{H}t}{\hbar}}\right)^{\dagger} = \hat{1} + \frac{i\hat{H}^{\dagger}t}{\hbar} + \frac{1}{2}\left(\frac{i\hat{H}^{\dagger}t}{\hbar}\right)^2 + \cdots \infty$$

$$= \hat{1} + \frac{i\hat{H}t}{\hbar} + \frac{1}{2}\left(\frac{i\hat{H}t}{\hbar}\right)^2 + \cdots \infty$$

(As $\hat{H}$ is a Hermitian operator, $\hat{H} = \hat{H}^{\dagger}$)

Therefore, LHS = RHS, or the time-evolution operator is a unitary operator.

(5) Since $\psi(x,t) = e^{\frac{-i\hat{H}t}{\hbar}}\psi(x,0)$, one may write $|\psi(x,t)|^2 = |\psi(x,0)|^2$. This indicates that the total probability density is conserved during the time evolution of the quantum system. Therefore, the time-evolution operator is a norm-preserving operator.

As a result, the normalization constant does not change during the dynamical evolution of the quantum system if the quantum dynamics is explored using the time-evolution operator. Note that the norm is preserved because the time-evolution operator is a unitary operator.

5.3 Numerical Implementation of the Time-Evolution Operator

Although a mathematical strategy to obtain a solution to the TDSE using the time propagator is quite straightforward, its numerical implementation is not an easy task, for reasons that will be revealed soon. As the time-evolution operator is an exponential operator, its numerical implementation requires a method of calculating the matrix exponential. This point is elaborated in the following.

Almost all currently prevailing numerical methods, which are used to solve the TDSE, make use of the grid representation of a continuous wavefunction. The discretization of the wavefunction on a position grid is done by dividing the entire position space into a uniform grid (see Chapter 4 and PythonChapter B for more details). When the initial (at $t = 0$) wavefunction $\psi(x,0)$ is represented on the x-grid, we get discretized values of the wavefunction:

$$\begin{aligned}
\psi(x_0,0) &= y_0 \\
\psi(x_1,0) &= y_1 \\
\psi(x_2,0) &= y_2 \\
&\vdots \\
\psi(x_{N-1},0) &= y_{N-1}
\end{aligned}$$

The above discretized wavefunction is conveniently represented by a column matrix:

$$\psi(x,0) = \begin{pmatrix} y_0 \\ y_1 \\ y_2 \\ \cdot \\ \cdot \\ \cdot \\ y_{N-1} \end{pmatrix}$$

Once the wavefunction is discretized on the x-grid, the next question arises: **What is the matrix representation of the time-evolution operator?** Or, in other words,

$$\psi(x,t) = e^{\frac{-i\hat{H}t}{\hbar}}\psi(x,0) = \begin{pmatrix} & & \\ & ? & \\ & & \end{pmatrix} \begin{pmatrix} y_0 \\ y_1 \\ y_2 \\ \cdot \\ \cdot \\ \cdot \\ y_{N-1} \end{pmatrix}$$

5.3.1 *Matrix representation of the time-evolution operator*

Finding an efficient and accurate technique to calculate the matrix exponential has been an open problem for many decades. The method is extraordinarily simplified if the matrix is diagonal because the exponential of a diagonal matrix can be obtained by exponentiating each diagonal element as follows:

$$\text{If } \mathbf{A} = \begin{pmatrix} a_{11} & 0 & 0 & 0 & 0 & 0 & 0 \\ 0 & a_{22} & 0 & 0 & 0 & 0 & 0 \\ 0 & 0 & a_{33} & 0 & 0 & 0 & 0 \\ 0 & 0 & 0 & a_{44} & 0 & 0 & 0 \\ 0 & 0 & 0 & 0 & a_{55} & 0 & 0 \\ 0 & 0 & 0 & 0 & 0 & a_{66} & 0 \\ 0 & 0 & 0 & 0 & 0 & 0 & a_{77} \end{pmatrix}, \text{ then}$$

$$e^{\mathbf{A}} = \begin{pmatrix} e^{a_{11}} & 0 & 0 & 0 & 0 & 0 & 0 \\ 0 & e^{a_{22}} & 0 & 0 & 0 & 0 & 0 \\ 0 & 0 & e^{a_{33}} & 0 & 0 & 0 & 0 \\ 0 & 0 & 0 & e^{a_{44}} & 0 & 0 & 0 \\ 0 & 0 & 0 & 0 & e^{a_{55}} & 0 & 0 \\ 0 & 0 & 0 & 0 & 0 & e^{a_{66}} & 0 \\ 0 & 0 & 0 & 0 & 0 & 0 & e^{a_{77}} \end{pmatrix}$$

As the time-evolution operator $e^{\frac{-i\hat{H}t}{\hbar}}$ contains the Hamiltonian operator, for the numerical implementation of the time-evolution operator, our target should be to present the Hamiltonian matrix in diagonal form under the grid representation. The Hamiltonian operator is a sum of the kinetic and potential terms (where we continue assuming that $V(x)$ is time-independent):

$$\hat{H} = \left[-\frac{\hbar^2}{2m}\frac{\partial^2}{\partial x^2} + V(x)\right] = \hat{T} + \hat{V}$$

Therefore, one can approximately express the time-evolution operator as a product of kinetic and potential factors **after discretizing the entire time interval** $[0, t]$ by a very short time step Δt:

$$e^{\frac{-i\hat{H}\Delta t}{\hbar}} = e^{\frac{-i(\hat{T}+\hat{V})\Delta t}{\hbar}} \approx e^{\frac{-i\hat{T}\Delta t}{\hbar}} e^{\frac{-i\hat{V}\Delta t}{\hbar}} \tag{5.4}$$

This is called the **split operator method**.[2] The reason behind discretizing the time by a very short time step Δt is elaborated in the following.

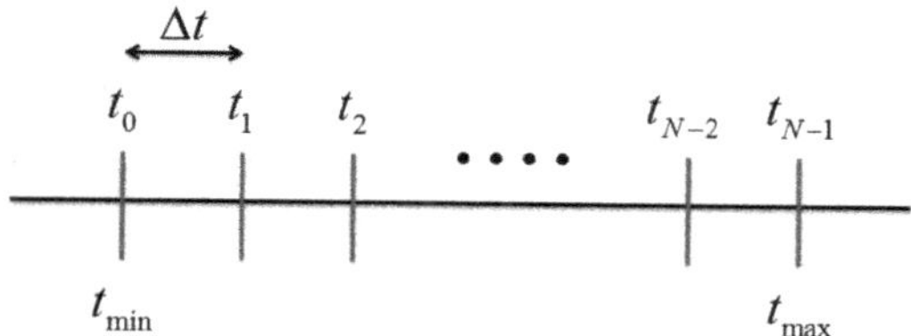

Figure 5.1. Discretized time frame.

If the entire time internal $[0, t]$ is discretized by a very short time step Δt (as shown in Figure 5.1), one may write each short time propagation as

$$\psi(x, 0 + \Delta t) = \hat{U}(\Delta t)\psi(x, 0) = e^{\frac{-i\hat{H}\Delta t}{\hbar}}\psi(x, 0)$$

$$\psi(x, 0 + 2\Delta t) = \hat{U}(\Delta t)\psi(x, 0 + \Delta t) = e^{\frac{-i\hat{H}\Delta t}{\hbar}}\psi(x, 0 + \Delta t)$$

$$\psi(x, 0 + 3\Delta t) = \hat{U}(\Delta t)\psi(x, 0 + 2\Delta t) = e^{\frac{-i\hat{H}\Delta t}{\hbar}}\psi(x, 0 + 2\Delta t)$$

$$\vdots$$

$$\begin{aligned}\psi(x, t) = \psi(x, 0 + N\Delta t) &= \hat{U}(\Delta t)\psi(x, 0 + [N - 1]\Delta t) \\ &= e^{\frac{-i\hat{H}\Delta t}{\hbar}}\psi(x, 0 + [N - 1]\Delta t)\end{aligned}$$

Or, collectively, one may write

$$\psi(x, t) = e^{\frac{-i\hat{H}\Delta t}{\hbar}} e^{\frac{-i\hat{H}\Delta t}{\hbar}} e^{\frac{-i\hat{H}\Delta t}{\hbar}} \cdots (N \text{ times}) \cdots \psi(x, 0)$$

Thus, the time propagator for the entire time interval can be expressed as the N-times product of the short time propagator:

$$e^{\frac{-i\hat{H}t}{\hbar}} = e^{\frac{-i\hat{H}\Delta t}{\hbar}} e^{\frac{-i\hat{H}\Delta t}{\hbar}} e^{\frac{-i\hat{H}\Delta t}{\hbar}} \cdots (N \text{ times}) \tag{5.5}$$

As mentioned earlier, the short time propagator can be approximately expressed as

$$e^{\frac{-i\hat{H}\Delta t}{\hbar}} \approx e^{\frac{-i\hat{T}\Delta t}{\hbar}} e^{\frac{-i\hat{V}\Delta t}{\hbar}}$$

In general, $e^{\hat{A}+\hat{B}}$ is not equal to $e^{\hat{A}}e^{\hat{B}}$, where $\hat{A}$ and $\hat{B}$ are non-commuting operators. We may prove this using the Taylor series expansion:

$$\begin{aligned} e^{\hat{A}+\hat{B}} &= \hat{1} + (\hat{A}+\hat{B}) + \frac{1}{2}(\hat{A}+\hat{B})^2 + \cdots\infty \\ &= \hat{1} + (\hat{A}+\hat{B}) + \frac{\hat{A}^2+\hat{B}^2}{2} + \frac{\hat{A}\hat{B}}{2} + \frac{\hat{B}\hat{A}}{2} + \cdots\infty \end{aligned} \tag{5.6}$$

and

$$\begin{aligned} e^{\hat{A}}e^{\hat{B}} &= \left[\hat{1} + \hat{A} + \frac{1}{2}\hat{A}^2 + \cdots\infty\right]\left[\hat{1} + \hat{B} + \frac{1}{2}\hat{B}^2 + \cdots\infty\right] \\ &= \hat{1} + (\hat{A}+\hat{B}) + \frac{\hat{A}^2+\hat{B}^2}{2} + \hat{A}\hat{B} + \cdots\infty \end{aligned} \tag{5.7}$$

Clearly, equations (5.6) and (5.7) are not equal. Therefore, one cannot decompose the exponential of the sum of two non-commuting operators into the product of two exponentials for each operator: $e^{\hat{A}+\hat{B}} \neq e^{\hat{A}}e^{\hat{B}}$, where $\hat{A}$ and $\hat{B}$ are non-commuting operators. As the kinetic and potential energy operators are non-commuting operators, the following inequality holds:

$$e^{\frac{-i\hat{H}t}{\hbar}} \neq e^{\frac{-i\hat{T}t}{\hbar}}e^{\frac{-i\hat{V}t}{\hbar}}$$

However, for a very short time propagation, one may approximately write

$$e^{\frac{-i\hat{H}\Delta t}{\hbar}} \approx e^{\frac{-i\hat{T}\Delta t}{\hbar}}e^{\frac{-i\hat{V}\Delta t}{\hbar}}$$

This approximate expression incorporates an error which is proportional to Δt^2. Therefore, by selecting a very short time step, Δt, one can reduce the error significantly. We first prove this argument as follows:

$$\begin{aligned} e^{\frac{-i(\hat{T}+\hat{V})\Delta t}{\hbar}} &= \hat{1} + (-i)(\hat{T}+\hat{V})\frac{\Delta t}{\hbar} + \frac{(-i)^2}{2}(\hat{T}+\hat{V})^2\frac{\Delta t^2}{\hbar^2} + \cdots\infty \\ &= \hat{1} + (-i)(\hat{T}+\hat{V})\frac{\Delta t}{\hbar} + \frac{(-i)^2}{2} \\ &\quad \times (\hat{T}^2 + \hat{V}^2 + \hat{T}\hat{V} + \hat{V}\hat{T})\frac{\Delta t^2}{\hbar^2} + \cdots\infty \end{aligned}$$

and

$$e^{\frac{-i\hat{T}\Delta t}{\hbar}} e^{\frac{-i\hat{V}\Delta t}{\hbar}} = \left[\hat{1} + (-i)\hat{T}\frac{\Delta t}{\hbar} + \frac{(-i)^2}{2}\hat{T}^2\frac{\Delta t^2}{\hbar^2} + \cdots\infty\right]$$
$$\times\left[\hat{1} + (-i)\hat{V}\frac{\Delta t}{\hbar} + \frac{(-i)^2}{2}\hat{V}^2\frac{\Delta t^2}{\hbar^2} + \cdots\infty\right]$$
$$= \hat{1} + (-i)(\hat{T} + \hat{V})\frac{\Delta t}{\hbar} + \frac{(-i)^2}{2}(\hat{T}^2 + \hat{V}^2)\frac{\Delta t^2}{\hbar^2}$$
$$- \hat{T}\hat{V}\frac{\Delta t^2}{\hbar^2} + \cdots\infty$$

To find the error, we may subtract the above two expressions:

$$e^{\frac{-i(\hat{T}+\hat{V})\Delta t}{\hbar}} - e^{\frac{-i\hat{T}\Delta t}{\hbar}} e^{\frac{-i\hat{V}\Delta t}{\hbar}}$$
$$= \left[\hat{1} + (-i)(\hat{T} + \hat{V})\frac{\Delta t}{\hbar} + \frac{(-i)^2}{2}(\hat{T}^2 + \hat{V}^2 + \hat{T}\hat{V} + \hat{V}\hat{T})\right.$$
$$\left.\frac{\Delta t^2}{\hbar^2} + \cdots\infty\right]$$
$$- \left[\hat{1} + (-i)(\hat{T} + \hat{V})\frac{\Delta t}{\hbar} + \frac{(-i)^2}{2}(\hat{T}^2 + \hat{V}^2)\frac{\Delta t^2}{\hbar^2}\right.$$
$$\left.-\hat{T}\hat{V}\frac{\Delta t^2}{\hbar^2} + \cdots\infty\right]$$

Or, considering up to a second-order error (higher-order errors are neglected), we get

$$e^{\frac{-i(\hat{T}+\hat{V})\Delta t}{\hbar}} - e^{\frac{-i\hat{T}\Delta t}{\hbar}} e^{\frac{-i\hat{V}\Delta t}{\hbar}} = -\frac{1}{2}(\hat{T}\hat{V} + \hat{V}\hat{T})\frac{\Delta t^2}{\hbar^2} + \hat{T}\hat{V}\frac{\Delta t^2}{\hbar^2}$$
$$= \frac{(\hat{T}\hat{V} - \hat{V}\hat{T})}{2}\frac{\Delta t^2}{\hbar^2} \tag{5.8}$$

So, second-order error $\propto \Delta t^2$, where higher-order terms can be neglected because Δt is selected to be very small.

Thus, the split operator approach using the product $e^{\frac{-i(\hat{T}+\hat{V})\Delta t}{\hbar}} \approx e^{\frac{-i\hat{T}\Delta t}{\hbar}} e^{\frac{-i\hat{V}\Delta t}{\hbar}}$ incorporates an error $\propto \Delta t^2$. This error can be further reduced if we take the following symmetrized product:

$$e^{\frac{-i(\hat{T}+\hat{V})\Delta t}{\hbar}} \approx e^{-i\frac{\hat{T}}{2}\frac{\Delta t}{\hbar}} e^{\frac{-i\hat{V}\Delta t}{\hbar}} e^{-i\frac{\hat{T}}{2}\frac{\Delta t}{\hbar}} \tag{5.9}$$

One can easily prove that the symmetrized product scheme of the split operator method incorporates an error $\propto \Delta t^3$ (prove it yourself). We will use this symmetrized product scheme to achieve a better approximation in the numerical implementation of the time propagator using the discretized time frame.

However, note that the final numerical implementation of the symmetrized split-operator-based time propagation requires us to present both the kinetic and potential energy matrices in diagonal form for the reason mentioned earlier. In the position space grid representation (as shown explicitly in Chapter 4), the potential energy operator can be represented by a diagonal matrix. Therefore, the exponential of a potential energy matrix can be easily evaluated: The exponential of a diagonal matrix is obtained by exponentiating each diagonal element. Thus, one can write the potential part of the propagator as

$$e^{\frac{-i\mathbf{V}\Delta t}{\hbar}} = \begin{pmatrix} e^{\frac{-iV(x_0)\Delta t}{\hbar}} & 0 & 0 & \cdots & 0 \\ 0 & e^{\frac{-iV(x_1)\Delta t}{\hbar}} & 0 & \cdots & 0 \\ 0 & 0 & e^{\frac{-iV(x_2)\Delta t}{\hbar}} & \cdots & 0 \\ \cdot & \cdot & \cdot & \cdots & \cdot \\ \cdot & \cdot & \cdot & \cdots & \cdot \\ \cdot & \cdot & \cdot & \cdots & \cdot \\ 0 & 0 & 0 & \cdots & e^{\frac{-iV(x_{N-1})\Delta t}{\hbar}} \end{pmatrix}$$

In a position space grid representation, the kinetic energy operator, on the other hand, features a tridiagonal matrix form (this is elaborately discussed in Chapter 4 using the finite difference method). Computing an exponential of a non-diagonal matrix is far from simple. One can, however, take advantage of the momentum space grid representation of the kinetic energy operator because this operator adopts a diagonal matrix form in momentum space grid for the reasons discussed in the following.

The kinetic energy operator has a differential form in position space: $\hat{T} = -\frac{\hbar^2}{2m}\frac{d^2}{dx^2}$. This differential form gives birth to the tridiagonal matrix representation of the kinetic energy operator in the position space grid. However, in momentum space, the kinetic energy operator does not have a differential form. In momentum space, the kinetic energy operator is just a function of momentum space: $\hat{T} = \frac{\hat{p}^2}{2m} = \frac{p^2}{2m} = \frac{f(p)}{2m}$. Therefore, just like the potential energy operator in position space grid representation, the kinetic energy operator can be represented by a diagonal matrix in momentum space grid representation. As a result, one can write the first kinetic part of the propagator as

$$e^{\frac{-i\mathbf{T}\Delta t}{2\hbar}} = \begin{pmatrix} e^{\frac{-iT(p_0)\Delta t}{2\hbar}} & 0 & 0 & \cdots & 0 \\ 0 & e^{\frac{-iT(p_1)\Delta t}{2\hbar}} & 0 & \cdots & 0 \\ 0 & 0 & e^{\frac{-iT(p_2)\Delta t}{2\hbar}} & \cdots & 0 \\ . & . & . & \cdots & . \\ . & . & . & \cdots & . \\ . & . & . & \cdots & . \\ 0 & 0 & 0 & \cdots & e^{\frac{-iT(p_{N-1})\Delta t}{2\hbar}} \end{pmatrix}$$

However, for the kinetic energy operator to operate on the wavefunction in momentum space, one has to first convert the wavefunction from its position space representation to its momentum space representation: $\psi(x) \rightarrow \psi(p)$.

Thus, the above discussion points to the following strategy for the numerical implementation of the symmetrized split operator method.

Analytical form of the short time propagator:

$$\psi(x, 0+\Delta t) = e^{-i\frac{\hat{T}}{2}\frac{\Delta t}{\hbar}} e^{\frac{-i\hat{V}\Delta t}{\hbar}} e^{-i\frac{\hat{T}}{2}\frac{\Delta t}{\hbar}} \psi(x, 0)$$

Numerical steps to be followed to implement the time propagator:

(1) Fourier transform the wavefunction from its position grid representation to momentum grid representation: $\psi(x) \rightarrow \psi(p)$, by making use of the efficient fast Fourier transform (FFT) algorithm.

(2) Carry out the first kinetic part of the propagation in momentum space:

$$\psi'(p, 0+\Delta t) = e^{-i\frac{\hat{T}}{2}\frac{\Delta t}{\hbar}} \psi(p, 0)$$

(3) Inverse Fourier transform the wavefunction from momentum space to position space:

$$\psi'(p, 0+\Delta t) \rightarrow \psi'(x, 0+\Delta t)$$

(4) Carry out the potential part of the propagation in position space:

$$\psi''(x, 0+\Delta t) = e^{-i\frac{\hat{V}\Delta t}{\hbar}} \psi'(x, 0+\Delta t)$$

(5) Fourier transform the wavefunction again from position space to momentum space:

$$\psi''(x, 0+\Delta t) \rightarrow \psi''(p, 0+\Delta t)$$

(6) Carry out the second kinetic part of the propagator in momentum space:

$$\psi'''(p, 0+\Delta t) = e^{-i\frac{\hat{T}}{2}\frac{\Delta t}{\hbar}} \psi''(p, 0+\Delta t)$$

(7) Finally, inverse Fourier transform the wavefunction from momentum space to position space:

$$\psi'''(p, 0+\Delta t) \rightarrow \psi'''(x, 0+\Delta t)$$

The above sequence is then repeated for each Δt, with ψ''' serving as the new initial wavefunction for the next time step.

Note that in position space grid representation, the initial wavefunction is represented by a column matrix:

$$\psi(x, 0) = \begin{pmatrix} \psi(x_0, 0) \\ \psi(x_1, 0) \\ \psi(x_2, 0) \\ \vdots \\ \psi(x_{N-1}, 0) \end{pmatrix}$$

(Continued)

(Continued)

After converting the discretized wavefunction from its position space to its momentum space representation, we obtain the following column matrix (with the same number of elements) in momentum space:

$$\psi(p,0) = \begin{pmatrix} \psi(p_0,0) \\ \psi(p_1,0) \\ \psi(p_2,0) \\ \vdots \\ \psi(p_{N-1},0) \end{pmatrix}$$

Subsequently, the $\psi'(p, 0 + \Delta t) = e^{-i\frac{\hat{T}}{2}\frac{\Delta t}{\hbar}}\psi(p, 0)$ routine can be directly performed by element-wise multiplication of the respective matrices as shown in the following:

$$\begin{pmatrix} e^{\frac{-iT(p_0)\Delta t}{2\hbar}} & 0 & 0 & \cdots & 0 \\ 0 & e^{\frac{-iT(p_1)\Delta t}{2\hbar}} & 0 & \cdots & 0 \\ 0 & 0 & e^{\frac{-iT(p_2)\Delta t}{2\hbar}} & \cdots & 0 \\ . & . & . & \cdots & . \\ . & . & . & \cdots & . \\ . & . & . & \cdots & . \\ 0 & 0 & 0 & \cdots & e^{\frac{-iT(p_{N-1})\Delta t}{2\hbar}} \end{pmatrix} \begin{pmatrix} \psi(p_0,0) \\ \psi(p_1,0) \\ \psi(p_2,0) \\ . \\ . \\ . \\ \psi(p_{N-1},0) \end{pmatrix}$$

$$= \begin{pmatrix} e^{\frac{-iT(p_0)\Delta t}{2\hbar}}\psi(p_0,0) \\ e^{\frac{-iT(p_1)\Delta t}{2\hbar}}\psi(p_1,0) \\ e^{\frac{-iT(p_2)\Delta t}{2\hbar}}\psi(p_2,0) \\ . \\ . \\ . \\ e^{\frac{-iT(p_{N-1})\Delta t}{2\hbar}}\psi(p_{N-1},0) \end{pmatrix}$$

Similar element-wise multiplication needs to be performed to implement the $\psi''(x, 0 + \Delta t) = e^{-i\frac{\hat{V}\Delta t}{\hbar}}\psi'(x, 0 + \Delta t)$ routine in position space.

Key Points to Remember

- $\psi(x,t) = e^{\frac{-i\hat{H}t}{\hbar}}\psi(x,0)$ is a solution to the TDSE if $\hat{H}$ is independent of time.
- Numerical implementation of the time propagator follows steps given below:

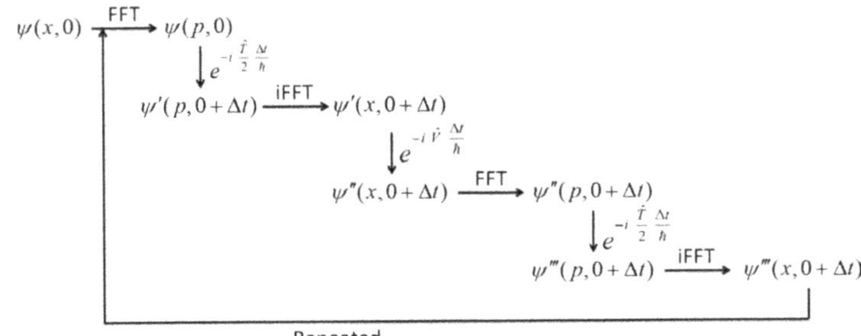

Notes, References, and Further Reading

1. I. N. Levine, Chapter 9: Perturbation theory. *Quantum Chemistry*, 7th edn. Pearson Education, Inc., New Jersey (2014).
2. (a) M. Feit, J. Fleck Jr. and A. Steiger, Solution of Schrödinger equation by a spectral method. *J. Comput. Phys.* 47, 412 (1982); (b) D. J. Tannor, Chapter 11: Numerical methods for solving the time-dependent Schrödinger equation. *Introduction to Quantum Mechanics a Time-Dependent Approach.* University Science Books, Sausalito (CA) (2007).

Exercise

5.1. Prove that the symmetrized product scheme of the split operator method, as expressed by $e^{\frac{-i(\hat{T}+\hat{V})\Delta t}{\hbar}} \approx e^{-i\frac{\hat{T}}{2}\frac{\Delta t}{\hbar}} e^{\frac{-i\hat{V}\Delta t}{\hbar}} e^{-i\frac{\hat{T}}{2}\frac{\Delta t}{\hbar}}$, incorporates an error which is proportional to Δt^3.

5.2. The norm of $\psi(x,t)$ at any time t is given by $||\psi(x,t)|| = \left[\int_{-\infty}^{+\infty} \psi^*(x,t)\,\psi(x,t)dx\right]^{\frac{1}{2}}$. By making use of the unitarity property of the time propagator, prove that $||\psi(x,t)|| = ||\psi(x,0)||$.

5.3. Prove that time evolution operator is a unitary operator.

5.4. The linear transformation which diagonalizes a Hermitian matrix is expressed by $\mathbf{U}^\dagger \mathbf{H} \mathbf{U} = \mathbf{\Lambda}$; where $\mathbf{U}$ is the unitary matrix that diagonalizes $\mathbf{H}$ to $\mathbf{\Lambda}$. Prove that $e^{\mathbf{H}} = \mathbf{U} e^{\mathbf{\Lambda}} \mathbf{U}^\dagger$.

5.5. prove that $\psi(x,t) = e^{\frac{-i\hat{H}t}{\hbar}} \psi(x,0)$ is a solution to the TDSE. Does it remain to be a solution if the Hamiltonian is time-dependent? Present your argument.

5.6. Find a simplified expression for $e^{\hat{A}+\hat{B}}$ if $\hat{A}$ and $\hat{B}$ commute.

PythonChapter E

Gaussian Wavepacket Dynamics

Highlights: *Gaussian Wavepacket Dynamics under Zero Potential and Linear Potential Employing Split-Operator Approach.*

E.1 Introduction

The goal of this PythonChapter is to finally accomplish a numerical solution to the TDSE to explore the quantum dynamics of a particle in one dimension. In Chapter 5, we have already understood that the split operator approach provides a numerical solution to the TDSE. In this chapter, we will implement this approach in Python programming.

E.2 Gaussian Wavepacket Dynamics under Zero Potential

Let us consider a free particle. This particle does not experience any potential, and its quantum dynamics is analytically solvable (refer to Chapter 3 for an analytical solution). In this section, we will make use of the split operator approach to explore the same dynamics numerically.

As we did before in Chapter 3, let us assume that the initial (at $t = 0$) wavefunction of the free particle is represented by a Gaussian wavepacket of the following form:

$$\psi_{w.p.}(x, 0) = e^{\frac{-x^2}{2}} e^{ik_0 x} \tag{E.1}$$

where k_0 is a real constant representing the angular wavenumber of the carrier wave of the wavepacket. In PythonChapter C, we have also understood that k_0 represents the central wavenumber of the wavepacket in Fourier space (or momentum space), and it is related to the velocity of the free particle. Furthermore, the initial wavepacket, as expressed in Equation (E.1), is centered at $x = 0$ in position space. The question which we wish to answer here is, *how do we explore the wavefunction* $\psi_{w.p.}(x,t)$ *numerically, given the initial wavefunction* $\psi_{w.p.}(x,0)$ *under zero potential?* We will make use of the split operator method to calculate $\psi_{w.p.}(x,t)$ numerically. The first step, however, is to normalize the initial wavefunction (note that in Chapter 2, we have realized that if we begin with a normalized wavefunction, TDSE preserves the normalization of the wavefunction for any later time).

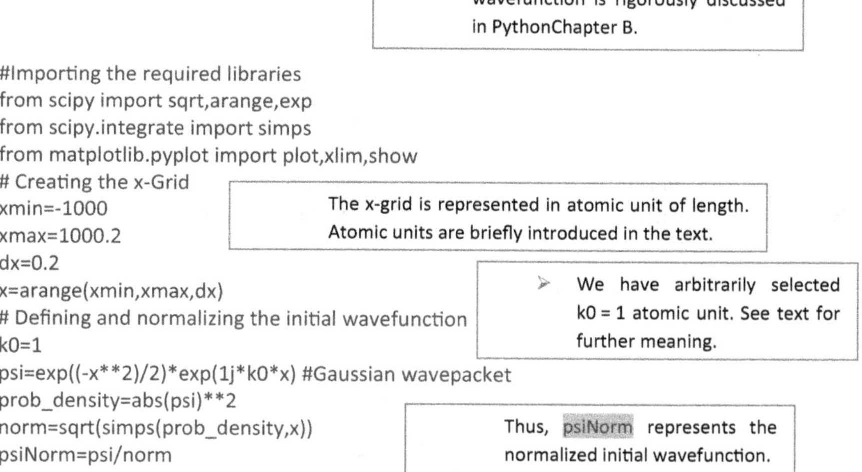

(a) Obtaining the normalized initial wavefunction:

> The procedure of normalizing a wavefunction is rigorously discussed in PythonChapter B.

```
#Importing the required libraries
from scipy import sqrt,arange,exp
from scipy.integrate import simps
from matplotlib.pyplot import plot,xlim,show
# Creating the x-Grid
xmin=-1000
xmax=1000.2
dx=0.2
x=arange(xmin,xmax,dx)
# Defining and normalizing the initial wavefunction
k0=1
psi=exp((-x**2)/2)*exp(1j*k0*x) #Gaussian wavepacket
prob_density=abs(psi)**2
norm=sqrt(simps(prob_density,x))
psiNorm=psi/norm
```

The x-grid is represented in atomic unit of length. Atomic units are briefly introduced in the text.

> We have arbitrarily selected k0 = 1 atomic unit. See text for further meaning.

Thus, psiNorm represents the normalized initial wavefunction.

In the above program, we have arbitrarily selected a limit of $(-1000, +1000)$ for the x-grid; however, note that care must be taken while selecting the limit of the x-grid to ensure that the wavefunction at any time must take zero value at the boundaries.

Recall that the momentum of the free particle is given by $\hbar k_0$ (refer to Chapter 3 for proof). In atomic units, as $\hbar = 1$, momentum is given by k_0. If the particle is considered to be an electron, the momentum of the particle can be expressed by $m_e v = k_0$. In atomic units, as $m_e = 1$, the speed of the electron is represented by $v = k_0$. Therefore, in the above program, by selecting $\boldsymbol{k0} = \mathbf{1}$, we assume that the particle (an electron) is traveling with 1 atomic unit of speed, which is equivalent to 2.188×10^6 m s^{-1} (in SI unit). Furthermore, the kinetic energy of the particle (an electron) is given by $\frac{1}{2}m_e v^2$ or, in atomic units, the kinetic energy of the particle (electron) is expressed as $\frac{1}{2}k_0^2$ (because in atomic units, $m_e = 1$ and $v = k_0$). We will make use of this expression of kinetic energy to implement the split operator approach in the next step.

Introduction to Atomic Units

The units which are frequently used in atomic and molecular quantum mechanical calculations are called atomic units.[1,2] To realize the atomic units (we closely follow the discussion by Szabo and Ostlund),[3] let us consider the time-independent Schrödinger equation for the hydrogen atom (in SI units):

$$\left[-\frac{\hbar^2}{2m_e}\nabla^2 - \frac{e^2}{4\pi\varepsilon_0 r}\right]\psi = E\psi$$

where $\hbar$ is the Plank's constant divided by 2π, m_e is the mass of the electron, and $-e$ is the charge on the electron. To convert the above time-independent Schrödinger equation to a dimensionless form, we may change the variables as $x, y, z \rightarrow \lambda x', \lambda y', \lambda z'$ and obtain

$$\left[-\frac{\hbar^2}{2m_e\lambda^2}\nabla'^2 - \frac{e^2}{4\pi\varepsilon_0\lambda r'}\right]\psi' = E\psi'$$

The constants in front of the kinetic and potential energy operators can be factored out if we choose $x, y, z \rightarrow \lambda x', \lambda y', \lambda z'$ such that

$$\frac{\hbar^2}{m_e\lambda^2} = \frac{e^2}{4\pi\varepsilon_0\lambda} = E_a$$

where $\frac{\hbar^2}{m_e\lambda^2} = \frac{e^2}{4\pi\varepsilon_0\lambda} = E_a$, which is the atomic unit of energy (also called the Hartree). By solving the relation for λ, we obtain $\lambda = \frac{4\pi\varepsilon_0\hbar^2}{m_e e^2}$. This exercise shows that λ is nothing but the Bohr radius a_0, which is the atomic unit of length.

(*Continued*)

(Continued)

Inserting $\frac{\hbar^2}{m_e \lambda^2} = \frac{e^2}{4\pi\varepsilon_0 \lambda} = E_a$ into the above time-independent Schrödinger equation, we obtain

$$E_a \left[-\frac{1}{2}\nabla'^2 - \frac{1}{r'} \right] \psi' = E\psi'$$

$$\text{or} \left[-\frac{1}{2}\nabla'^2 - \frac{1}{r'} \right] \psi' = \frac{E}{E_a}\psi'$$

$$\text{or} \left[-\frac{1}{2}\nabla'^2 - \frac{1}{r'} \right] \psi' = E'\psi', \text{ where } E' = \frac{E}{E_a}$$

The above equation is called the time-independent Schrödinger equation for the hydrogen atom in atomic units. As already realized, the use of atomic units greatly simplifies the quantum mechanical equations and expressions. For example, the Hamiltonian operator of a free electron, expressed by $-\frac{\hbar^2}{2m_e}\frac{\partial^2}{\partial x^2}$, becomes simply $-\frac{1}{2}\frac{\partial^2}{\partial x^2}$ in atomic units.

A list of atomic units and their corresponding SI equivalents is given in Table E.1.[4]

Table E.1.

Physical Property or Constant	**Atomic Unit (a.u.)**	**SI Equivalent**
Mass	1 a.u. of mass = mass of an electron, m_e	9.1094×10^{-31} kg
Charge	1 a.u. of charge = charge of a proton, e	1.6022×10^{-19} C
Angular Momentum	Planck constant divided by 2π, $\hbar$	1.0546×10^{-34} J · s
Length	Bohr Radius, $a_0 = \dfrac{4\pi\varepsilon_0\hbar^2}{m_e e^2}$	5.2918×10^{-11} m
Energy	$\dfrac{e^2}{4\pi\varepsilon_0 a_0} = E_a$	4.3597×10^{-18} J
Permittivity	$4\pi\varepsilon_0$	1.1127×10^{-10} $C^2J^{-1}m^{-1}$
Time		2.41888×10^{-17} s

Guiding Questions

E.1. Show that the speed of an electron in the first Bohr orbit is $\frac{e^2}{4\pi\varepsilon_0\hbar} = 2.188 \times 10^6$ m s^{-1}. This is the atomic unit of speed.

E.2. Express the speed of light in atomic units.

(b) The time evolution of the Gaussian wavepacket

We have already realized in Chapter 5 that a recipe for implementing the split operator method, which propagates the wavefunction from an initial time $t = 0$ to a final time t, is given by

$$\psi(x,t) = \left[\prod_{N} e^{-i\frac{\hat{T}}{2}\frac{\Delta t}{\hbar}} e^{-i\hat{V}\frac{\Delta t}{\hbar}} e^{-i\frac{\hat{T}}{2}\frac{\Delta t}{\hbar}}\right] \psi(x,0) \tag{E.2}$$

where $\prod \cdots$ implies an N-times product. Recall (from Chapter 5) that for each term in the product, the potential part of the propagator is executed in position space, and the kinetic part of the propagator is carried out in momentum space (or k-space). In atomic units, which consider $\hbar = 1$, and for a free particle (for which we consider $\hat{V} = 0$ because the particle does not experience any potential), the above split operator recipe can be expressed as

$$\psi(x,t) = \left[\prod_{N} e^{-i\hat{T}\Delta t}\right] \psi(x,0) \tag{E.3}$$

Alternatively, in explicit numerical steps, it is summarized as follows:

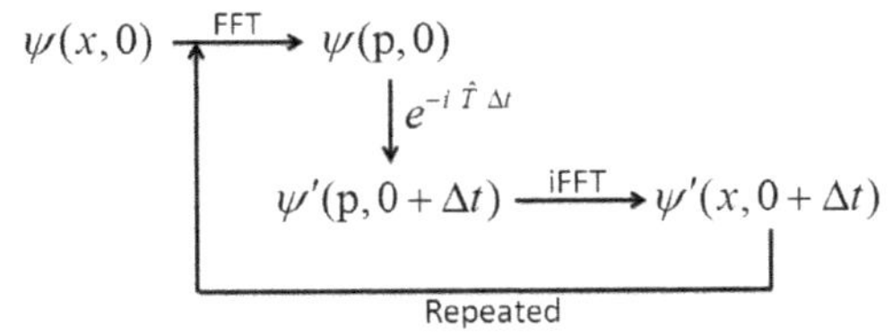

$\psi(x,0) \xrightarrow{\text{FFT}} \psi(\text{p},0)$

$\downarrow e^{-i\hat{T}\Delta t}$

$\psi'(\text{p},0+\Delta t) \xrightarrow{\text{iFFT}} \psi'(x,0+\Delta t)$

Repeated

where $\psi(x,0)$ is the initial wavefunction and $\psi'(x,0+\Delta t)$ is the new wavefunction at time $(0 + \Delta t)$. To implement the product form as given in Equation (E.3), the above sequence is repeated with the help of a **for loop** for every Δt, with ψ' serving as the new initial wavefunction for the next iteration of the loop. We must select a sufficiently small Δt to achieve higher accuracy in this computation. Furthermore, note that the kinetic energy operator on the k-grid is represented by (in atomic units) $\hat{T} = 0.5 \times k^2$ (a quadratic function in the k-space). Based on these results, one may now easily implement the split operator method to determine the $\psi_{w.p.}(x,t)$.

```
#Importing the required libraries
from scipy import sqrt,arange,exp,pi
from scipy.integrate import simps
from scipy.fftpack import fftfreq,fft,ifft
from matplotlib.pyplot import plot,xlim,show
#Creating the x-Grid
xmin=-1000
xmax=1000.2
dx=0.2
x=arange(xmin,xmax,dx)
#Defining and normalizing the initial wavefunction
k0=1
psi=exp((-x**2)/2)*exp(1j*k0*x) #Gaussian wavepacket
prob_density=abs(psi)**2
norm=sqrt(simps(prob_density,x))
psiNorm=psi/norm
prob_density_initial=abs(psiNorm)**2
#Time propagation (implementing split operator)
N=len(x)
k=2*pi*fftfreq(N,dx) #Creating the k-Grid
KE_k=(0.5*k**2) #Defining the kinetic energy
Nt=100 #Total number of time steps
dt=0.1 #Time step size in atomic unit
for i in arange(Nt):
        #fft of wavefunction to k-space
        psiNorm_k=fft(psiNorm)
        #Time evolution due to KE part of the propagator
        psiNorm_k_KE=exp(-1.0j*KE_k*dt)*psiNorm_k
        #iFFT of the wavefunction to x-space
        psiNorm=ifft(psiNorm_k_KE)
prob_density_final=abs(psiNorm)**2
#Plotting the initial and final probability densities
plot(x,prob_density_final)
plot(x,prob_density_initial)
xlim(-20,40)
show()
```

In the above program, final time is defined by **Nt*dt=100*0.1=10** atomic unit. Executing the above program, one obtains the initial and final probability density distributions for the free particle, as shown in Figure E.1. This exercise clearly demonstrates the fact that the free-particle wavepacket moves and spreads over time. Note that the graph, as shown in Figure E.1, obtained by executing the above program requires further formatting for better presentation (such as axis labeling, giving a title, etc.). This task is left for readers.

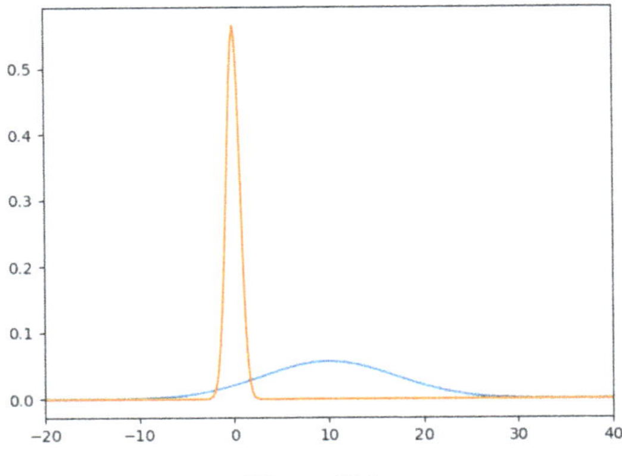

Figure E.1.

Self-Study

One can change **Nt** to obtain a different final time. Note that by selecting **dt=0.1** atomic unit, we make sure that each loop implements a very short time propagation. Record the changes if **dt** is selected to be very large.

(c) Monitoring the spread and testing the Ehrenfest theorem

Spread: A Gaussian function (or distribution) is characterized by the mean (average) and width (spread) of its distribution (details are discussed in Chapter 3). For example, the Gaussian function expressed by the form Ae^{-ax^2} exhibits the average of the distribution at $x = 0$ and has a width of Δx. Here, the width (spread of the distribution) is expressed **as the full width at half-maximum (at 50% of the maximum)** of the probability density distribution $A^2e^{-2ax^2}$. Furthermore, we have realized in Chapter 3 that $a = \frac{2\ln 2}{\Delta x^2}$.

In a different context, the spread of a Gaussian distribution is also conveniently expressed as the standard deviation (square root of the variance) of the distribution, which is defined by

$$\sigma = \sqrt{\langle \hat{x}^2 \rangle - \langle \hat{x} \rangle^2}$$

where $\langle \hat{x}^2 \rangle = \int_{-\infty}^{+\infty} |\psi(x)|^2 x^2 dx$ and $\langle \hat{x} \rangle = \int_{-\infty}^{+\infty} |\psi(x)|^2 x dx$, by making use of the expression for the expectation value in quantum mechanics (see Chapter 2 for these definitions). When a Gaussian function is defined in terms of its standard deviation, it adopts the form $Ae^{-\frac{x^2}{4\sigma^2}}$. Graphically, the standard deviation σ corresponds to **the half-width at about 60% of the maximum** of the probability density distribution $Ae^{-\frac{x^2}{2\sigma^2}}$. Therefore, care must be taken to be clear about the parameter which is used to define a Gaussian function.

As the standard deviation of the probability distribution can be calculated directly from the corresponding wavefunction, we may easily explore the time evolution of the spread of the wavepacket using the standard deviation of the probability density distribution.

Ehrenfest theorem: Ehrenfest theory (see Chapter 2) suggests that the expectation value of position of the particle follows classical mechanics. We may decipher the time evolution of the expectation value of position of the particle as time progresses numerically.

In the following exercise, we will explore the time evolution of both the spread of the wavepacket and the expectation value of position of the particle. For this exercise, we have to execute the program with different final times (by selecting different values of ***Nt***).

```
#Importing the required libraries
from scipy import sqrt,arange,exp,pi
from scipy.integrate import simps
from scipy.fftpack import fftfreq,fft,ifft
from matplotlib.pyplot import plot,xlim,show
#Creating the x-grid
xmin=-1000
xmax=1000.2
dx=0.2
x=arange(xmin,xmax,dx)
#Defining and normalizing the initial wavefunction
k0=1
psi=exp((-x**2)/2)*exp(1j*k0*x) #Gaussian wavepacket
prob_density=abs(psi)**2
norm=sqrt(simps(prob_density,x))
psiNorm=psi/norm
prob_density_initial=abs(psiNorm)**2
#Time propagation (implementing split operator)
```

```
N=len(x)
k=2*pi*fftfreq(N,dx) #Creating the k-grid
KE_k=(0.5*k**2) #Defining the kinetic energy
Nt=100 #Total number of time steps
dt=0.1 #Time step size in atomic unit
for i in range(Nt):
        t=(i+1)*dt
        #fft of wavefuncition to k-space
        psiNorm_k=fft(psiNorm)
        #time evolution due to KE part of propagator
        psiNorm_k_KE=exp(-1.0j*KE_k*dt)*psiNorm_k
        #ifft to x-space
        psiNorm=ifft(psiNorm_k_KE)
prob_density_final=abs(psiNorm)**2
avg1=simps(prob_density_final*x,x)
avg2=simps(prob_density_final*x*x,x)
std=sqrt(avg2-avg1* avg1)
print("Final Time=%f"%Nt*dt)
print("<x>=%f"%avg1)
print("width=%f"%std)
```

Upon execution of the above program, the expectation value of position and the spread of the distribution can be obtained for different given final times, as typically exemplified in Table E.2. Note that here the standard deviation of the probability density distribution is used to define the spread of the distribution (not using FWHM).

One may easily plot the expectation value of position and the spread (as measured by the standard deviation) of the distribution as functions of final time, as shown in Figure E.2.

It is evident from Figure E.2 that the expectation value of position (blue line) changes linearly with time, corroborating the fact that the particle under the influence of zero potential must travel with a constant speed, and as a result, the trajectory created by the expectation value must be linear with time.

On the other hand, the spread of the distribution monotonically increases with time, resulting in the spreading of the wavepacket. In other words, the probability of finding the particle becomes distributed over a large range (in position) as time progresses. This spreading is a purely quantum mechanical behavior of a particle undergoing quantum dynamics.

Table E.2.

Final Time (a.u.)	**<x>(a.u.)**	σ**(a.u.)**
0.0	0.707107	0.00
0.2	0.72111	0.20
0.5	0.790569	0.50
1.0	1.000000	1.00
1.5	1.274755	1.50
2.0	1.581139	2.00
3.0	2.236068	3.00
4.0	2.915476	4.00
6.0	4.301163	6.00
10.0	7.106335	10.0

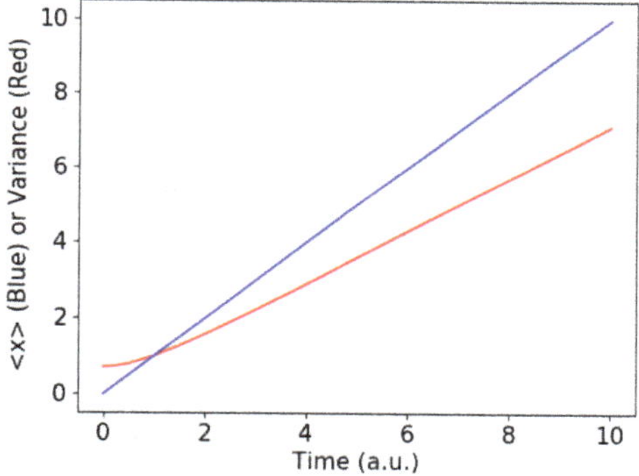


```
#Importing the required libraries
from matplotlib.pyplot import
plot,xlim,show,xlabel,ylabel,tick_params

x=[0.0,0.2,0.5,1.0,1.5,2.0,3.0,4.0,6.0,10.0]
width=[0.707107,0.72111,0.790569,1.0,1.274755,1.5811
39,2.236068,2.915476,4.301163,7.106335]
expectation_value=[0,0.2,0.5,1.0,1.5,2.0,3.0,4.0,6.0,10.0]

plot(x,width,color='red')
plot(x,expectation_value,color='blue')
xlabel('Time (a.u.)',fontsize=15)
ylabel('<x> (Blue) or Std. Deviation (Red)',fontsize=15)
tick_params(labelsize=15)
show()
```

Figure E.2.

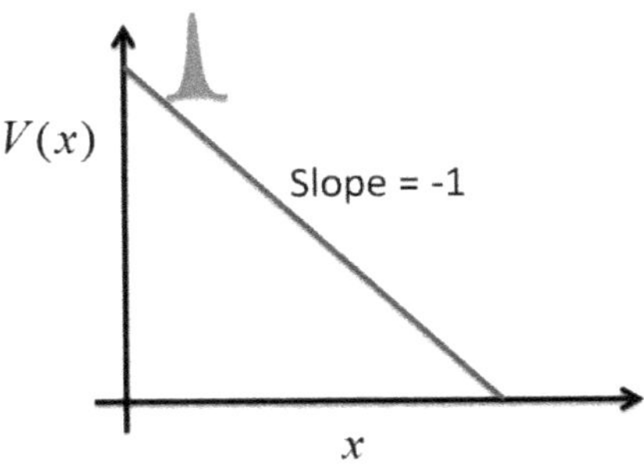

Figure E.3.

E.3 Gaussian Wavepacket Dynamics under Linear Potential

To explore the wavepacket dynamics under a linear potential, we will directly make use of both the kinetic and potential parts of the propagator, as given in atomic unit: $\psi(x,t) = [\prod_{N} e^{-i\frac{\hat{T}}{2}\Delta t} e^{-i\hat{V}\Delta t} e^{-i\frac{\hat{T}}{2}\Delta t}]$ $\psi(x,0)$. Note that the potential part of the propagator needs to be executed in position space, while the kinetic part of the propagator needs to be carried out in momentum space (or k-space). For simplicity, assume that the linear potential is expressed by $\hat{V}(x) = -x$, as depicted in Figure E.3.

Explicit numerical steps for the current problem are summarized as follows:

$\psi(x,0) \xrightarrow{\text{FFT}} \psi(p,0)$

$\downarrow e^{-i\frac{\hat{T}}{2}\Delta t}$

$\psi'(p,0+\Delta t) \xrightarrow{\text{iFFT}} \psi'(x,0+\Delta t)$

$\downarrow e^{-i\hat{V}\Delta t}$

$\psi''(x,0+\Delta t) \xrightarrow{\text{FFT}} \psi''(p,0+\Delta t)$

$\downarrow e^{-i\frac{\hat{T}}{2}\Delta t}$

$\psi'''(p,0+\Delta t) \xrightarrow{\text{iFFT}} \psi'''(x,0+\Delta t)$

Repeated

Here, $\psi(x,0)$ is the initial wavefunction and $\psi'''(x,0+\Delta t)$ is the new wavefunction at time $(0+\Delta t)$. To achieve propagation for the final time t, the above numerical steps need to be repeated with the help of a **for loop** for every Δt, with ψ''' serving as the new initial wavefunction for the next iteration. Python implementation is given as follows.

```
#Importing the required libraries
from scipy import sqrt,arange,exp,pi
from scipy.integrate import simps
from scipy.fftpack import fftfreq,fft,ifft
from matplotlib.pyplot import plot,xlim,show
#Creating the x-grid in atomic unit
xmin=-1000
xmax=1000.2
dx=0.2
x=arange(xmin,xmax,dx)
#Defining and normalizing the initial wavefunction
k0=1
psi=(exp((-x**2)/2))*(exp(1j*k0*x)) #Gaussian wavepacket
prob_density=abs(psi)**2
norm=sqrt(simps(prob_density,x))
psiNorm=psi/norm
prob_density_initial=abs(psiNorm)**2
#Defining the potential
v=-x
#Time propagation (implementing split operator method)
N=len(x)
k=2*pi*fftfreq(N,dx) #creating k-grid
KE_k=(0.5*k**2) #Defining kinetic energy
Nt=100 #Total Number of time steps
dt=0.1 #Time step size in atomic unit
for i in arange(Nt):
        #fft of wavefuncition to k-space
        psiNorm_k=fft(psiNorm)
        #Time evolution due to 1st KE part of the propagator
        psiNorm_k_KE1=exp(-1.0j*KE_k*dt/2.0)*psiNorm_k
        #ifft of wavefunction to x-space
        psiNorm_k_KE1_x=ifft(psiNorm_k_KE1)
        #Time evolution due to PE part of propagator
        psiNorm_k_KE1_x_V=exp(-1.0j*v*dt)*psiNorm_k_KE1_x
        #fft of wavefuncition to k-space again
        psiNorm_k_KE1_x_V_k=fft(psiNorm_k_KE1_x_V)
        #Time evolution due to 2nd KE part of the Propagator
        psiNorm_k_KE1_x_V_k_KE2=exp(-1.0j*KE_k*dt/2.0)*psiNorm_k_KE1_x_V_k
        #ifft to x-space
        psiNorm=ifft(psiNorm_k_KE1_x_V_k_KE2)
prob_density_final=abs(psiNorm)**2
#Plotting the initial and final probability densities
plot(x,prob_density_final)
plot(x,prob_density_initial)
xlim(-50,100)
show()
```

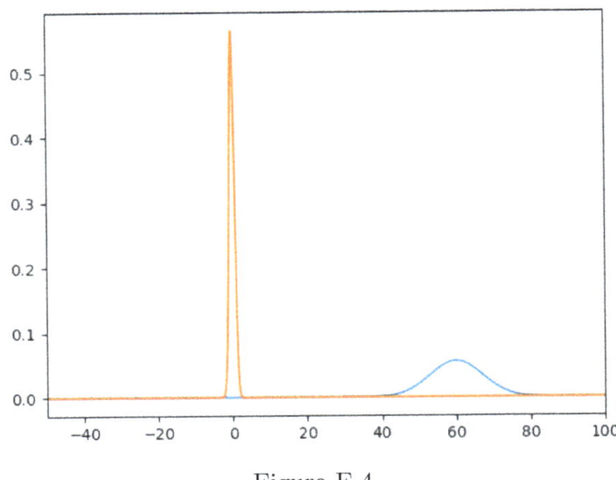

Figure E.4.

Execution of the above Python program renders the initial and final probability density distributions for the particle experiencing a linear potential, as shown in Figure E.4.

Self-Study

(a) Compare the wavepacket dynamics under linear potential with those under zero potential (free particle).

(b) Plot the expectation value of position and the standard deviation of the probability density distribution of a particle experiencing a linear potential as functions of time. Compare the results with those of a free particle.

Key Points to Remember

- In atomic units, numerical steps to implement split operator approach are given by

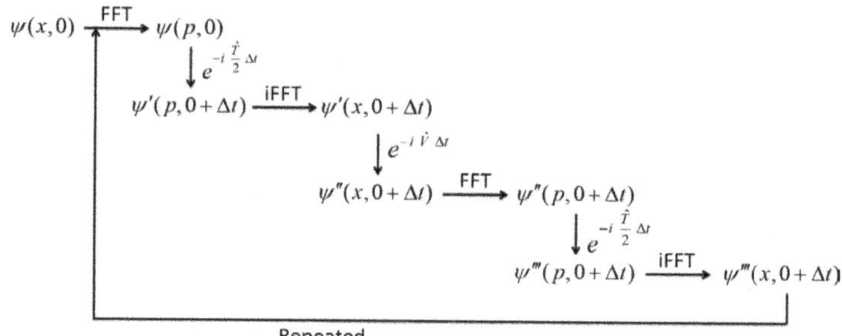

- Short-time propagation is repeated with the help of a **for loop** in Python Implementation.

Notes, References, and Further Reading

1. D. R. Hartree, *The Wave Mechanics of an Atom with a Non-Coulomb Central Field, Part 1, Theory and Methods.* Mathematical Proceedings of the Cambridge Philosophical Society, Vol. 24, pp. 89–110 (1928).
2. H. Shull and G. G. Hall, Atomic units. *Nature*, 184, 1559–1560 (1954).
3. A. Szabo and N. S. Ostlund, *Modern Quantum Chemistry: Introduction to Advanced Electronic Structure Theory*, 1st edn. Dover Publications Inc., New York (1996).
4. E. Tiesinga, P. J. Mohr, D. B. Newell and B. N. Taylor, CODATA recommended value of fundamental physical constants: 2018. *J. Phys. Chem. Ref. Data*, 50, 033105 (2021).

Exercises

E.1. Propagate a Gaussian wavepacket under a quadratic potential (e.g., quantum harmonic oscillator), $V(x) = \frac{1}{2}m\omega^2x^2$, where m and ω are the mass of the particle and the angular frequency of the oscillation, respectively. Plot the probability density at the initial time and at some later time. Plot the expectation value of position and the standard deviation of the probability density distribution as functions of time.

E.2. Propagate a Gaussian wavepacket under a Morse potential (quantum anharmonic oscillator), $V(x) = D_e[1 - e^{ax}]^2$, where D_e and a are the potential well depth and the characteristic parameter which controls the width of the potential well, respectively. Plot the probability density at the initial time and at some later time. Plot the expectation value of position and the standard deviation of the probability density distribution as functions of time.

E.3. Propagate a Gaussian wavepacket under a soft-Coulomb potential, $V(x) = \frac{1}{\sqrt{1+x^2}}$. Plot the probability density at the initial time and at some later time. Plot the expectation value of position and the standard deviation of the probability density distribution as functions of time.

Chapter 6

Application of Wavepacket Dynamics: Illuminating Examples

Highlights: *Role of Wavepacket Dynamics in Chemistry, Dynamics of Molecular Photophysics and Photochemistry, Dynamics on Single Potential Energy Curve, Dynamics on Multiple Potential Energy Curves, Time-Dependent Approach to Spectroscopy, Wavepacket Correlation Function and its Fourier Transform, Time-Dependent Viewpoint of Quantum Scattering, Phase Shift, Attosecond Time Delay in Photoionization.*

6.1 Introduction

Through the previous five theory chapters and five PythonChapters, we have developed an in-depth theoretical understanding of the wavepacket dynamics in one dimension and its numerical implementation using Python programming. From the research front, here, an obvious question arises: "**What is the role of wavepacket dynamics in chemistry?**" In this chapter (which is the concluding chapter of this book), the role of wavepacket dynamics in chemistry is presented very briefly. Wavepacket dynamics plays an important role in understanding photochemistry at the molecular level. The presentation of this chapter cannot be made as rigorous as that for the other chapters because our task here is to *go over it more quickly* to obtain a bird's-eye view without delving deeper into the research front. However, the language of the presentation is brought down to the level of graduate students so that this chapter serves the purpose of this textbook and raises curiosity in the chemistry student's mind.

6.2 Molecular Photophysics and Photochemistry

The Born–Oppenheimer approximation (BOA),[1] which separates the electronic and nuclear motions of a molecule, is the keystone to the visualization of chemical processes. This approximation enables us to picture molecules as a set of nuclei moving over a potential energy surface created by the faster moving electrons. Despite its central role in chemical theory, breakdown of the BOA is very common in the photochemistry and photophysics of polyatomic molecules,[2] for which a large number of energetically close-lying electronic states and many degrees of freedom exist. Specific examples of the breakdown of the BOA include charge transfer, electronic quenching, spin forbidden reactions, and nature's two most basic processes: the initial radiationless energy transport step in photosynthesis and the cis–trans isomerization that initiates the process of vision.

The breakdown of BOA leads to the formation of a conical intersection between electronic potential energy surfaces.[2] A conical intersection is a point at which two potential energy surfaces cross. The potential energy depends linearly on the nuclear coordinates near this interaction, and this is why the confluence is referred to as a conical intersection since the local topography near the intersection looks like a double cone, as illustrated in Figure 6.1.

Conical intersections are now firmly established as key features in the photophysical and photochemical processes of polyatomic molecules.[2] They provide an efficient pathway for radiationless transition between electronic states. When a conical intersection is involved in a chemical or physical process, the nuclei move on more than one potential energy surface. The probability of transitions from an upper state to a lower state is high if the nuclei encounter a conical intersection, at which point the two potential energy surfaces are energetically degenerate. Thus, conical intersections provide pathways for ultrafast internal conversions (ICs) or intersystem crossings (ISCs).

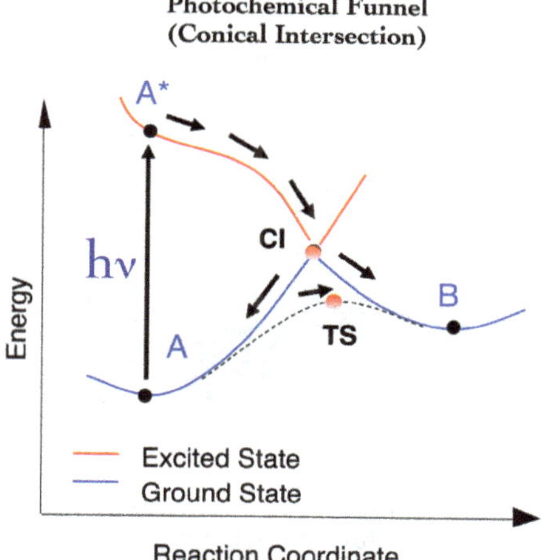

Figure 6.1. A generic illustration of a conical intersection between two potential energy surfaces. In this figure, **CI** refers to the conical intersection and **TS** refers to the transition state. Reproduced with permission from Ref. 2(d). Copyright (2011) The Royal Society of Chemistry (RSC) on behalf of the Centre National de la Recherche Scientifique (CRNS) and the RSC.

In general, a number of intrinsic issues are associated with chemical dynamics through a conical intersection: All these issues can have a significant influence on understanding photophysics and photochemistry at the molecular level. Electronic excitation promotes a molecule from its ground state to an electronically excited state. As illustrated in Figure 6.2, the nuclei in the electronic ground state are assumed to have their positions characterized by point $\boldsymbol{p'}$ in the reaction coordinate space. Electronic excitation takes place so fast that the nuclei do not have enough time to move. Thus, the positions of the nuclei in the excited electronic state are identical to those in the ground state (this is called the classical Franck–Condon rule). The quantum mechanical analogue of the classical Franck–Condon rule is that the probability of the nuclei making a transition from one potential energy surface to another is greatest for those places at which the square of the overlap between the vibrational wavefunction of the initial and final states, $q_{v'v''} \eqqcolon |\langle(\psi_{v'}(R)|\psi_{v''}(R)\rangle|^2$, called the Franck–Condon factor, is the greatest. Thus, the **FC** point in

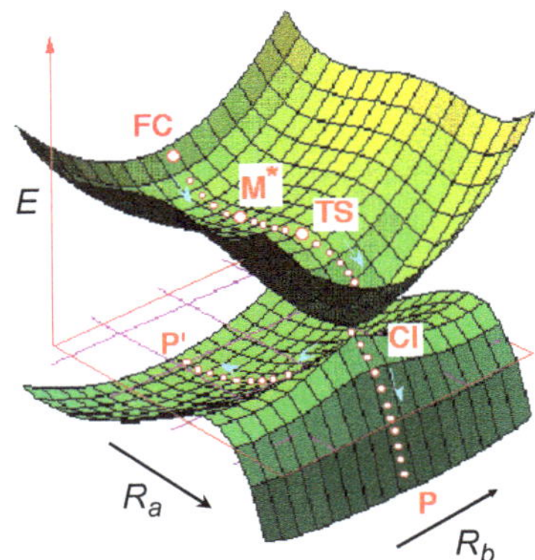

Figure 6.2. Generic view of chemistry through a conical intersection. A conical intersection can mediate successful photochemistry by directing the molecule to the product side, following transition through it. Thanks are due to Professor Mike Robb of Imperial College London for providing this figure.

Figure 6.2 shows the very essence of the Franck–Condon rule — a vertical transition.

The nuclear configuration at the FC point of the upper excited state may differ quite significantly from the nearest potential energy minimum $\boldsymbol{M}^*$ in the excited state PES. In a few tens (or hundreds) of femtoseconds, the system "slides down" (evolves) from **FC** to the neighborhood of $\boldsymbol{M}^*$, transforming its potential energy into kinetic energy of the nuclei. If $\boldsymbol{M}^*$ is separated from the conical intersection (**CI**) by a barrier, as soon as the molecule during vibration overcomes the barrier to the **TS**, by going over it or by tunneling through it, it will be driven by the new force to the conical intersection "attractor" (often with almost 100% probability). At (or near) the conical intersection, the molecule undergoes a nonradiative transition to the ground state. If the relative momentum of the nuclei is large enough, the system slides down along the pathway toward **P**, the nearest minimum for the product. If the momentum is small, the system may go back to $\mathbf{P}'$, returning to the original molecular configuration. The actual direction or path taken by the molecule, in general, depends on the relative steepness of the two paths.

The total energy, of course, has to be conserved. In an isolated gas-phase environment, the system finds a way to redistribute its electronic excitation energy to its own molecular vibrational, rotational, and translational degrees of freedom. In a condensed-phase environment, the system also finds a way to dissipate its energy to the environment, i.e., a part of the electronic excitation energy is transferred to the neighboring solvent molecules. Consequently, a transition of a molecular system through a conical intersection from an upper excited electronic state to the ground (or a lower) electronic surface often leads to molecular fragmentation or rearrangement.

The development of femtosecond (fs) pulsed lasers has made possible the observation of transient species (in the course of a photochemical or photophysical process), which have very short lifetimes (often, surviving only for several hundred femtoseconds or a few picoseconds). With respect to the lifetimes, the transient species formed near an avoided crossing or a conical intersection in photo-induced processes may quite resemble an "activated complex" or "transition state" in thermally activated processes (frequently invoked in the context of "transition state theory"). In the following sections, ultrafast relaxation dynamics through an avoided crossing and through a conical intersection are presented to demonstrate the role of wavepacket dynamics in photochemistry and photophysics.

Avoided Crossing versus Conical Intersection

In thermal reactions, the molecular structure associated with the transition state often provides information on the expected products. Similarly, in a photochemical reaction, the geometry associated with a conical intersection is often found to be related to the expected photoproducts. For diatomic molecules, with only one degree of freedom, electronic states with the same symmetry do not cross, yielding a non-crossing rule. More details of the origin of non-crossing rules are discussed in Ref. 3, which suggests that conical intersections may not exist frequently in diatomic molecules. Instead, an avoided crossing is more frequent for diatomic species. However, for polyatomic molecules, two electronic states, even with the same symmetry, may intersect and form a conical intersection.

6.2.1 *Ultrafast dynamics through avoided crossing of NaI*

Electronically ground and first excited states of a sodium iodide (NaI) molecule are described with two potential energy curves: one

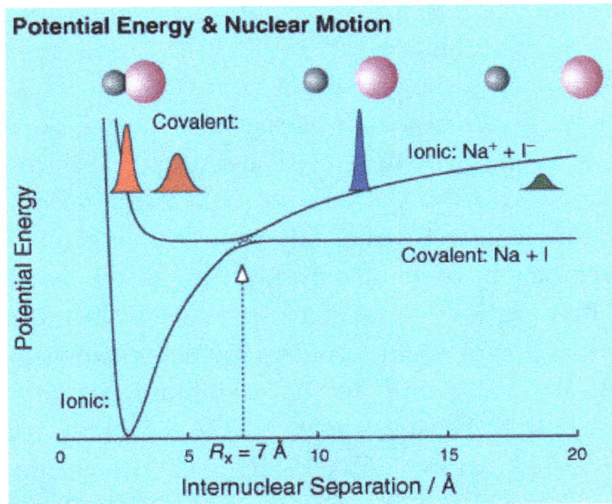

Figure 6.3. The ground electronic and first electronically excited potential energy curves of sodium iodide (NaI). Reprinted with permission from Ref. 4. Copyright (2001) American Chemical Society.

"ionic" and another "covalent," as depicted in Figure 6.3.[4] Both curves create an avoided crossing at an internuclear separation of approximately 7 Å. In the ground state, the NaI molecule exhibits an ionic electronic character (represented as Na^+I^-, in which an ionic bond is formed between elements), and the same molecule features a covalent electronic character (represented as NaI, in which a covalent bond is formed between elements) in the first electronically excited state. However, because of the difference in the ionization potentials of Na and Na^+ and the electron affinity of I and I^-, the dissociation limit for (Na^+I^- (ionic infinite separation) is greater than that of (Na + I) (covalent infinite separation).

NaI is a heteronuclear diatomic molecule, and it does not possess a center of symmetry. Hence, the symmetry argument (as elaborately discussed in Ref. 3) does not help two electronic potential energy curves to cross. This renders an avoided crossing. Because of this avoided crossing at an internuclear distance of approximately 7 Å, "electron harpooning" occurs when the molecule is bound in the upper electronic state. Consequently, the electron is transferred from one atom to another during vibration on the upper excited curve, as the electronic character changes from covalent to ionic in the bound upper curve. However, if there is a nonadiabatic transition from the upper state to the lower state through the avoided-crossing region, which ultimately results in dissociation, the molecule does not need to change its electronic character. It always stays covalent.

In the fs pump–probe spectroscopy experiment, the pump pulse at λ_1 (center wavelength) excites the molecule from the ionic (Na^+I^-) ground state to the covalent (NaI) excited state. A short laser pulse is composed of a wide range of frequency components, which excite the many vibrational states of NaI simultaneously, while the molecule undergoes electronic excitation.[3] Consequently, the electronically excited complex exists as a superposition of many vibrational states, which renders a localized vibrational wavepacket. An illustrative example of such coherent excitation of many vibrational states is depicted in Figure 6.4, taking diatomic harmonic potential wells. This example does not necessarily depict the NaI problem; rather, it gives a general sense of vibrational wavepacket creation by fs pulse-excitation.

There are two limiting possibilities following an electronic excitation, as shown in Figure 6.3: (a) The vibrational wavepacket of the electronically excited NaI may get trapped on the adiabatic potential energy curve, in which case, the wavepacket oscillates between the covalent and ionic potential energy curves; or (b) the vibrational wavepacket may dissociate, which is shown as a movement of the wavepacket toward a very long interatomic separation on the dissociative covalent curve after undergoing a nonadiabatic transition through the avoided crossing. These two possibilities should exhibit entirely different temporal behaviors, and if there is trapping (in the upper bound curve), the details of the nature of the curves and the strength of the nonadiabatic coupling near the avoided crossing can be deciphered from the fs pump–probe spectroscopy.

In fs pump–probe spectroscopy, the dynamics of the vibrational wavepacket (created on the electronically excited state) is probed by a second probe pulse (this probe excitation essentially creates molecular ions which are detected by a mass spectrometer) with a frequency that corresponds to the absorption frequency of perturbed

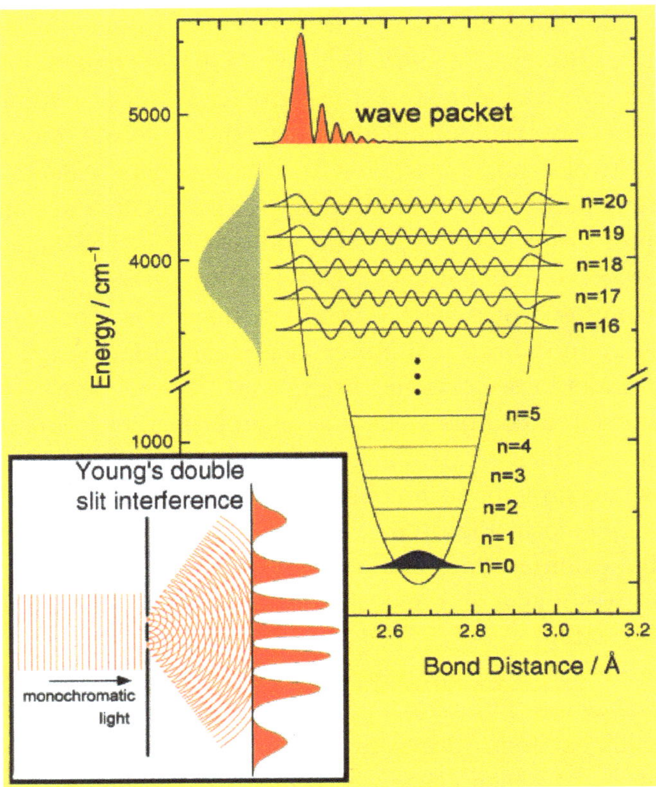

Figure 6.4. Ground and excited electronic states of a diatomic molecule are shown under the harmonic oscillator approximation (i.e., potential energy curves are bound). Stationary wave functions associated with vibrational states are also shown. A short pulse (the spectrum of the pulse is shown in gray color along the left vertical axis) coherently excites many vibrational states. This coherent excitation leads to the creation of a vibrational wavepacket (schematically shown in the top panel) in the upper excited state. The vibrational wavepacket is the result of quantum interference, similar to the interference of light observed in Thomas Young's double-slit experiment (bottom left panel). Reprinted with permission from Ref. 4. Copyright (2001) American Chemical Society.

Na, i.e., Na absorbs the probe light when it is part of the complex (or in other words, when Na and I attain a specific inter-atomic distance). Hence, the excitation frequency depends on a specific (long) Na······I distance, so that an absorption is obtained each time the wavepacket returns to that separation.

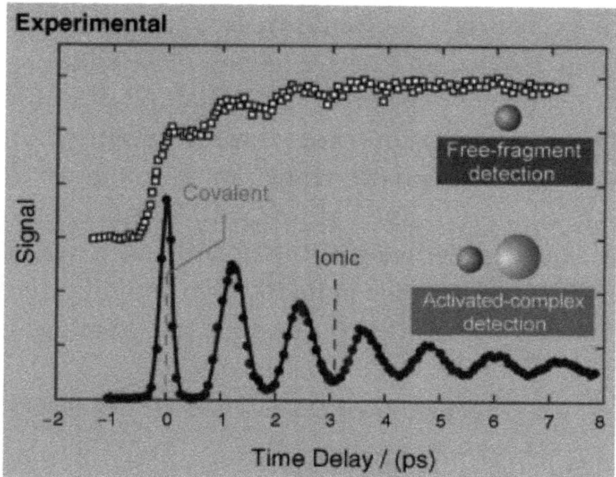

Figure 6.5. The transient ionization spectrum of quasi-bound complex (transition-state configuration) of NaI, and the photoproduct Na as a function of pump–probe time delay. Reprinted with permission from Ref. 4. Copyright (2001) American Chemical Society.

In a seminal experiment, performed by Zewail and co-workers, the fs pump pulse was selected at 310 nm and the probe pulse was selected from the wavelength range of 560–630 nm. In this probe pulse wavelength range, only 589 nm is on resonance with the free Na D-line. The other probe wavelengths are off-resonance with the free Na atom; however, they are on-resonance with perturbed Na, i.e., bound Na with I at a specific large internuclear distance.

The observed parent ion intensity as a function of pump–probe delay time is plotted in Figure 6.5, which shows striking oscillatory behavior that recurs in about 1.25 ps, featuring the wavepacket oscillation with about that period. This period corresponds to a frequency of $27\,\mathrm{cm}^{-1}$. The decline in intensity shows the rate at which the excited state species dissociates (via a nonadiabatic jump from the upper curve to the lower curve through avoided crossing). Furthermore, the complex survives for about 10 oscillations, and the oscillations are damped in 10 ps. Therefore, the average probability of a nonadiabatic transition through avoided crossing is estimated to be 0.1.

For NaBr (which is a molecule with a structure very similar to that of the NaI molecule), the frequency of oscillation is found to be similar in magnitude; however, severe damping is observed: NaBr barely survives one oscillation (not shown in Figure 6.5). Therefore, the probability of nonadiabatic transition through avoided crossing for NaBr must be more than that for NaI. This conclusion is consistent with the mass dependency of the nonadiabatic coupling term: A larger mass exhibits a smaller nonadiabatic coupling term, resulting in a slow nonadiabatic transition through the crossing.[3]

Question to Raise Curiosity

Can you write a Python program to explore the wavepacket dynamics of NaI system numerically if the potential energy curves are known? Try it yourself.

6.2.2 *Ultrafast dynamics through conical intersection: Photophysics of vision*

Vision is one of the most important biological processes in our body. Much of the sensory information about our world comes to us through our vision. How do we see? What is the initial step toward vision? These are scientific questions which we have asked for a long time.

Vision is the result of the conversion of light energy into an electrochemical impulse. The impulse is transmitted through neurons to the brain, where signals from all the visual receptors are interpreted. Vision is initiated when a photon is absorbed by a pigment called *rhodopsin*,[5] which is located in the rods of the retina (see Figure 6.6). This pigment consists of an organic molecule, retinal, in association with a protein named opsin. After optical excitation, the 11-cis-retinal prosthetic group of rhodopsin is converted to an all-trans primary photoproduct in an efficient and barrierless isomerization reaction (see Figure 6.7). This change in shape of retinal apparently gives the signal to opsin to undergo a sequence of dark (thermal) reactions involved in triggering neural excitation.

The absorption spectra of the photo-intermediates of rhodopsin help one identify the intermediates very easily because their absorption maxima appear at distinctly different wavelengths (see

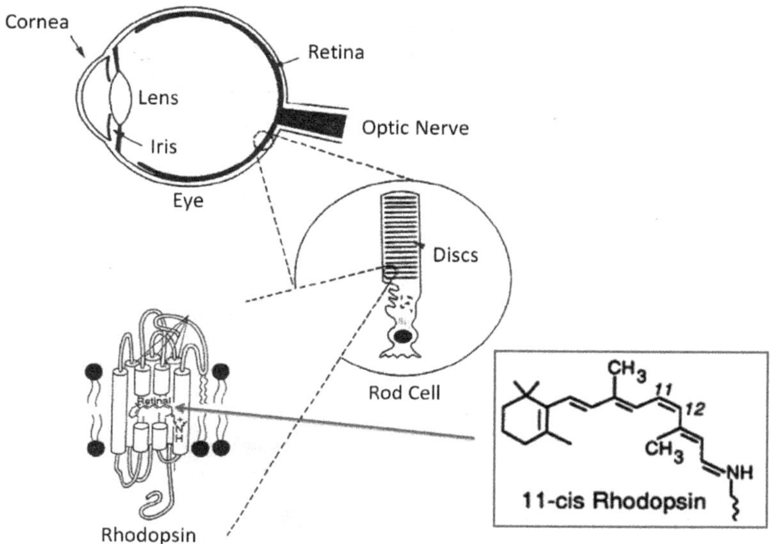

Figure 6.6. Molecular structure of our eye: Rhodopsin is a photoreceptive molecule which comprises a light-absorbing part (retinal) and a protein (opsin). Rhodopsin is embedded within the light-sensitive rod and cone cells in the retina, which is responsible for our vision. For more detail, see text.

Figure 6.8).[7] Figure 6.8 shows that while the absorption maximum of rhodopsin (11-cis) appears near 500 nm, the absorption of photo-intermediates (11-trans) shifts to longer wavelengths (called the red shift). More specifically, the absorption maximum of photorhodopsin appears at 570 nm. This difference in absorption spectra can be efficiently utilized in transient absorption spectroscopy to monitor the formation dynamics of the photoproduct by recording the time evolution of the red-shifted absorption of the photoproduct.

The fs transient absorption measurements of the cis–trans isomerization of the visual pigment rhodopsin explore the dynamics of the first step in vision. The formation time for the rhodopsin photoproduct is found to be around 200 fs, which occurs via cis–trans isomerization and which is identified as the initial photochemistry of vision.[6] The fs transient absorption spectra of rhodopsin following a photoexcitation by a 35 fs pulse centered at 500 nm (pump or excitation pulse) are presented in Figure 6.9 (left figure). A broadband

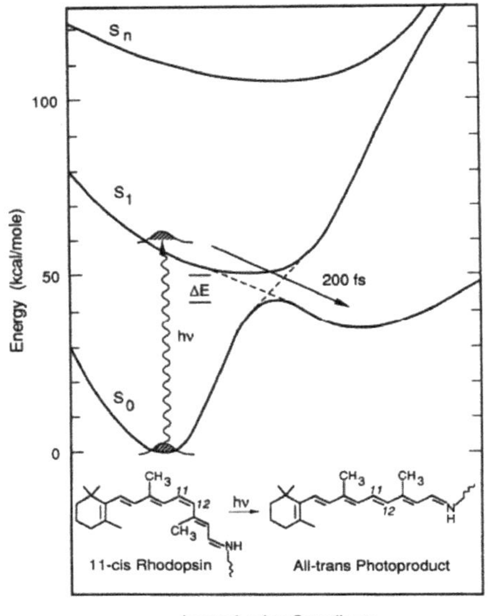

Figure 6.7. Schematic ground and excited state potential energy curves are depicted to demonstrate the mechanism of the 11-cis → 11-trans isomerization in rhodopsin. Dashed lines, connecting the S_1 and S_0 potential energy curves, represent the nonadiabatic surfaces, over which isomerization occurs. Reproduced with permission from Ref. 6. Copyright (1993) National Academy of Sciences, USA.

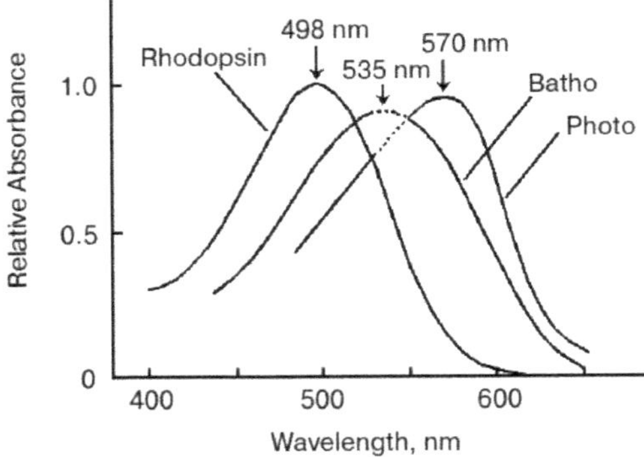

Figure 6.8. Absorption spectra of bovine rhodopsin, bathorhodopsin, and photorhodopsin. Reproduced with permission from Ref. 7.

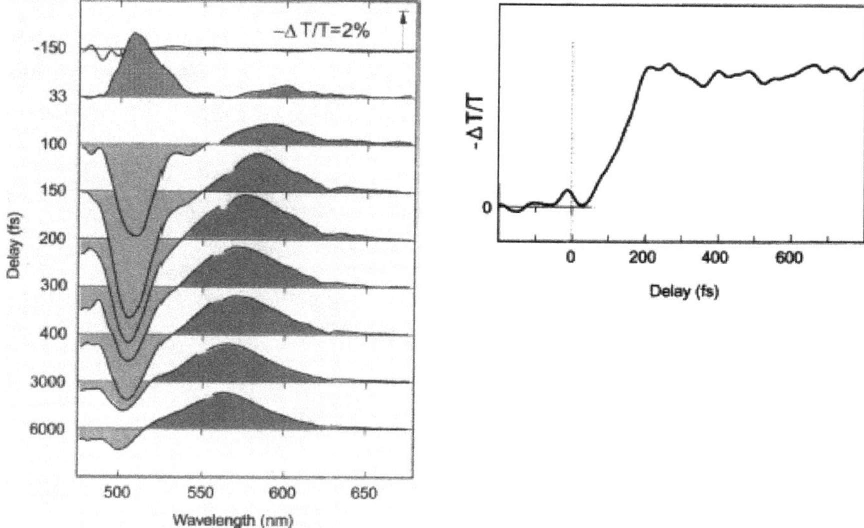

Figure 6.9. Left: Differential absorption spectra of rhodopsin at various time delays of the probe pulse after photoexcitation with a 35 fs pump pulse at 500 nm. Right: Transient absorption measurement of rhodopsin at 570 nm probe wavelength after excitation at 500 nm. Reproduced with permission from Ref. 6. Copyright (1993) National Academy of Sciences, USA.

10 fs probe pulse is used to cover the spectral range of 490–670 nm. At the 33 fs delay time, the difference absorption spectrum consists of very weak absorption between 550 and 620 nm. Note that the photoproduct absorbs in this range. The absorption of the photoproduct uniformly grows in strength, reaching its largest amplitude at 200 fs after the excitation by the pump pulse. Although the maximum of the photoproduct absorption shifts from 575 to 565 nm between 200 fs and 6 ps, there is little change in the integrated area or shape of this absorption feature. The difference absorbance observed at the single-probe wavelength of 570 nm is shown in Figure 6.9 (right figure), which illustrates that the difference absorption signal is at the baseline at time zero and does not fully develop until $\approx$200 fs after the excitation by the pump pulse.

Measurement of the full difference absorption spectrum of the rhodopsin photoproduct confirms that the first step in vision occurs at only 200 fs. The rapid disappearance (≈50 fs) of the Franck–Condon excited state absorption near 500 nm and the rapid appearance of (within 200 fs) a photoproduct absorption band extending from 530 to 620 nm support the rapidity of the isomerization process.

Question to Raise Curiosity

Can you write a Python program to explore the wavepacket dynamics of photoinduced cis–trans isomerization of the ethylene molecule numerically if the potential energy curves are known? Try it yourself.

6.2.3 *Illuminating computational exercises*

6.2.3.1 *Wavepacket dynamics on single potential energy curve*

In this exercise, we explore the dissociation dynamics of Ar_2^+ following electronic excitation from its ground electronic state (named as $A^2\sum_u^+$) to its first electronically excited state (named as $D^2\sum_g^+$). The first electronically excited state ($D^2\sum_g^+$) of Ar_2^+ is dissociative in nature (as shown in Figure 6.10), and as a result, Ar_2^+ species dissociates immediately following its electronic excitation from the ground electronic state $A^2\sum_u^+$ to the excited electronic state $D^2\sum_g^+$.

To explore the dissociation dynamics of Ar_2^+ on the excited potential energy curve, we first need to obtain the respective potential energy curve on which the dynamics will evolve. As Ar_2^+ is a diatomic species, one can obtain the respective potential energy curve as a function of the internuclear coordinate R (representing the bond length) using *ab initio* calculations, which can be performed easily using any currently available quantum chemistry software package, such as Gaussian or Molpro.[8]

In the past, Schwendner *et al.* have obtained an analytical fit to the *calculated potential energy* points for each electronic state of Ar_2^+ computed by Naumkin *et al.*[9] Their results are directly used for the present exercise. The potential energy curves for the ground electronic state $A^2\sum_u^+$ and electronically excited state $D^2\sum_g^+$ of Ar_2^+ are denoted as $V_0(R)$ and $V_1(R)$, respectively. Their analytical

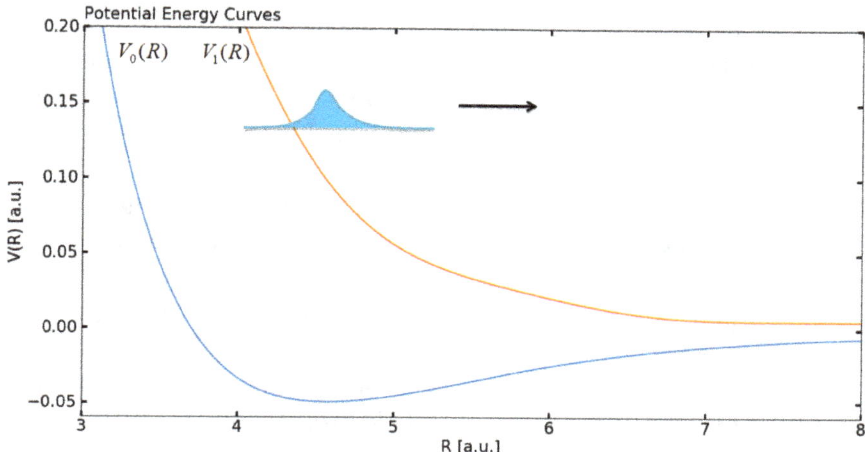

Figure 6.10. Potential energy curves (in atomic units) for the $V_1(D^2\sum_g^+)$ and $V_0(A^2\sum_u^+)$ states of Ar_2^+ are plotted using the analytical expressions given in the text and Ref. 9. A wavepacket is placed on the dissociative state $V_1(D^2\sum_g^+)$ for illustration.

expressions are given as follows:

$$\begin{aligned}V_0(R) &= -0.00234146 + 0.0000819646\, e^{-(R-9.2)^2} \\ &\quad + 0.000165879\, e^{-(R-7.8)^2} - 0.000584086\, e^{-(R-6.5)^2} \\ &\quad - 0.0055544\, e^{-(R-5.0)^2} - 0.002371\, e^{-(R-3.5)^2} \\ &\quad + 0.041821\left[e^{-1.74(R-4.535)} - 2e^{-0.87(R-4.535)}\right]\end{aligned}$$

$$\begin{aligned}V_1(R) &= 0.0032635 + 0.00109782\, e^{-0.5(R-7.5)^2} - 0.00487849\, e^{-(R-6.5)^2} \\ &\quad + 0.013737\, e^{-(R-5.5)^2} \\ &\quad - 0.0120447\, e^{-(R-5.0)^2} + 0.0380757\, e^{-(R-3.5)^2} \\ &\quad + 0.360969\, e^{-1.27(R-3.45)}\end{aligned}$$

Here, both R and the potential energy are expressed in atomic units. These two curves are depicted in Figure 6.10.

Note that the $V_0(R)$ and $V_1(R)$ curves of the Ar_2^+ species remain well-separated in energy. As a result, the wavepacket, which is created on the $V_1(R)$ curve following the electronic excitation of the

Ar_2^+ species, can be legitimately assumed to **evolve on a single potential energy curve**, $V_1(R)$. This is an important realization before we implement the split-operator method (introduced in Chapter 5) to propagate the wavepacket on the dissociative $V_1(R)$ curve because, thus far, without perhaps noticing it, we have learned the split-operator method, which can only be employed if the dynamics evolves only on a single potential energy curve. The split-operator approach to exploring the dynamics involving multiple (one-dimensional) potential energy curves requires special attention. This will be discussed in the later section.

One can now explore the dissociation dynamics of Ar_2^+ following electronic excitation from its ground electronic state to its first electronically excited state using the following steps (the execution of the present exercise using Python is left for the reader's self-study):

(a) First, using the grid representation, find the normalized ground vibrational state wavefunction of Ar_2^+ on its ground electronic state potential energy curve. Find the equilibrium bond length of Ar_2^+.

(b) Next, assume that electronic excitation instantaneously places the ground vibrational wavefunction of Ar_2^+ on the electronically excited potential energy curve, retaining the equilibrium bond distance (manifesting a well-known concept of vertical excitation during the electronic excitation process). Therefore, the ground vibrational state wavefunction may serve as the initial wavepacket on the electronically excited state curve of Ar_2^+. Propagate the wavepacket until the center of the wavepacket moves up to an internuclear separation of 8 a.u. (the asymptotic region of the excited state potential energy curve). In the present exercise, we assume that when the wavepacket moves to the asymptotic region of the excited state potential energy curve, it manifests the dissociation of the Ar_2^+ species.

6.2.3.2 *Wavepacket dynamics on multiple potential energy curves*

Thus far (in Chapter 5 and PythonChapter E), employing the time-evolution operator, we have directly monitored the time evolution

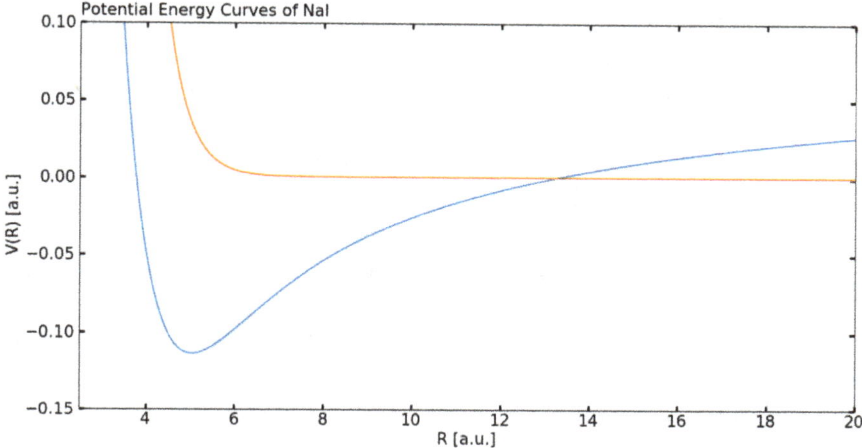

Figure 6.11. Potential energy curves (in atomic units) for the ground and electronically excited states of NaI are plotted using the analytical expressions given in the text and Ref. 10.

of the wavefunction using the relation (which is also a solution to the TDSE): $\psi(x,t) = e^{\frac{-i\hat{H}t}{\hbar}}\psi(x,0)$, where the initial state of the quantum system is represented by $\psi(x,0)$. Note that **this relation is strictly valid for dynamics evolving on a single potential energy curve.** The primary objective of the present exercise is to construct a simple theoretical framework for dynamics evolving on multiple (one-dimensional) potential energy curves.

We have apprehended that the ground electronic state potential energy curve and the first electronically excited potential energy curve of Ar_2^+ species remain well separated in energy, as evident in Figure 6.10. However, this scenario may change for other species. For example, the ground electronic state and the first excited electronic state potential energy curves of NaI species cross each other, as shown in Figure 6.11. Note that this figure does not portray avoided crossing, as depicted in Figure 6.3, for the same curves. This difference originates from the representation of the respective curves. Figure 6.3 makes use of an **adiabatic** representation of the electronic states, while Figure 6.11 considers a **diabatic** representation of the electronic states. For further details, see Piela's quantum chemistry book.[1]

Once a wavepacket is created on the first excited electronic state potential energy curve of NaI, it travels on the same potential energy curve until it reaches the crossing point where the wavepacket splits. While some part of the wavepacket continues traveling on the excited electronic state potential energy curve, another part starts traveling on the ground electronic state potential energy curve. To visualize a wavepacket splitting near or at the crossing of two potential energy curves, our first task is to construct the **time evolution operator for two states.**

If two states do not cross and remain isolated for the internuclear coordinate over which the dynamics evolves, one can express the time evolution of the wavefunction at each state by the respective time-evolution operator as follows:

$$\psi_1(x,t) = e^{\frac{-i\hat{H}_1 t}{\hbar}} \psi_1(x,0)$$

$$\psi_2(x,t) = e^{\frac{-i\hat{H}_2 t}{\hbar}} \psi_2(x,0)$$

Note that as the two states have different potential energy curves, and consequently their Hamiltonians are different. In matrix form, these two equations can be jointly expressed as

$$\begin{pmatrix} \psi_1(x,t) \\ \psi_2(x,t) \end{pmatrix} = \begin{pmatrix} e^{\frac{-i\hat{H}_1 t}{\hbar}} & \psi_1(x,0) \\ e^{\frac{-i\hat{H}_2 t}{\hbar}} & \psi_2(x,0) \end{pmatrix} = \begin{pmatrix} e^{\frac{-i\hat{H}_1 t}{\hbar}} & 0 \\ 0 & e^{\frac{-i\hat{H}_2 t}{\hbar}} \end{pmatrix} \begin{pmatrix} \psi_1(x,0) \\ \psi_2(x,0) \end{pmatrix}$$

$$= \exp\left[-\frac{i}{\hbar} \begin{pmatrix} \hat{H}_1 & 0 \\ 0 & \hat{H}_2 \end{pmatrix} t \right] \begin{pmatrix} \psi_1(x,0) \\ \psi_2(x,0) \end{pmatrix}$$

Therefore, the time-evolution operator for two isolated states (which do not cross) can be expressed through the following form:

$$\exp\left[-\frac{i}{\hbar} \begin{pmatrix} \hat{H}_1 & 0 \\ 0 & \hat{H}_2 \end{pmatrix} t \right]$$

where the Hamiltonian matrix $\begin{pmatrix} \hat{H}_1 & 0 \\ 0 & \hat{H}_2 \end{pmatrix}$ is diagonal. Note that we have made use of the fact that if a matrix is diagonal, its exponential is obtained simply by exponentiating each diagonal term (see Chapter 5 for details). If the two potential energy curves cross, the Hamiltonian matrix does not remain diagonal for the reason given in the following Illuminating Discussion. Readers are suggested to go over the discussion before proceeding further. As the Hamiltonian matrix does not remain diagonal when the two potential energy curves cross, one immediate concern pertaining to the time-evolution operator for two coupled states arises: **How do we obtain the exponential of a Hermitian (or Hamiltonian) matrix if it is not diagonal?**

Illuminating Discussion

In quantum mechanics, we often work on a two-level system which serves as a model for a physical or chemical system either having energetically close-lying two states, which are quite different from all other energy states, or having two states getting coupled, ignoring all other energy levels (the quantum mechanics of a two-level system is discussed in adequate detail in Ref. 11). For example, consider a quantum system (described by the Hamiltonian $\hat{H}_0$) with two eigenstates specified by $|\phi_1\rangle$ and $|\phi_2\rangle$ with corresponding energies E_1 and E_2, respectively, as shown in the following figure.

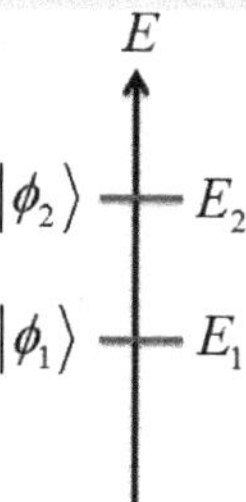

It is obvious that $|\phi_1\rangle$ and $|\phi_2\rangle$ are orthonormal, i.e., $\langle\phi_1|\phi_1\rangle = \langle\phi_2|\phi_2\rangle = 1$ and $\langle\phi_1|\phi_2\rangle = \langle\phi_2|\phi_1\rangle = 0$, and E_1 and E_2 are two eigenvalues of the Hamiltonian $\hat{H}_0$. Therefore, the Hamiltonian matrix has a diagonal form:

$$\mathbf{H_0} = \begin{pmatrix} E_1 & 0 \\ 0 & E_2 \end{pmatrix}$$

(*Continued*)

(Continued)

Consequently, the corresponding eigenstates can be represented in matrix form as

$$|\phi_1\rangle = \begin{pmatrix} 1 \\ 0 \end{pmatrix} \text{ and } |\phi_2\rangle = \begin{pmatrix} 0 \\ 1 \end{pmatrix}$$

Note that $|\phi_1\rangle$ and $|\phi_2\rangle$ are discrete isolated states of the quantum system defined by the Hamiltonian $\hat{H}_0$. They are not coupled.

Next, due to the reason to be revealed later, we assume that $|\phi_1\rangle$ and $|\phi_2\rangle$ states, which were eigenstates of $\hat{H}_0$, suddenly start interacting through a coupling operator $\hat{V}$. The exact nature of this coupling operator does not need to be understood immediately. The total Hamiltonian for the quantum system in which two states are coupled can be represented as

$$\hat{H} = \hat{H}_0 + \hat{V}$$

or, in matrix form,

$$\mathbf{H} = \mathbf{H_0} + \mathbf{V} = \begin{pmatrix} E_1 & 0 \\ 0 & E_2 \end{pmatrix} + \begin{pmatrix} 0 & V_{12} \\ V_{21} & 0 \end{pmatrix} = \begin{pmatrix} E_1 & V_{12} \\ V_{21} & E_2 \end{pmatrix}$$

Here, note that the coupling operator has the form $\mathbf{V} = \begin{pmatrix} 0 & V_{12} \\ V_{21} & 0 \end{pmatrix}$, which possesses only off-diagonal elements. This indicates that coupling does not allow interaction of a state with itself; it just allows interaction between two different states.

Before we proceed further, we must understand the physical meaning of the above matrix formulation of the total Hamiltonian. It indicates that, in the absence of coupling, the $|\phi_1\rangle$ state (with energy E_1) and the ϕ_2 state (with energy E_2) are possible two eigenstates of the quantum system defined by the uncoupled Hamiltonian $\hat{H}_0$. As a result, if the quantum system is placed in one of these two states, it remains there forever. Coupling, however, mixes these two states, definitely yielding two different eigenstates, $|\phi_+\rangle$ (with eigenstate energy E_+) and $|\phi_-\rangle$ (with eigenstate energy E_-), of the total Hamiltonian $\hat{H}$. Therefore, $|\phi_1\rangle$ and $|\phi_2\rangle$ are not eigenstates of the coupled Hamiltonian $\hat{H}$. They are eigenstates of the uncoupled Hamiltonian $\hat{H}_0$. With this consensus, we will first calculate the two eigenstate energies E_+ and E_- of the total Hamiltonian $\hat{H}$ in terms of E_1, E_2 and the matrix elements (V_{12} and V_{21}) of $\hat{V}$.

(Continued)

(*Continued*)

We have noted that in the $[|\phi_1\rangle, |\phi_2\rangle]$ basis, the total Hamiltonian (of the coupled system) adopts the following matrix form:

$$\hat{H} = \begin{pmatrix} E_1 & V_{12} \\ V_{21} & E_2 \end{pmatrix}$$

where $V_{12} = V_{21}^* = \langle\phi_1|\hat{V}|\phi_2\rangle$. One can easily diagonalize the above matrix to obtain eigenvalues with the help of the characteristic equation:

$$\begin{vmatrix} E_1 - E & V_{12} \\ V_{21} & E_2 - E \end{vmatrix} = 0$$

$$\text{or } (E_1 - E)(E_2 - E) - |V_{12}|^2 = 0$$

$$\text{or } E^2 - (E_1 + E_2)E + (E_1E_2 - |V_{12}|^2) = 0$$

$$\text{or } E = \frac{(E_1 + E_2) \pm \sqrt{(E_1 + E_2)^2 - 4(E_1E_2 - |V_{12}|^2)}}{2}$$

$$\text{or } E = \frac{E_1 + E_2}{2} \pm \sqrt{\left(\frac{E_1 - E_2}{2}\right)^2 + |V_{12}|^2}$$

Thus, the eigenvalues of the total Hamiltonian (representing the coupled system) are

$$E_+ = \frac{E_1 + E_2}{2} + \sqrt{\left(\frac{E_1 - E_2}{2}\right)^2 + |V_{12}|^2}$$

$$\text{and } E_- = \frac{E_1 + E_2}{2} - \sqrt{\left(\frac{E_1 - E_2}{2}\right)^2 + |V_{12}|^2}$$

For the mathematical convenience, we may define following two terms:

$$E_{\text{avg}} = \left(\frac{E_1 + E_2}{2}\right), \text{ referring to the average energy of the two states}$$

$$\Delta E = \left(\frac{E_1 - E_2}{2}\right), \text{ referring to the energy separation of the two states}$$

As a result, we may re-express the eigenvalues of the total Hamiltonian as

$$E_+ = E_{\text{avg}} + \sqrt{\Delta E^2 + |V_{12}|^2}$$

$$\text{and } E_- = E_{\text{avg}} - \sqrt{\Delta E^2 + |V_{12}|^2}$$

(*Continued*)

(Continued)

In the above derivation, it has been assumed that the parameter on which the energy of each eigenstate (for either uncoupled or coupled) depends remains fixed. If the parameter is varied, then the energy level will change, and one may conveniently plot the energy of the two states (coupled and uncoupled) as a function of energy separation ΔE as follows.

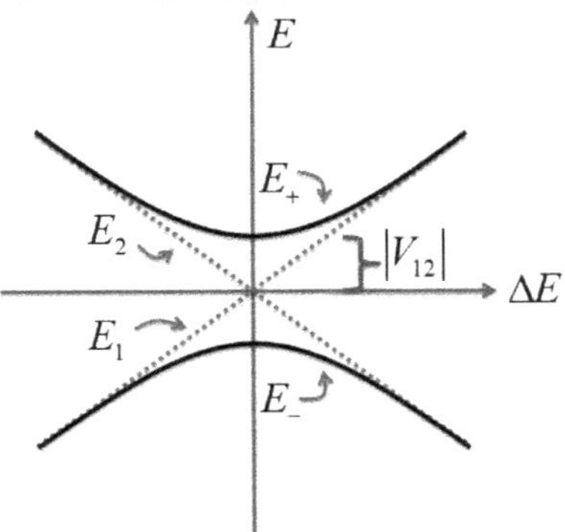

Here, $\Delta E = \left(\frac{E_1-E_2}{2}\right)$, or $E_1 = E_2 + 2\Delta E$ and $E_2 = E_1 - 2\Delta E$, in the absence of coupling, the variation of E_1 and E_2 as a function of ΔE gives two straight lines with positive and negative slopes, respectively (as plotted with dotted lines). This scenario changes in the presence of coupling. To realize the scenario in the presence of coupling, let us inspect carefully the values of E_+ and E_- for large magnitudes of ΔE (i.e., $\Delta E \gg |V_{12}|$):

$$E_+ \approx E_{\text{avg}} + \Delta E = E_1$$

$$\text{and } E_- \approx E_{\text{avg}} - \Delta E = E_2$$

Therefore, for large magnitudes of ΔE, the E_+ and E_- curves meet the E_1 and E_2 lines, respectively. In other words, the asymptotes of the E_+ and E_- curves meet the E_1 and E_2 lines, respectively. For $\Delta E = 0$, on the other hand, $E_+ = E_{\text{avg}} + |V_{12}|$ and $E_- = E_{\text{avg}} - |V_{12}|$. Therefore, the minimum separation between the E_+ and E_- curves is $2|V_{12}|$. The E_+ and E_- curves describe two symmetrical branches of a hyperbola. These findings directly indicate that in the absence of coupling, the two energy states $|\phi_1\rangle$ and $|\phi_2\rangle$ cross at $\Delta E = 0$. Due to the effect of coupling, two new states ($|\phi_+\rangle$ and $|\phi_-\rangle$) repel each other, giving birth to the non-crossing rule. $|\phi_+\rangle$ and $|\phi_-\rangle$ states do not cross at all.

(Continued)

(Continued)

In the present analysis, our next task will be to find out the matrix form of the $|\phi_+\rangle$ and $|\phi_-\rangle$ states. The procedure, which will be followed to find $|\phi_+\rangle$ and $|\phi_-\rangle$, is a general procedure of obtaining eigenvectors for any (2×2) Hermitian matrix. For this purpose, a great simplification of the mathematical procedure can be obtained if we introduce two angles, θ and φ. They are defined as follows.

Since $V_{12} = V_{21}^*$ (off-diagonal coupling terms), we may present the coupling terms using a real part V and an imaginary part $e^{-i\varphi}$ as follows:

$$V_{12} = V\, e^{-i2\varphi}$$

$$\text{and } V_{21} = Ve^{i2\varphi}$$

$$\text{Then, the coupling matrix is } \mathbf{V} = \begin{pmatrix} 0 & Ve^{-i2\varphi} \\ Ve^{i2\varphi} & 0 \end{pmatrix}$$

Note that instead of $e^{-i\varphi}$, we have taken $e^{i2\varphi}$ for mathematical convenience to be realized soon. On the other hand, the magnitude of the energy separation $(\Omega = \sqrt{\Delta E^2 + V^2})$ with respect to the average energy (E_{avg}) may be presented by an angle of 2θ as depicted in the following figure (note that instead of θ, we have taken 2θ again for mathematical convenience to be realized soon).

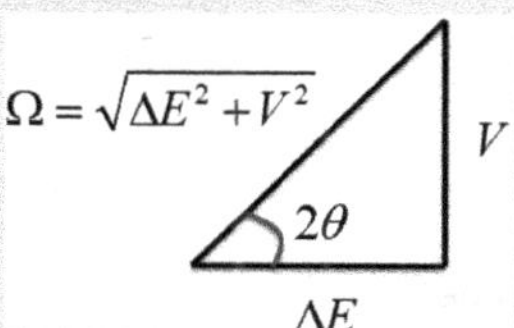

As a result, $\tan 2\theta = \frac{V}{\Delta E}$, and the range of θ is restricted by $0 \le \theta \le \frac{\pi}{4}$ (this limit geometrically defines two extreme situations: one with $\Delta E = 0$ and another with $V = 0$).

Thus, we have the following Hamiltonian matrix for the coupled system in the (θ, φ) representation:

$$\mathbf{H} = \begin{pmatrix} E_1 & \Delta E \tan 2\theta e^{-i2\varphi} \\ \Delta E \tan 2\theta e^{i2\varphi} & E_2 \end{pmatrix}$$

The eigenvalues of the total Hamiltonian can also be presented using the (θ, φ) representation as follows:

(Continued)

(Continued)

$$E_+ = E_{\text{avg}} + \sqrt{\Delta E^2 + \Delta E^2 \tan^2 2\theta}$$
$$= E_{\text{avg}} + \Delta E \sqrt{1 + \tan^2 2\theta}$$
$$= E_{\text{avg}} + \Delta E \sec 2\theta$$
$$\text{and } E_- = E_{\text{avg}} - \Delta E \sec 2\theta$$
$$(\text{since, } 1 + \tan^2 2\theta = \sec^2 2\theta)$$

On the other hand, the eigenvectors of the total Hamiltonian can be presented using the (θ, φ) representation as follows:

As $E_{\text{avg}} = \left(\frac{E_1+E_2}{2}\right)$ and $\Delta E = \left(\frac{E_1-E_2}{2}\right)$, we have $E_1 = E_{\text{avg}} + \Delta E$ and $E_2 = E_{\text{avg}} - \Delta E$. Thus, the eigenvalue equation for the eigenvalue $E_+ = E_{\text{avg}} + \Delta E \sec 2\theta$ can be written as

$$\begin{pmatrix} E_{\text{avg}} + \Delta E & \Delta E \tan 2\theta e^{-i2\varphi} \\ \Delta E \tan 2\theta e^{i2\varphi} & E_{\text{avg}} - \Delta E \end{pmatrix} \begin{pmatrix} a \\ b \end{pmatrix} = (E_{\text{avg}} + \Delta E \sec 2\theta) \begin{pmatrix} a \\ b \end{pmatrix}$$

$$\text{or } aE_{\text{avg}} + a\Delta E + b\Delta E \tan 2\theta e^{-2i\varphi} = aE_{\text{avg}} + a\Delta E \sec 2\theta$$
$$\text{or } a + b \tan 2\theta e^{-2i\varphi} = a \sec 2\theta$$
$$\text{or } a - a \sec 2\theta = -b \tan 2\theta e^{-2i\varphi}$$
$$\text{or } a \left[1 - \frac{1}{\cos 2\theta}\right] = -b \frac{\sin 2\theta}{\cos 2\theta} e^{-2i\varphi}$$
$$\text{or } a \left[\frac{\cos 2\theta - 1}{\cos 2\theta}\right] = -b \frac{\sin 2\theta}{\cos 2\theta} e^{-2i\varphi}$$
$$\text{or } -2a \sin^2\theta = -2b \sin\theta \cos\theta e^{-2i\varphi} \; (\text{as, } 2\sin^2\theta = (1 - \cos 2\theta) \text{ and } \sin 2\theta = 2 \sin\theta \cos\theta)$$
$$\text{or } a \sin\theta e^{i\varphi} = b \cos\theta e^{-i\varphi}$$
$$\text{or } \frac{a}{b} = \frac{\cos\theta e^{-i\varphi}}{\sin\theta e^{i\varphi}}$$

Thus, corresponding to the eigenvalue $E_+ = E_{\text{avg}} + \Delta E \sec 2\theta$, the eigenvector is (check it yourself)

$$|\phi_+\rangle = \begin{pmatrix} \cos\theta e^{-i\varphi} \\ \sin\theta e^{i\varphi} \end{pmatrix}$$

(Continued)

(Continued)

Interestingly, note that $|\phi_+\rangle$ is already normalized:

$$\sqrt{\langle\phi_+|\phi_+\rangle} = \sqrt{(\cos\theta e^{i\varphi} \quad \sin\theta e^{-i\varphi})\begin{pmatrix}\cos\theta e^{-i\varphi}\\ \sin\theta e^{i\varphi}\end{pmatrix}} = \sqrt{\cos^2\theta + \sin^2\theta} = 1$$

Following a similar derivation, one can easily show that corresponding to the eigenvalue $E_- = E_{\text{avg}} - \Delta E \sec 2\theta$, the eigenvector is (check it yourself)

$$|\phi_-\rangle = \begin{pmatrix}-\sin\theta e^{-i\varphi}\\ \cos\theta e^{i\varphi}\end{pmatrix}$$

Furthermore, as $\Omega^2 = \Delta E^2 + V^2$, or considering $4\Omega^2 = (E_1 - E_2)^2 + 4V^2 = D$, we get $\Omega = \frac{\sqrt{D}}{2}$, where $D = (E_1 - E_2)^2 + 4V^2$. As a result, we obtain

$$E_+ = \left(\frac{E_1 + E_2}{2} + \frac{\sqrt{D}}{2}\right) \text{ and } E_- = \left(\frac{E_1 + E_2}{2} - \frac{\sqrt{D}}{2}\right)$$

Important realization

Based on the above discussion, we obtain an important realization that the general unitary matrix which diagonalizes any (2×2) Hermitian (and, consequently, Hamiltonian) matrix $\mathbf{H} = \begin{pmatrix}E_1 & V_{12}\\ V_{21} & E_2\end{pmatrix}$ is given by

$$\mathbf{U} = \begin{pmatrix}-\sin\theta\, e^{-i\varphi} & \cos\theta\, e^{-i\varphi}\\ \cos\theta\, e^{i\varphi} & \sin\theta\, e^{i\varphi}\end{pmatrix}$$

where the coupling terms are represented by a product of real and imaginary parts, $V_{12} = V\,e^{-i2\varphi}$, $V_{21} = V\,e^{+i2\varphi}$, and $\theta = \frac{1}{2}\tan^{-1}\left(\frac{V}{\Delta E}\right)$. Here, $\Delta E = \left(\frac{E_1 - E_2}{2}\right)$, and the range of θ is restricted by $0 \le \theta \le \frac{\pi}{4}$. Therefore, the linear transformation which diagonalizes any (2×2) Hermitian matrix is expressed as $\mathbf{U}^\dagger\mathbf{H}\mathbf{U} = \mathbf{\Lambda}$, where $\mathbf{\Lambda} = \begin{pmatrix}E_- & 0\\ 0 & E_+\end{pmatrix}$, $E_- = E_{\text{avg}} - \Delta E \sec 2\theta$, $E_+ = E_{\text{avg}} + \Delta E \sec 2\theta$, and $E_{\text{avg}} = \left(\frac{E_1+E_2}{2}\right)$. E_+ and E_- can also be expressed as $E_+ = \left(\frac{E_1+E_2}{2} + \frac{\sqrt{D}}{2}\right)$ and $E_- = \left(\frac{E_1+E_2}{2} - \frac{\sqrt{D}}{2}\right)$, respectively, with $D = (E_1 - E_2)^2 + 4V^2$.

Self-study

Prove that $\mathbf{U} = \begin{pmatrix}-\sin\theta\, e^{-i\varphi} & \cos\theta\, e^{-i\varphi}\\ \cos\theta\, e^{i\varphi} & \sin\theta\, e^{i\varphi}\end{pmatrix}$ is a unitary matrix (calculate its adjoint and show that it is equal to its inverse) that diagonalizes the (2×2)

(Continued)

(Continued)

Hamiltonian matrix $\mathbf{H} = \begin{pmatrix} E_1 & \Delta E \tan 2\theta e^{-i2\varphi} \\ \Delta E \tan 2\theta e^{i2\varphi} & E_2 \end{pmatrix}$. Show that $\mathbf{U}^\dagger\mathbf{H}\mathbf{U} = \mathbf{\Lambda}$, where $\mathbf{\Lambda} = \begin{pmatrix} E_- & 0 \\ 0 & E_+ \end{pmatrix}$, $E_+ = E_{\text{avg}} + \Delta E \sec 2\theta$ and $E_- = E_{\text{avg}} - \Delta E \sec 2\theta$. For this exercise, use the following trigonometric relations: $\sin 2\theta = 2\sin\theta\cos\theta, \cos 2\theta = \cos^2\theta - \sin^2\theta$, and $\sin^2\theta + \cos^2\theta = 1$.

6.2.4 *Revisiting the exponential of a Hermitian matrix*

Recall, from Chapter 4, that the linear transformation which diagonalizes a Hermitian matrix is expressed by

$$\mathbf{U}^\dagger\mathbf{H}\mathbf{U} = \mathbf{\Lambda} \tag{6.1}$$

where $\mathbf{U}$ is the unitary matrix that diagonalizes the Hermitian matrix $\mathbf{H}$ to $\mathbf{\Lambda}$. Multiply equation (6.1) by $\mathbf{U}$ from the left-hand side:

$$\mathbf{U}\mathbf{U}^\dagger\mathbf{H}\mathbf{U} = \mathbf{U}\mathbf{\Lambda}$$

As $\mathbf{U}\mathbf{U}^\dagger = \mathbf{1}$ (check it yourself), we get $\mathbf{H}\mathbf{U} = \mathbf{U}\mathbf{\Lambda}$

Next, multiply the above equation by $\mathbf{U}^\dagger$ from the right-hand side:

$$\mathbf{H}\mathbf{U}\mathbf{U}^\dagger = \mathbf{U}\mathbf{\Lambda}\mathbf{U}^\dagger$$

As $\mathbf{U}\mathbf{U}^\dagger = \mathbf{1}$ (check it yourself), we can write

$$\mathbf{H} = \mathbf{U}\mathbf{\Lambda}\mathbf{U}^\dagger \tag{6.2}$$

We may now use the above relationship to obtain the exponential of the Hermitian matrix $\mathbf{H}$ using Taylor series expansion:

$$e^{\mathrm{H}} = \mathbf{1} + \mathbf{H} + \frac{1}{2!}\mathbf{H}^2 + \frac{1}{3!}\mathbf{H}^3 + \cdots\infty = \mathbf{1} + \mathbf{H} + \frac{1}{2!}\mathbf{H}\mathbf{H} + \frac{1}{3!}\mathbf{H}\mathbf{H}\mathbf{H} + \cdots\infty$$

or, inserting equation (6.2) into the above equation, we obtain

$$e^{\mathbf{H}} = \mathbf{1} + \mathbf{U\Lambda U^{\dagger}} + \frac{1}{2!}\mathbf{U\Lambda U^{\dagger}U\Lambda U^{\dagger}} + \frac{1}{3!}\mathbf{U\Lambda U^{\dagger}U\Lambda U^{\dagger}U\Lambda U^{\dagger}} + \cdots\infty$$

As $\mathbf{U^{\dagger}U} = \mathbf{1}$, we may write

$$e^{\mathbf{H}} = \mathbf{1} + \mathbf{U\Lambda U^{\dagger}} + \frac{1}{2!}\mathbf{U\Lambda^2 U^{\dagger}} + \frac{1}{3!}\mathbf{U\Lambda^3 U^{\dagger}} + \cdots\infty$$

$$\text{or } e^{\mathbf{H}} = \mathbf{U}\left[\mathbf{1} + \mathbf{\Lambda} + \frac{1}{2!}\mathbf{\Lambda}^2 + \frac{1}{3!}\mathbf{\Lambda}^3 + \cdots\infty\right]\mathbf{U}^{\dagger}$$

$$\text{or } e^{\mathbf{H}} = \mathbf{U}e^{\mathbf{\Lambda}}\mathbf{U}^{\dagger} \tag{6.3}$$

In Chapter 5, we have noted that calculating the exponential of a matrix is not a straight forward job. If the matrix is diagonal, however, the exponential of that matrix can be easily obtained by exponentiating each diagonal element. In the above discussion, on the other hand, we have apprehended another perspective of obtaining the exponential of a Hermitian matrix using equation (6.3).

Thus, based on the above perspective, one can obtain the exponential of a Hermitian matrix, which is not diagonal, if the unitary matrix, which diagonalizes the Hermitian matrix, is known (analytically or numerically).

As for any (2×2) Hermitian (and, consequently, Hamiltonian) matrix $\mathbf{H} = \begin{pmatrix} E_1 & V_{12} \\ V_{21} & E_2 \end{pmatrix}$, one can obtain analytically the general unitary matrix $\mathbf{U} = \begin{pmatrix} -\sin\theta\, e^{-i\varphi} & \cos\theta\, e^{-i\varphi} \\ \cos\theta\, e^{i\varphi} & \sin\theta\, e^{i\varphi} \end{pmatrix}$, which can diagonalize the Hermitian matrix, an analytical form of the exponential of a (2×2) Hermitian matrix can be obtained. In the next part of the present exercise, we go over the algebra, closely following Schwendner *et al.*[9]

As realized already, the time-evolution operator for two coupled states can be expressed using the following form, considering short-time propagation (need for short-time propagation is discussed in

Chapter 5):

$$
\begin{aligned}
\exp\left[-\frac{i}{\hbar}\begin{pmatrix}\hat{H}_1 & \hat{H}_{12}\\ \hat{H}_{21} & \hat{H}_2\end{pmatrix}\Delta t\right] &= \exp\left[-\frac{i}{\hbar}\begin{pmatrix}T+\hat{V}_1 & \hat{V}_{12}\\ \hat{V}_{21} & T+\hat{V}_2\end{pmatrix}\Delta t\right]\\
&\approx \exp\left[-\frac{i}{\hbar}\begin{pmatrix}\hat{T} & 0\\ 0 & T\end{pmatrix}\frac{\Delta t}{2}\right]\exp\left[-\frac{i}{\hbar}\begin{pmatrix}\hat{V}_1 & \hat{V}_{12}\\ \hat{V}_{21} & \hat{V}_2\end{pmatrix}\Delta t\right]\\
&\quad\times \exp\left[-\frac{i}{\hbar}\begin{pmatrix}\hat{T} & 0\\ 0 & T\end{pmatrix}\frac{\Delta t}{2}\right] + O(\Delta t^3)
\end{aligned}
$$

where the off-diagonal terms represent the coupling terms and the diagonal elements represent the respective Hamiltonian without coupling. Note that we have made use of the symmetrized product form of the time-evolution operator. It is clear from the above expression that complications arise since the potential energy matrix is not diagonal any more. Therefore, using equations (6.1) and (6.3), we further reexpress the potential part of the two-state time-evolution operator as follows

$$
\begin{aligned}
&\exp\left[-\frac{i}{\hbar}\begin{pmatrix}V_1 & V_{12}\\ V_{21} & V_2\end{pmatrix}\Delta t\right]\\
&= \mathbf{U}\exp\left[-\frac{i}{\hbar}\mathbf{U}^\dagger\begin{pmatrix}V_1 & V_{12}\\ V_{21} & V_2\end{pmatrix}\mathbf{U}\Delta t\right]\mathbf{U}^\dagger\\
&= \mathbf{U}\exp\left[-\frac{i}{\hbar}\begin{pmatrix}\lambda_1 & 0\\ 0 & \lambda_2\end{pmatrix}\Delta t\right]\mathbf{U}^\dagger = \mathbf{U}\begin{pmatrix}e^{-\frac{i}{\hbar}\lambda_1\Delta t} & 0\\ 0 & e^{-\frac{i}{\hbar}\lambda_2\Delta t}\end{pmatrix}\mathbf{U}^\dagger\\
&= \begin{pmatrix}-\sin\theta\, e^{-i\varphi} & \cos\theta\, e^{-i\varphi}\\ \cos\theta\, e^{i\varphi} & \sin\theta\, e^{i\varphi}\end{pmatrix}\begin{pmatrix}e^{-\frac{i}{\hbar}\lambda_1\Delta t} & 0\\ 0 & e^{-\frac{i}{\hbar}\lambda_2\Delta t}\end{pmatrix}\\
&\quad\times\begin{pmatrix}-\sin\theta\, e^{i\varphi} & \cos\theta\, e^{-i\varphi}\\ \cos\theta\, e^{i\varphi} & \sin\theta\, e^{-i\varphi}\end{pmatrix}
\end{aligned}
$$

Here, the coupling terms are represented by a product of real and imaginary parts, $V_{12} = V\, e^{-i2\varphi}$ and $V_{21} = V\, e^{i2\varphi}$. Furthermore, as already noted in the Illuminating Discussion section, $\tan 2\theta = \left(\frac{V}{\Delta V}\right)$, where the range of θ is restricted by $0 \leq \theta \leq \frac{\pi}{4}$, and $\Delta V = \left(\frac{V_1 - V_2}{2}\right)$, $\lambda_1 = \left(\frac{V_1+V_2}{2} - \frac{\sqrt{D}}{2}\right)$, $\lambda_2 = \left(\frac{V_1+V_2}{2} + \frac{\sqrt{D}}{2}\right)$, and $D = (V_1 - V_2)^2 + 4V^2$.

For further reducing the above expression, first, assume that $\varphi = 0$, i.e., the coupling terms are real (justification of this assumption will be given soon). Then, using matrix multiplication rules, the potential part of the two-state time-evolution operator can be expressed as

$$\begin{aligned}
&\exp\left[-\frac{i}{\hbar}\begin{pmatrix} V_1 & V_{12} \\ V_{21} & V_2 \end{pmatrix}\Delta t\right] \\
&= \begin{pmatrix} -\sin\theta\, e^{-\frac{i}{\hbar}\lambda_1\Delta t} & \cos\theta\, e^{-\frac{i}{\hbar}\lambda_2\Delta t} \\ \cos\theta\, e^{-\frac{i}{\hbar}\lambda_1\Delta t} & \sin\theta\, e^{-\frac{i}{\hbar}\lambda_2\Delta t} \end{pmatrix} \begin{pmatrix} -\sin\theta & \cos\theta \\ \cos\theta & \sin\theta \end{pmatrix} \\
&= \begin{pmatrix} \sin^2\theta\, e^{-\frac{i}{\hbar}\lambda_1\Delta t} + \cos^2\theta\, e^{-\frac{i}{\hbar}\lambda_2\Delta t} & -\sin\theta\cos\theta\, e^{-\frac{i}{\hbar}\lambda_1\Delta t} + \sin\theta\cos\theta\, e^{-\frac{i}{\hbar}\lambda_2\Delta t} \\ -\sin\theta\cos\theta\, e^{-\frac{i}{\hbar}\lambda_1\Delta t} + \sin\theta\cos\theta\, e^{-\frac{i}{\hbar}\lambda_2\Delta t} & \cos^2\theta\, e^{-\frac{i}{\hbar}\lambda_1\Delta t} + \sin^2\theta\, e^{-\frac{i}{\hbar}\lambda_2\Delta t} \end{pmatrix}
\end{aligned} \tag{6.4}$$

One can further simplify each term of the matrix given in equation (6.4) using $\lambda_1 = \left(\frac{V_1+V_2}{2} - \frac{\sqrt{D}}{2}\right)$ and $\lambda_2 = \left(\frac{V_1+V_2}{2} + \frac{\sqrt{D}}{2}\right)$, where $D = (V_1 - V_2)^2 + 4V^2$, $\cos 2\theta = \frac{V_1 - V_2}{\sqrt{D}}$, and $\sin 2\theta = \frac{2V}{\sqrt{D}}$.

The *11-element* of the matrix in equation (6.4) is

$$\begin{aligned}
&\sin^2\theta e^{-\frac{i}{\hbar}\lambda_1\Delta t} + \cos^2\theta\, e^{-\frac{i}{\hbar}\lambda_2\Delta t} \\
&\quad = \sin^2\theta\, e^{-\frac{i}{\hbar}\left(\frac{V_1+V_2}{2} - \frac{\sqrt{D}}{2}\right)\Delta t} + \cos^2\theta\, e^{-\frac{i}{\hbar}\left(\frac{V_1+V_2}{2} + \frac{\sqrt{D}}{2}\right)\Delta t} \\
&\quad = e^{-\frac{i}{\hbar}\left(\frac{V_1+V_2}{2}\right)\Delta t}\left[\sin^2\theta\, e^{\frac{i}{\hbar}\frac{\sqrt{D}}{2}\Delta t} + \cos^2\theta\, e^{-\frac{i}{\hbar}\frac{\sqrt{D}}{2}\Delta t}\right] \\
&\quad = e^{-\frac{i}{\hbar}\left(\frac{V_1+V_2}{2}\right)\Delta t}\frac{1}{2}\left[(1 - \cos 2\theta)e^{\frac{i}{\hbar}\frac{\sqrt{D}}{2}\Delta t} + (1 + \cos 2\theta)e^{-\frac{i}{\hbar}\frac{\sqrt{D}}{2}\Delta t}\right]
\end{aligned}$$

$$= e^{-\frac{i}{\hbar}\left(\frac{V_1+V_2}{2}\right)\Delta t}\frac{1}{2}\left[\left(e^{\frac{i}{\hbar}\frac{\sqrt{D}}{2}\Delta t}+e^{-\frac{i}{\hbar}\frac{\sqrt{D}}{2}\Delta t}\right)\right.$$
$$\left.+\cos 2\theta\left(e^{-\frac{i}{\hbar}\frac{\sqrt{D}}{2}\Delta t}-e^{\frac{i}{\hbar}\frac{\sqrt{D}}{2}\Delta t}\right)\right]$$
$$= e^{-\frac{i}{\hbar}\left(\frac{V_1+V_2}{2}\right)\Delta t}\left[\cos\left(\frac{\sqrt{D}}{2\hbar}\Delta t\right)-i\cos 2\theta\sin\left(\frac{\sqrt{D}}{2\hbar}\Delta t\right)\right]$$
$$= e^{-\frac{i}{\hbar}\left(\frac{V_1+V_2}{2}\right)\Delta t}\left[\cos\left(\frac{\sqrt{D}}{2\hbar}\Delta t\right)+i\frac{V_2-V_1}{\sqrt{D}}\sin\left(\frac{\sqrt{D}}{2\hbar}\Delta t\right)\right]$$

The *12-element* of the matrix in equation (6.4) is

$$-\sin\theta\ \cos\theta e^{-\frac{i}{\hbar}\lambda_1\Delta t}+\sin\theta\cos\theta\, e^{-\frac{i}{\hbar}\lambda_2\Delta t}$$
$$= \sin\theta\cos\theta\left[e^{-\frac{i}{\hbar}\lambda_2\Delta t}-e^{-\frac{i}{\hbar}\lambda_1\Delta t}\right]$$
$$= \sin\theta\cos\theta\left[e^{-\frac{i}{\hbar}\left(\frac{V_1+V_2}{2}+\frac{\sqrt{D}}{2}\right)\Delta t}-e^{-\frac{i}{\hbar}\left(\frac{V_1+V_2}{2}-\frac{\sqrt{D}}{2}\right)\Delta t}\right]$$
$$= e^{-\frac{i}{\hbar}\left(\frac{V_1+V_2}{2}\right)\Delta t}\sin\theta\cos\theta\left[e^{-\frac{i}{\hbar}\frac{\sqrt{D}}{2}\Delta t}-e^{\frac{i}{\hbar}\frac{\sqrt{D}}{2}\Delta t}\right]$$
$$= e^{-\frac{i}{\hbar}\left(\frac{V_1+V_2}{2}\right)\Delta t}\sin 2\theta(-i)\sin\left(\frac{\sqrt{D}}{2\hbar}\Delta t\right)$$
$$= e^{-\frac{i}{\hbar}\left(\frac{V_1+V_2}{2}\right)\Delta t}\frac{2V}{\sqrt{D}}(-i)\sin\left(\frac{\sqrt{D}}{2\hbar}\Delta t\right)$$

Similarly, the *21- and 22-elements* can be simplified to obtain a final expression for equation (6.4) as follows (check it yourself):

$$\exp\left[-\frac{i}{\hbar}\begin{pmatrix}V_1 & V_{12}\\ V_{21} & V_2\end{pmatrix}\Delta t\right]$$
$$=\begin{pmatrix} e^{-\frac{i}{\hbar}\left(\frac{V_1+V_2}{2}\right)\Delta t}\left[\cos\left(\frac{\sqrt{D}}{2\hbar}\Delta t\right)+i\frac{V_2-V_1}{\sqrt{D}}\sin\left(\frac{\sqrt{D}}{2\hbar}\Delta t\right)\right] & e^{-\frac{i}{\hbar}\left(\frac{V_1+V_2}{2}\right)\Delta t}\frac{2V}{\sqrt{D}}(-i)\sin\left(\frac{\sqrt{D}}{2\hbar}\Delta t\right)\\ e^{-\frac{i}{\hbar}\left(\frac{V_1+V_2}{2}\right)\Delta t}\frac{2V}{\sqrt{D}}(-i)\sin\left(\frac{\sqrt{D}}{2\hbar}\Delta t\right) & e^{-\frac{i}{\hbar}\left(\frac{V_1+V_2}{2}\right)\Delta t}\left[\cos\left(\frac{\sqrt{D}}{2\hbar}\Delta t\right)+i\frac{V_1-V_2}{\sqrt{D}}\sin\left(\frac{\sqrt{D}}{2\hbar}\Delta t\right)\right]\end{pmatrix} \tag{6.5}$$

6.2.5 *Numerical implementation of the two-state time-evolution operator*

Thus, finally, we obtain an analytical form for the two-state short-time propagation as follows:

$$\begin{pmatrix} \psi_1(x,\Delta t) \\ \psi_2(x,\Delta t) \end{pmatrix} \approx \begin{pmatrix} e^{-\frac{i}{\hbar}T\frac{\Delta t}{2}} & 0 \\ 0 & e^{-\frac{i}{\hbar}T\frac{\Delta t}{2}} \end{pmatrix} \begin{pmatrix} 11 & 12 \\ 21 & 22 \end{pmatrix} \times \begin{pmatrix} e^{-\frac{i}{\hbar}T\frac{\Delta t}{2}} & 0 \\ 0 & e^{-\frac{i}{\hbar}T\frac{\Delta t}{2}} \end{pmatrix} \begin{pmatrix} \psi_1(x,0) \\ \psi_2(x,0) \end{pmatrix} \tag{6.6}$$

where the $\begin{pmatrix} 11 & 12 \\ 21 & 22 \end{pmatrix}$ matrix represents the potential part of the two-state time-evolution operator, as given in equation (6.5). Closely following the scheme presented in Chapter 5, one can now sketch numerical steps to implement the above short-time propagation:

$$\begin{pmatrix} \psi_1(x,\Delta t) \\ \psi_2(x,\Delta t) \end{pmatrix} \equiv \Im^{-1} \begin{pmatrix} e^{-\frac{i}{\hbar}T\frac{\Delta t}{2}} & 0 \\ 0 & e^{-\frac{i}{\hbar}T\frac{\Delta t}{2}} \end{pmatrix} \Im \begin{pmatrix} 11 & 12 \\ 21 & 22 \end{pmatrix} \times \Im^{-1} \begin{pmatrix} e^{-\frac{i}{\hbar}T\frac{\Delta t}{2}} & 0 \\ 0 & e^{-\frac{i}{\hbar}T\frac{\Delta t}{2}} \end{pmatrix} \Im \begin{pmatrix} \psi_1(x,0) \\ \psi_2(x,0) \end{pmatrix}$$

where $\Im$ and $\Im^{-1}$ denote the Fourier transform and inverse Fourier transform, respectively. Furthermore, the execution of the above numerical steps is carried out from right to left. Explicitly, one can express each step as follows:

$$\begin{aligned} \begin{pmatrix} \psi_1(x,\Delta t) \\ \psi_2(x,\Delta t) \end{pmatrix} &\equiv \Im^{-1} \begin{pmatrix} e^{-\frac{i}{\hbar}T\frac{\Delta t}{2}} & 0 \\ 0 & e^{-\frac{i}{\hbar}T\frac{\Delta t}{2}} \end{pmatrix} \Im \begin{pmatrix} 11 & 12 \\ 21 & 22 \end{pmatrix} \\ &\quad \times \Im^{-1} \begin{pmatrix} e^{-\frac{i}{\hbar}T\frac{\Delta t}{2}} & 0 \\ 0 & e^{-\frac{i}{\hbar}T\frac{\Delta t}{2}} \end{pmatrix} \Im \begin{pmatrix} \psi_1(x,0) \\ \psi_2(x,0) \end{pmatrix} \\ &\equiv \Im^{-1} \begin{pmatrix} e^{-\frac{i}{\hbar}T\frac{\Delta t}{2}} & 0 \\ 0 & e^{-\frac{i}{\hbar}T\frac{\Delta t}{2}} \end{pmatrix} \Im \begin{pmatrix} 11 & 12 \\ 21 & 22 \end{pmatrix} \\ &\quad \times \Im^{-1} \begin{pmatrix} e^{-\frac{i}{\hbar}T\frac{\Delta t}{2}} & 0 \\ 0 & e^{-\frac{i}{\hbar}T\frac{\Delta t}{2}} \end{pmatrix} \begin{pmatrix} \psi_1(p,0) \\ \psi_2(p,0) \end{pmatrix} \end{aligned}$$

$$\equiv \Im^{-1} \begin{pmatrix} e^{-\frac{i}{\hbar}T\frac{\Delta t}{2}} & 0 \\ 0 & e^{-\frac{i}{\hbar}T\frac{\Delta t}{2}} \end{pmatrix} \Im \begin{pmatrix} 11 & 12 \\ 21 & 22 \end{pmatrix}$$

$$\times \Im^{-1} \begin{pmatrix} e^{-\frac{i}{\hbar}T\frac{\Delta t}{2}} \psi_1(p,0) \\ e^{-\frac{i}{\hbar}T\frac{\Delta t}{2}} \psi_2(p,0) \end{pmatrix}$$

$$\equiv \Im^{-1} \begin{pmatrix} e^{-\frac{i}{\hbar}T\frac{\Delta t}{2}} & 0 \\ 0 & e^{-\frac{i}{\hbar}T\frac{\Delta t}{2}} \end{pmatrix} \Im \begin{pmatrix} 11 & 12 \\ 21 & 22 \end{pmatrix} \Im^{-1} \begin{pmatrix} \psi_1'(p,\Delta t) \\ \psi_2'(p,\Delta t) \end{pmatrix}$$

$$\equiv \Im^{-1} \begin{pmatrix} e^{-\frac{i}{\hbar}T\frac{\Delta t}{2}} & 0 \\ 0 & e^{-\frac{i}{\hbar}T\frac{\Delta t}{2}} \end{pmatrix} \Im \begin{pmatrix} 11 & 12 \\ 21 & 22 \end{pmatrix} \begin{pmatrix} \psi_1'(x,\Delta t) \\ \psi_2'(x,\Delta t) \end{pmatrix}$$

$$\equiv \Im^{-1} \begin{pmatrix} e^{-\frac{i}{\hbar}T\frac{\Delta t}{2}} & 0 \\ 0 & e^{-\frac{i}{\hbar}T\frac{\Delta t}{2}} \end{pmatrix} \Im \begin{pmatrix} 11\psi_1'(x,\Delta t) + 12\psi_2'(x,\Delta t) \\ 21\psi_1'(x,\Delta t) + 22\psi_2'(x,\Delta t) \end{pmatrix}$$

$$\equiv \Im^{-1} \begin{pmatrix} e^{-\frac{i}{\hbar}T\frac{\Delta t}{2}} & 0 \\ 0 & e^{-\frac{i}{\hbar}T\frac{\Delta t}{2}} \end{pmatrix} \Im \begin{pmatrix} \psi_{1,11}''(x,\Delta t) + \psi_{2,12}''(x,\Delta t) \\ \psi_{1,21}''(x,\Delta t) + \psi_{2,22}''(x,\Delta t) \end{pmatrix}$$

$$\equiv \Im^{-1} \begin{pmatrix} e^{-\frac{i}{\hbar}T\frac{\Delta t}{2}} & 0 \\ 0 & e^{-\frac{i}{\hbar}T\frac{\Delta t}{2}} \end{pmatrix} \begin{pmatrix} \psi_{1,11}''(p,\Delta t) + \psi_{2,12}''(p,\Delta t) \\ \psi_{1,21}''(p,\Delta t) + \psi_{2,22}''(p,\Delta t) \end{pmatrix}$$

$$\equiv \Im^{-1} \begin{pmatrix} e^{-\frac{i}{\hbar}T\frac{\Delta t}{2}} \left[\psi_{1,11}''(p,\Delta t) + \psi_{2,12}''(p,\Delta t)\right] \\ e^{-\frac{i}{\hbar}T\frac{\Delta t}{2}} \left[\psi_{1,21}''(p,\Delta t) + \psi_{2,22}''(p,\Delta t)\right] \end{pmatrix}$$

$$\equiv \Im^{-1} \begin{pmatrix} \left[\psi_{1,11}'''(p,\Delta t) + \psi_{2,12}'''(p,\Delta t)\right] \\ \left[\psi_{1,21}'''(p,\Delta t) + \psi_{2,22}'''(p,\Delta t)\right] \end{pmatrix}$$

$$\equiv \begin{pmatrix} \left[\psi_{1,11}'''(x,\Delta t) + \psi_{2,12}'''(x,\Delta t)\right] \\ \left[\psi_{1,21}'''(x,\Delta t) + \psi_{2,22}'''(x,\Delta t)\right] \end{pmatrix}$$

where, in atomic unit, the elements of the two-state potential part of the propagator are given by

$$\mathit{11-element}: e^{-i\left(\frac{V_1+V_2}{2}\right)\Delta t}\left[\cos\left(\frac{\sqrt{D}}{2}\Delta t\right) + i\frac{V_2 - V_1}{\sqrt{D}}\sin\left(\frac{\sqrt{D}}{2}\Delta t\right)\right]$$

$$\mathit{12-element\ or\ 21-element}: e^{-i\left(\frac{V_1+V_2}{2}\right)\Delta t}\frac{2V}{\sqrt{D}}(-i)\sin\left(\frac{\sqrt{D}}{2}\Delta t\right)$$

$$22 - element : e^{-i(\frac{V_1+V_2}{2})\Delta t}\left[\cos\left(\frac{\sqrt{D}}{2}\Delta t\right) + i\frac{V_1 - V_2}{\sqrt{D}}\sin\left(\frac{\sqrt{D}}{2}\Delta t\right)\right]$$

In the above numerical steps, $\psi_1(x,0)$ and $\psi_2(x,0)$ are the initial wavefunctions associated with states 1 and 2, respectively. On the other hand, $[\psi'''_{1,11}(x,\Delta t) + \psi'''_{2,12}(x,\Delta t)]$ and $[\psi'''_{1,21}(x,\Delta t) + \psi'''_{2,22}(x,\Delta t)]$ are new wavefunctions associated with states 1 and 2, respectively, at time $(0 + \Delta t)$. As we did before (in PythonChapter E), to accomplish the time propagation for the final time t, we have to repeat the above numerical steps using a for loop for every Δt, with $[\psi'''_{1,11}(x,\Delta t)+\psi'''_{2,12}(x,\Delta t)]$ and $[\psi'''_{1,21}(x,\Delta t)+\psi'''_{2,22}(x,\Delta t)]$ serving as the new initial wavefunction (associated with states 1 and 2, respectively) for the next iteration.

For Python implementation, we need to know the analytical forms of the potential energy curves of NaI (as depicted in Figure 6.11). The ground state potential energy curve of NaI is given by[10]

$$V_0(R) = \left[A_2 + \left(\frac{B_2}{R}\right)^8\right]e^{-\frac{R}{\rho}} - \frac{1}{R} - \frac{1}{2}\frac{(\lambda_1 + \lambda_2)}{R^4} - \frac{C}{R^6} - \frac{2\lambda_1\lambda_2}{R^7} + \Delta E \tag{6.7}$$

which has several important terms. Readers are referred to Ref. 10 for illuminating discussions on the terms used to define the ground state potential of NaI. Briefly, the charge–charge interaction term R^{-1} comes from the fact that ground state NaI exhibits ionic character. The dipole–dipole interaction terms R^{-4} and R^{-7} come from the polarizabilities of the ions. The term R^{-6} originates from the van der Waals attraction between two ions. Finally, the exponential repulsion term $e^{-\frac{R}{\rho}}$ makes the potential well behaved at small R. On the other hand, the lowest-lying electronically excited state potential energy curve of NaI has a simple repulsive form given by

$$V_1(R) = A_1 e^{[-B_1(R-R_1)]} \tag{6.8}$$

Table 6.1. Values for the coefficients used in equations (6.7)–(6.9) to define the potential energy curves of NaI. Note that these values are given in Ref. 10(a) in SI units. They have been converted to atomic units for the present exercise.

Coefficients	**Values (in atomic units)**
A_2	101.43
B_2	3.00
ρ	0.6593
λ_1	2.753
λ_2	43.39
C	18.91
ΔE	0.0761
A_1	0.0299
B_1	2.163
R_1	5.102
A_{12}	0.002
B_{12}	0.194
R_x	13.095

Finally, the coupling potential is considered to be of the following form:

$$V = A_{12}e^{[-B_{12}(R-R_x)^2]} \tag{6.9}$$

which is nothing but a Gaussian centered around the crossing point R_x.

The values of the respective coefficients, which define all the above-mentioned potentials of NaI, are given in Table 6.1. The lowest-lying electronically excited state potential energy curve of NaI is quite steep at short R and becomes almost flat at large R (as evident in Figure 6.11).

Final Python implementation for the two-state wavepacket dynamics of NaI is given in the following.

```
#Importing the required libraries
from scipy import sqrt,arange,exp,pi,cos,sin
from scipy.integrate import simps
from scipy.fftpack import fftfreq,fft,ifft
from matplotlib.pyplot import plot,xlim,ylim,show

#Creating the x-grid in atomic unit
xmin=-50
xmax=500.1
dx=0.1
x=arange(xmin,xmax,dx)

#Defining and normalizing the initial wavefunctions
k0=0
psi0=0*x #No webpacket in ground state
psi1=((exp((-(x-4)**2)/5))*(exp(1j*k0*x))) #Gaussian wavepacket in the excited state
prob_density1=abs(psi1)**2
norm1=sqrt(simps(prob_density1,x))
psiNorm1=psi1/norm1
prob_density_initial1=abs(psiNorm1)**2
psiNorm0=psi0

#Defining the potential of each state
v1=0.0299*exp(-2.163*(x-5.102))
v0=((101.43+(3/x)**8)*exp(-x/0.6593))-(1/x)-((0.5*46.143)/(x**4))-(18.91/(x**6))-
((2*2.753*43.39)/(x**7))+0.0761

#Defining the coupling potential
v=0.02*exp(-0.194*(x-13.095))

#Defining the potential part of two-state propagator
D=((v0-v1)**2)+(4*v**2)
dt=0.1 #Time step size in atomic unit
v11=exp(-1.0j*((v1+v0)/2)*dt)*(cos(0.5*sqrt(D)*dt)+1.0j*((v1-v0)/sqrt(D))*sin(0.5*sqrt(D)*dt))
v12=-1.0j*exp(-1.0j*((v1+v0)/2)*dt)*(2*v/sqrt(D))*sin(0.5*sqrt(D)*dt)
v21=v12
v22=exp(-1.0j*((v1+v0)/2)*dt)*(cos(0.5*sqrt(D)*dt)+1.0j*((v0-v1)/sqrt(D))*sin(0.5*sqrt(D)*dt))

#Two-state time propagation
N=len(x)
k=2*pi*fftfreq(N,dx) #Creating k-grid
KE_k=(0.5*k**2) #Defining kinetic energy
Nt=100 #Total number of time steps
for i in arange(Nt):
        #fft of psi1 and psi0 to k-space
        psiNorm1_k=fft(psiNorm1)
        psiNorm0_k=fft(psiNorm0)
        #Time evolution of psi1 and psi0 due to 1st KE part
        psiNorm1_k_KE1=exp(-1.0j*KE_k*dt/2.0)*psiNorm1_k
```

> ➢ **Note that A_{12} (featuring the strength of the coupling) is intentionally taken to be 0.02 a.u. (instead of 0.002 a.u., as given in the table) to induce significant population transfer in short time for quick visualization of the outcome.**

```
        psiNorm0_k_KE1=exp(-1.0j*KE_k*dt/2.0)*psiNorm0_k
        #ifft of psi1 and psi0 to x-space
        psiNorm1_k_KE1_x=ifft(psiNorm1_k_KE1)
        psiNorm0_k_KE1_x=ifft(psiNorm0_k_KE1)
        #Time evolution of psi1 and psi0 due to PE part
        psiNorm1_k_KE1_x_V=(v22*psiNorm1_k_KE1_x)+(v21*psiNorm0_k_KE1_x)
        psiNorm0_k_KE1_x_V=(v11*psiNorm0_k_KE1_x)+(v12*psiNorm1_k_KE1_x)
        #fft of psi1 and psi0 to k-space again
        psiNorm1_k_KE1_x_V_k=fft(psiNorm1_k_KE1_x_V)
        psiNorm0_k_KE1_x_V_k=fft(psiNorm0_k_KE1_x_V)
        #Time evolution of psi1 and psi0 due to 2nd KE part
        psiNorm1_k_KE1_x_V_k_KE2=exp(-1.0j*KE_k*dt/2.0)*psiNorm1_k_KE1_x_V_k
        psiNorm0_k_KE1_x_V_k_KE2=exp(-1.0j*KE_k*dt/2.0)*psiNorm0_k_KE1_x_V_k
        #ifft of psi1 and psi0 to x-space
        psiNorm1=ifft(psiNorm1_k_KE1_x_V_k_KE2)
        psiNorm0=ifft(psiNorm0_k_KE1_x_V_k_KE2)
prob_density_final1=abs(psiNorm1)**2
prob_density_final0=abs(psiNorm0)**2

#Plotting the initial and final probability densities
plot(x,prob_density_final0,color='red',linewidth=3)
plot(x,prob_density_final1,color='blue',linewidth=3)
plot(x,prob_density_initial1,linestyle='--',color='grey',linewidth=3)
xlim(0,30)
ylim(0,0.16)
show()
```

Upon execution of the above program for two different **Nt (50 and 100)**, the probability density distribution functions associated with the two states can be obtained. They are plotted in Figure 6.12, which evidently shows that population is transferred from the excited electronic state to the ground electronic state as time progresses.

Self-Study

(a) Improve the formatting of all graphs given in Figure 6.12 for better presentation.

(b) Find the expectation value of the position and spread of the distribution for each wavepacket as a function of time. Plot them. The spread of the distribution can be represented by the standard deviation of the probability density distribution.

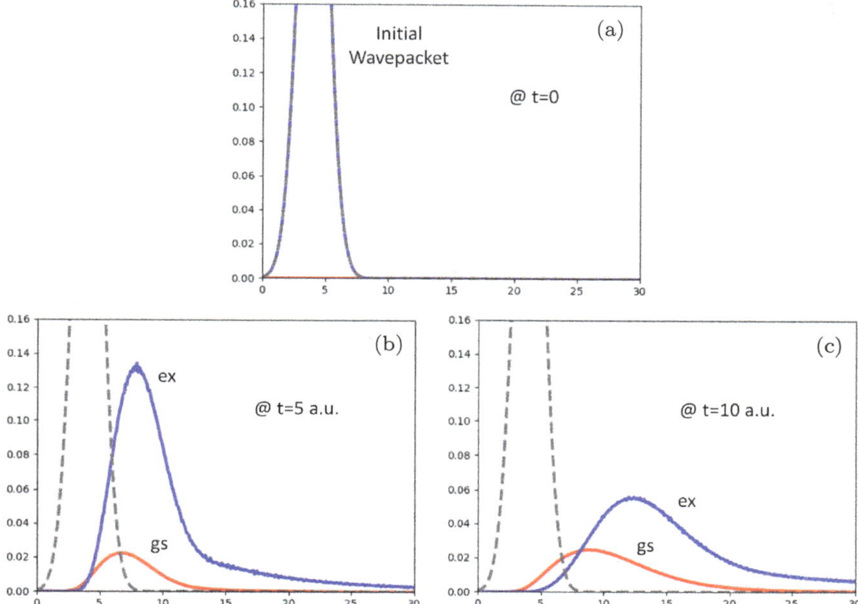

Figure 6.12. (a) At $t = 0$, initial wavepacket appears only on the excited state (zero contribution of ground state wavepacket); subsequent decay of the excited state wavepacket and rise of ground state wavepacket (b) at $t = 5$ a.u. and (c) 10 a.u.

(c) Change the value of the coupling strength (A_{12}) from 0.02 a.u. to 0.002 a.u. Record your observation.

6.3 Time-Dependent Approach to Spectroscopy

We have already realized that the time-independent Schrödinger equation (TISE) $\hat{H}(x)\psi_n(x) = E_n\psi_n(x)$ for a quantum system provides stationary eigenstate energies (E_n) and corresponding eigenstate wavefunctions ($\psi_n(x)$). These eigenstate wavefunctions are orthonormal, i.e., $\langle\psi_m|\psi_n\rangle = \delta_{mn} = 1$ when $m = n$ and $\delta_{mn} = 0$ when $m \neq n$. Thus, solutions to the TISE provide the spectral information (frequency ω at which spectral transition between two eigenstates should occur) of the quantum system, as schematically depicted in

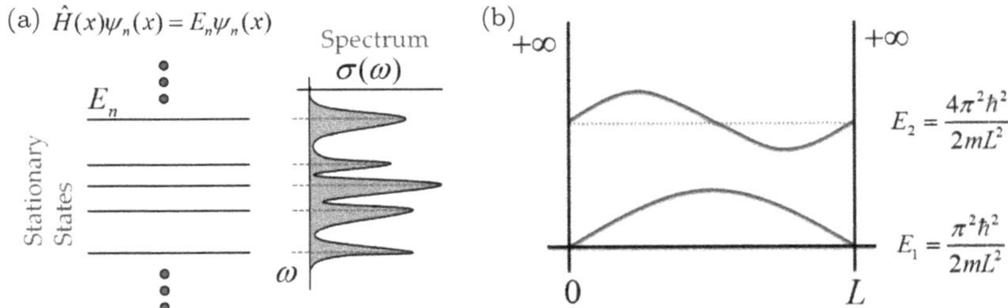

Figure 6.13. (a) Time-independent Schrödinger equation reveals the spectral information of a quantum system by exploring the (stationary) eigenstates. (b) Lowest-lying eigenstates of our most familiar quantum system, particle in one-dimensional box, are depicted.

Figure 6.13(a). The spectrum of a quantum system can be experimentally verified using energy- or frequency-resolved spectroscopy.

For example, the nth (stationary) eigenstate wavefunction of a particle in a one-dimensional box is given by (see Figure 6.13(b))

$$\psi_n(x) = \sqrt{\frac{2}{L}} \sin\left(\frac{\pi x n}{L}\right) \tag{6.10}$$

and the corresponding eigenstate energy is given by

$$E_n = \frac{n^2\pi^2\hbar^2}{2mL^2} \tag{6.11}$$

These are obtained by solving the corresponding TISE with the Hamiltonian $\hat{H}(x;L) = \left[-\frac{\hbar^2}{2m}\frac{d^2}{dx^2} + V(x;L)\right]$, where $V(x;L) = 0$ when $0 \leq x \leq L$ and $V(x;L) = +\infty$ otherwise define the potential of the particle. As the eigenstate energies of the particle are known using equation (6.11), one can predict the frequency ω (or photon energy) at which a spectral transition between two eigenstates may occur for the particle in the one-dimensional box. Note that the relative intensity of each transition depends on many factors and cannot be explored by only knowing the stationary eigenstates. The TISE provides the only possible eigenstates of a quantum system. As a result, the frequency ω (or photon energy) at which a spectral transition between two eigenstates may occur can be derived from the TISE.

On the contrary to the above time-independent picture, in this section, we will realize that it will be possible to determine the eigenvalues and eigenstates for any one-dimensional (we continue the discussion based on the one-dimensional problem) potential directly from solutions to the time-dependent Schrödinger equation (TDSE), $i\hbar\frac{\partial\psi(x,t)}{\partial t} = \hat{H}(x)\psi(x,t)$. This time-dependent approach to obtaining the spectral information of a quantum system plays an important role in theoretical spectroscopy. To apprehend the time-dependent approach to spectroscopy, we will first become familiar with a function called the wavepacket correlation function.[12]

6.3.1 *Wavepacket correlation function*

We have understood in Chapter 3 that, for a given initial wavepacket, $\psi(x,0)$, an analytical or numerical solution to the TDSE provides the final wavepacket, $\psi(x,t)$, for any later time t. By correlating the initial and final wavepackets, one can construct a new function at time t, called the wavepacket correlation function $c(t)$:

$$c(t) = \langle\psi(x,0)|\psi(x,t)\rangle = \int_{-\infty}^{+\infty} \psi^*(x,0)\psi(x,t)dx \tag{6.12}$$

Recall that $\psi(x,t)$ (the wavepacket at any later time t) can be expanded using the basis of orthonormal (stationary) eigenstates of the quantum system:

$$\psi(x,t) = \sum_n c_n(t)\,\psi_n(x)e^{-i\frac{E_n t}{\hbar}} \tag{6.13}$$

$$\text{So, at } t = 0, \psi(x,0) = \sum_m c_m(0)\,\psi_m(x) \tag{6.14}$$

Here, note that two separate indices (namely m and n) are selected to represent, respectively, the initial and final wavepackets for mathematical convenience.

Inserting equations (6.13) and (6.14) into equation (6.12), we get

$$
\begin{aligned}
c(t) &= \int_{-\infty}^{+\infty} \psi^*(x,0)\,\psi(x,t)dx \\
&= \int_{-\infty}^{+\infty} \sum_m c_m^*(0)\,\psi_m^*(x) \sum_n c_n(t)\,\psi_n(x)e^{-i\frac{E_n t}{\hbar}}\,dx \\
\text{or } c(t) &= \sum_m \sum_n c_m^*(0)\,c_n(t)\,e^{-i\frac{E_n t}{\hbar}} \int_{-\infty}^{+\infty} \psi_m^*(x)\,\psi_n(x)dx \\
\text{or } c(t) &= \sum_m \sum_n c_m^*(0)\,c_n(t)\,e^{-i\frac{E_n t}{\hbar}}\,\delta_{mn}
\end{aligned}
$$

By making use of the orthonormalization condition of the bases, i.e., $\delta_{mn} = 1$ when $m = n$ and $\delta_{mn} = 0$ when $m \neq n$, we get the final form of the wavepacket correlation function as

$$
c(t) = \sum_n c_n^*(0)c_n(t)e^{-i\frac{E_n t}{\hbar}} = \sum_n c_n^*(0)c_n(t)e^{-i\omega_n t} \tag{6.15}
$$

6.3.2 *Fourier transform of wavepacket correlation function*

For the reason to be revealed soon, next, take the Fourier transform of the wavepacket correlation function:

$$
\sigma(\omega) = \int_{-\infty}^{+\infty} c(t)e^{i\omega t}dt
$$

Inserting equation (6.15) into the above expression, we get

$$
\begin{aligned}
\sigma(\omega) &= \int_{-\infty}^{+\infty} \sum_n c_n^*(0)c_n(t)e^{-i\omega_n t}\,e^{i\omega t}dt \\
\text{or } \sigma(\omega) &= \sum_n c_n^*(0)c_n(t) \int_{-\infty}^{+\infty} e^{-i\omega_n t}e^{i\omega t}dt
\end{aligned} \tag{6.16}
$$

Different Representations of the Fourier Transform

Note that in the above representation of the Fourier transform, we have used the following form as the forward Fourier transform:

$$E(\omega) = \int_{-\infty}^{+\infty} E(t) e^{i\omega t} dt$$

As a result, its inverse Fourier transform is given by $E(t) = \frac{1}{2\pi}\int_{-\infty}^{+\infty} E(\omega) e^{-i\omega t} d\omega$.

On the other hand, if the forward transform is taken to be

$$E(\omega) = \int_{-\infty}^{+\infty} E(t) e^{-i\omega t} dt$$

and the corresponding inverse transform takes the following form:

$$E(t) = \frac{1}{2\pi}\int_{-\infty}^{+\infty} E(\omega) e^{i\omega t} d\omega$$

Recall from Chapter 3 that the delta function is represented as $\delta(\omega - \omega') = \frac{1}{2\pi}\int_{-\infty}^{+\infty} e^{i\omega t} e^{-i\omega' t} dt$. In the present context, using the present notations, therefore, we may write

$$\int_{-\infty}^{+\infty} e^{i\omega t} e^{-i\omega_n t} dt = 2\pi\, \delta\left(\omega - \omega_n\right) \tag{6.17}$$

Thus, inserting equation (6.17) into equation (6.16), we obtain

$$\sigma(\omega) = 2\pi \sum_n c_n^*(0)\, c_n(t)\, \delta(\omega - \omega_n) \tag{6.18}$$

Here, ω_n defines the nth stationary state of the quantum system as $\omega_n = \frac{E_n}{\hbar}$. Therefore, the $\sigma(\omega)$ function has a value at every $\omega = \omega_n$. Consequently, $\sigma(\omega)$ provides the spectral information of the quantum system.

The above exercise demonstrates that the **Fourier transform of the wavepacket correlation function carries the spectral information of the quantum system.**

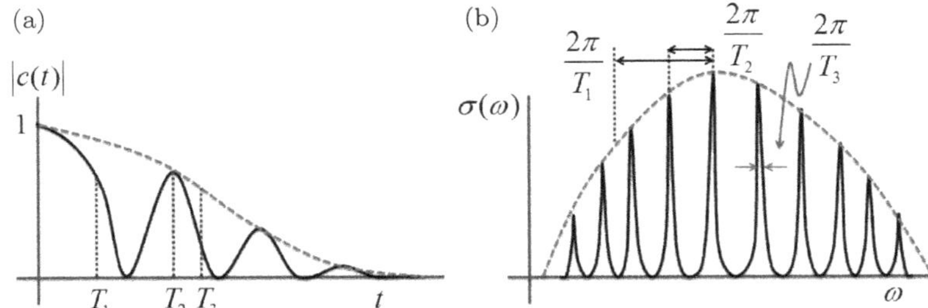

Figure 6.14. (a) Three characteristic timescales associated with the correlation function are schematically illustrated. (b) The same is depicted in the corresponding spectrum. See Ref. 12.

6.3.3 *Meaning of the wavepacket correlation function from time-domain perspective*

A typical example of a wavepacket correlation function $c(t)$ is given in Figure 6.14(a). At $t = 0$, $c(t) = \langle\psi(0)|\psi(0)\rangle = 1$, (as the initial wavepacket is taken to be normalized). Subsequent temporal evolution of $|c(t)|$ can be characterized using three timescales in the time domain:

(a) The first corresponds to the initial decay of $|c(t)|$ with a characteristic timescale, T_1. This decay occurs because the wavepacket moves away from its initial location. The integral $\langle\psi(0)|\psi(t)\rangle$ can be viewed as the extent of overlap between the initial and final wavepackets. Consequently, as the wavepacket moves away from its initial location in position space, its overlap with the initial wavepacket decays.

(b) After one period of motion, the wavepacket may return to its initial location in position space, rendering a recurrence of $|c(t)|$. This recurrence time is characterized by the timescale T_2. Figure 6.14(a) demonstrates, for example, three recurrences in $|c(t)|$.

(c) The recurrence in $|c(t)|$ may also decay with the third timescale T_3. The origin of this decay may be anharmonicity, coupling of the wavepacket motion to other degrees of freedom, irreversible decay due to dissociation (or predissociation), and radiative or nonradiative decay.

6.3.4 *Meaning of the wavepacket correlation function from frequency-domain perspective*

Figure 6.14(b) shows an example of the Fourier transform of $c(t)$. The Fourier transformed structure of $c(t)$ carries a very close correspondence with the time-domain structure of $c(t)$.

(a) The first decay in time-domain $|c(t)|$, which is characterized by the timescale T_1, appears as a width (a spectral width features a decay in the time-domain structure) of the spectral envelope, as given by $\frac{2\pi}{T_1}$. Note that the spectral width is inversely proportional to the corresponding decay time constant.
(b) A second decay in the time domain $|c(t)|$, characterized by the timescale T_3, appears as a width of each spectral feature in the frequency domain, as given by $\frac{2\pi}{T_3}$.
(c) Finally, recurrence in the time domain, characterized by the timescale T_2, corresponds to the spacing (which represents a recurrence of the spectral feature in the frequency domain) between adjacent spectral features, as given by $\frac{2\pi}{T_2}$.

6.3.5 *Fourier transform of the wavepacket*

Once we obtain the eigenstate energies (using the Fourier transform of the wavepacket correlation function), we may obtain the eigenstate wavefunction corresponding to each eigenstate energy using a hitherto unstated fact that **an eigenstate is nothing but a superposition of wavepackets**. We elaborate on this point in the following.

We have already understood that action of the time-evolution operator on the initial wavepacket renders the final wavepacket at any later time t:

$$\psi(x,t) = e^{-\frac{i\hat{H}t}{\hbar}}\psi(x,0) \tag{6.19}$$

Next, consider the following integral (physical meaning of this integral will be revealed soon):

$$\int_{-\infty}^{+\infty} \psi(x,t)e^{\frac{iE_n t}{\hbar}}dt$$

Inserting equation (6.19) into the above integral, we obtain

$$\int_{-\infty}^{+\infty} \psi(x,t)e^{\frac{iE_n t}{\hbar}}\,dt = \int_{-\infty}^{+\infty} e^{-\frac{i\hat{H}t}{\hbar}}\psi(0,t)e^{\frac{iE_n t}{\hbar}}\,dt$$

Next, expand the initial wavepacket $\psi(0,t)$ in the orthonormal basis (stationary eigenstates of $\hat{H}$) and rewrite the above integral as

$$\int_{-\infty}^{+\infty} \psi(x,t)e^{\frac{iE_n t}{\hbar}}\,dt = \int_{-\infty}^{+\infty} e^{-\frac{i\hat{H}t}{\hbar}}\psi(0,t)e^{\frac{iE_n t}{\hbar}}\,dt$$

$$= \textstyle\int_{-\infty}^{+\infty} e^{-\frac{i\hat{H}t}{\hbar}} \sum_m c_m(0)\psi_m(x)e^{\frac{iE_n t}{\hbar}}\,dt$$

$$\text{or} \int_{-\infty}^{+\infty} \psi(x,t)e^{\frac{iE_n t}{\hbar}}\,dt = \sum_m c_m(0)\int_{-\infty}^{+\infty} e^{-\frac{i\hat{H}t}{\hbar}}\psi_m(x)e^{\frac{iE_n t}{\hbar}}\,dt$$

As $e^{-\frac{i\hat{H}t}{\hbar}}\psi_m(x) = e^{-\frac{iE_m t}{\hbar}}\psi_m(x)$ (prove it yourself using the Taylor series expansion of the time-evolution operator), we may write

$$\int_{-\infty}^{+\infty} \psi(x,t)e^{\frac{iE_n t}{\hbar}}\,dt = \sum_m c_m(0)\int_{-\infty}^{+\infty} e^{-\frac{iE_m t}{\hbar}}\psi_m(x)e^{\frac{iE_n t}{\hbar}}\,dt$$

$$\text{or} \int_{-\infty}^{+\infty} \psi(x,t)e^{\frac{iE_n t}{\hbar}}\,dt = \sum_m c_m(0)\psi_m(x)\int_{-\infty}^{+\infty} e^{i\omega_n t}e^{-i\omega_m t}dt$$

or, using our familiar integral representation of the Dirac δ-function, we may write

$$\int_{-\infty}^{+\infty} \psi(x,t)e^{\frac{iE_n t}{\hbar}}\,dt = 2\pi\sum_m c_m(0)\,\delta(\omega_n-\omega_m)\psi_m(x) = 2\pi c_n(0)\,\psi_n(x)$$

$$\text{Therefore,} \int_{-\infty}^{+\infty} \psi(x,t)e^{\frac{iE_n t}{\hbar}}\,dt \propto \psi_n(x) \tag{6.20}$$

The above exercise provides a very useful perspective on the integral $\int_{-\infty}^{+\infty} \psi(x,t)e^{\frac{iE_n t}{\hbar}}\,dt$, which is the Fourier transform of the moving wavepacket at the eigenvalue energy E_n. This integral gives the eigenstate wavefunction ψ_n of $\hat{H}$ corresponding to the eigenvalue E_n.

6.3.6 *An illuminating computational example*

In PythonChapter D, we have gone over a numerical solution to the TISE of a one-dimensional quantum harmonic oscillator (which is also an analytically solvable problem). In atomic units ($\hbar = 1$) and considering $k = m = 1$ (defining the conditions of the problem), we have obtained the vibrational energies expressed by $\mathrm{E}_v = \left(v + \frac{1}{2}\right)$ a.u., where v represents the vibrational quantum number. As a result, we have obtained the energies of the three lowest-lying eigenstates of the quantum harmonic oscillator as 0.5, 1.5, and 2.5 a.u.

Following the time-domain approach, as discussed in earlier sections, one can make use of the wavepacket correlation function to explore the eigenstate energies of a one-dimensional quantum harmonic oscillator. In this approach, one has to *start with a Gaussian wavepacket centered at the minimum of the potential and gradually displace the wavepacket until the spectrum of the wavepacket contains the first three lowest-lying eigenstate energies.* For its Python implementation, one can use the split-operator approach (for wavepacket propagation, as demonstrated in PythonChapter E), compute the time correlation function (as defined in equation (6.12)), and perform the Fourier transform of the wavepacket correlation function (using the **fft** algorithm as implemented in **SciPy**) to determine the three lowest-lying eigenstate energies of the quantum harmonic oscillator.

Python implementation

```
#Import the libraries
from scipy import exp,sqrt,pi,arange,conj
from scipy.integrate import simps
from scipy.fftpack import fftfreq,fft,ifft
from matplotlib.pyplot import plot,xlim,ylim,show
#Create the x-grid
xmin=-100
xmax=+100.1
dx=0.02
x=arange(xmin,xmax,dx)
#Define and normalize the initial wavefunction
k0=0.0
x0=2.0
psi=exp((-(x-x0)**2)/2)*exp(1j*k0*x) #Gaussian wavepacket
```

```
prob_density=abs(psi)**2
Norm=sqrt(simps(prob_density,x))
psiNorm=psi/Norm
psiNorm_initial=psiNorm
prob_density_intial=abs(psiNorm_initial)**2
#Defining the harmonic potential
v=0.5*(x**2)
#Time propagation
N=len(x)
k=2*pi*fftfreq(N,dx) #Creating k-grid
KE_k=(0.5*k**2)
Nt=1500 #Total number of time steps
dt=0.1 #Time step-size in atomic unit
t_arr=[] #Empty array for final time
wcf_arr=[] #Empty array for wcf
wcf_arr.append(simps(abs(psiNorm_initial)**2,x))
t_arr.append(0)
for i in arange(Nt):
        #fft of the wavefunction to k-space
        psiNorm_k=fft(psiNorm)
        #Time evolution due to first kinetic part of the propagator
        psiNorm_k_KE1=exp(-1.0j*KE_k*dt/2.0)*psiNorm_k
        #ifft of wavefunction to x space
        psiNorm_k_KE1_x=ifft(psiNorm_k_KE1)
        #Time evolution due to PE part of propagator
        psiNorm_k_KE1_x_V=exp(-1.0j*v*dt)*psiNorm_k_KE1_x
        #fft of the wavefunction to k-space again
        psiNorm_k_KE1_x_V_k=fft(psiNorm_k_KE1_x_V)
        #Time evolution due to second kinetic part of the propagator
        psiNorm_k_KE1_x_V_k_KE2=exp(-1.0j*KE_k*dt/2.0)*psiNorm_k_KE1_x_V_k
        #ifft of wavefunction to x space
        psiNorm=ifft(psiNorm_k_KE1_x_V_k_KE2)
        #Defining wavepacket correlation function
        wcf=simps((conj(psiNorm_initial)*psiNorm),x)
        wcf_arr.append(wcf)
        t=(i+1)*dt
        t_arr.append(t)
#Fourier Transform the wcf
n=len(t_arr)
omega=2*pi*fftfreq(n,dt)
sigma=fft(wcf_arr)
#Plotting spectrum
plot(omega*(-1),abs(sigma))
xlim(0,5)
show()
```

> A Gaussian wavepacket centered at 2.0 atomic unit of length is considered.

> Harmonic potential is considered $V(x) = 0.5x^2$ (in atomic unit).

> An empty array (an array which does not have any element) can be created simply by Python's `arr=[]` construct. Later, elements can be inserted into the array using the `arr.append()` construct.

Execution of the above Python program renders the following graph:

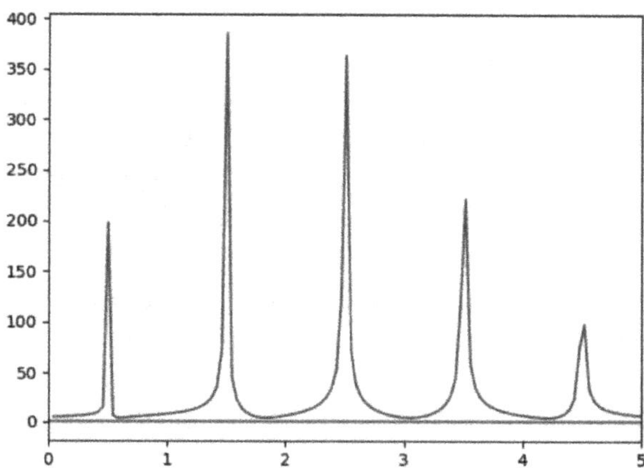

This graph displays the Fourier transformed spectrum of the wavepacket correlation function, with peaks at 0.5, 1.5, 2.5 a.u., and so on.

Self-Study

(a) Improve the formatting of the graphs for better presentation.
(b) Change the center of the initial wavepacket from `x0=2.0` to `x0=1.0`. Record your observation.
(c) Numerically find wavefunctions associated with the eigenvalues 0.5, 1.5, and 2.5 a.u. using equation (6.20).
(d) Plot the wavepacket correlation function as time progresses.

Additional Exercise

Make use of wavepacket dynamics, the time correlation function, and its Fourier transform to determine the three lowest-lying eigenstate energies and wavefunctions of diatomic species under the quantum anharmonic oscillator approximation. Assume that $\mu = k = 1$ a.u. Furthermore, consider a Morse potential $V(x) = D_e(1 - e^{-ax})^2$, where $D_e = 1.5\,\text{eV}$ and $a = 1\,\text{Å}^{-1}$ for the quantum anharmonic oscillator.

6.4 Time-Dependent Viewpoint of Quantum Scattering

The scattering phenomenon is the focus of many experimental and theoretical studies in physical chemistry. Specific examples include: (a) the scattering of molecules at the surface — relevant to heterogeneous catalysis; (b) cross-molecular beam experiments — relevant to chemical reactivity; (c) chemical reaction induced by collision — relevant to chemical dynamics, and (d) time delay measurements — relevant to attosecond phenomena and spectroscopy. Wavepacket dynamics plays an important role in exploring these phenomena. This is why we will go over the quantum scattering problem and its wavepacket dynamics perspective in adequate detail.[13]

6.4.1 *Bound and scattering states of a particle in one dimension*

A time-independent potential $V(x)$ can give rise to two physical states depending on the particle's total energy, E: bound and (unbound) scattering states, as shown in Figure 6.15. When the total energy of a particle is less than the potential $V(x)$ at infinity on both sides ($V(+\infty)$ and $V(-\infty)$), the particle remains stuck inside the potential feature, and its physical state is called the **bound state** (the harmonic oscillator potential creates a bound state, as depicted in Figure 6.15(a)). When the total energy of the particle exceeds the

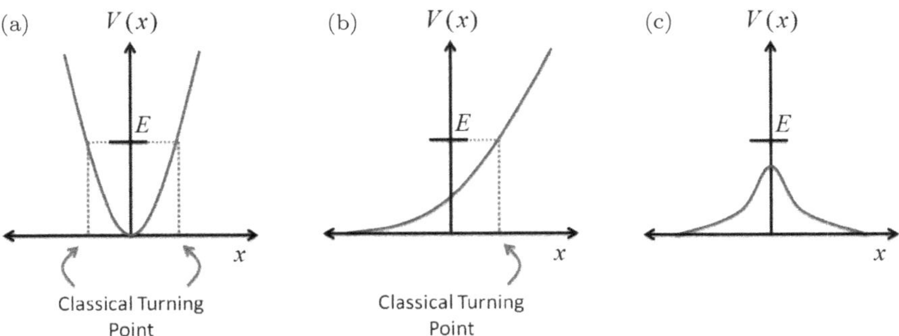

Figure 6.15. (a) Harmonic oscillator potential represents a bound state; frames (b) and (c) give typical examples of scattering states.

potential $V(x)$ at infinity on one side ($V(+\infty)$ or $V(-\infty)$) or both sides ($V(+\infty)$ and $V(-\infty)$), the particle can come from infinity, and it can slow down or speed up under the influence of the potential, and finally, it can return to infinity. This is called the particle's **scattering state or unbound state** (as schematically depicted in Figures 6.15(b) and 6.15(c)).

Elastic Scattering: When the total translational kinetic energy of the particle remains unchanged.

6.4.2 *One-dimensional elastic scattering of a particle*

Let us first consider the one-dimensional scattering of a free particle of mass m and energy $E(>0)$ by a step potential. As shown in Figure 6.16, in a step potential, the potential energy $V(x)$ remains constant (say V_0) and vanishes beyond a distance, say $x = 0$, which is mathematically expressed as

$$V(x) = 0 \text{ when } x \geq 0$$
$$V(x) = V_0 \text{ when } x < 0$$

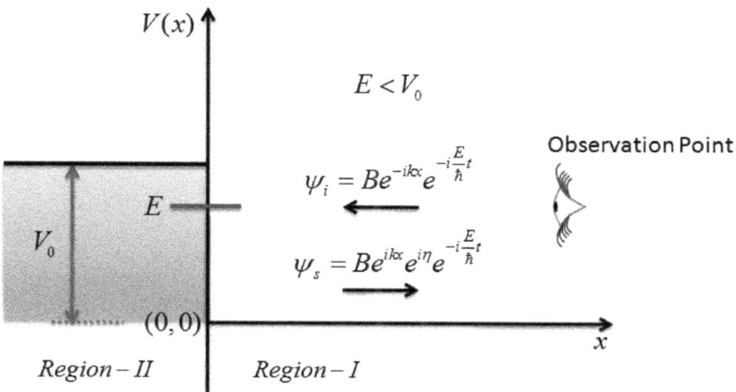

Figure 6.16. A free particle is scattered by a step potential. The observation point is selected to be quite away from the point of scattering. This is why we obtain a particular form of solution (called the asymptotic form giving rise to the stationary scattering state) for the given scattering problem.

We assume that the particle is incident from the right (coming from $x = +\infty$) on the potential with kinetic energy $E < V_0$.

Classical mechanical picture: Under the influence of a step potential, classically, if the particle moves from the right to the left with kinetic energy $E < V_0$, the particle will be fully reflected by the wall of the potential. Therefore, a classical particle always remains in $Region - I$ (this region is defined in Figure 6.16).

Quantum mechanical picture: The quantum mechanical consequence of the one-dimensional scattering is very intriguing. To realize the quantum mechanical consequence, we have to carefully inspect the two regions defined in Figure 6.16. The TISE can be written for both regions as follows:

$Region - I$: As $V(x) = 0$ for $x \geq 0$, we may write

$$-\frac{\hbar}{2m}\frac{d^2}{dx^2}\psi_I(x) = E\,\psi_I(x)$$

$Region - II$: As $V(x) = V_0$ for $x < 0$, we may write

$$\left[-\frac{\hbar}{2m}\frac{d^2}{dx^2} + V_0\right]\psi_{II}(x) = E\,\psi_{II}(x)$$

Both equations represent a free particle. Note that a particle which experiences a constant potential (space- and time-independent potential) is also considered a free particle because the particle does not experience any force due to the absence of the gradient of the potential. Therefore, the general solution for both regions can be written as follows (as understood in Chapter 3):

$$Region - I : \psi_I = Ae^{ik_Ix} + Be^{-ik_Ix}, \text{ where } k_I = \sqrt{\frac{2mE}{\hbar^2}}$$

$$Region - II : \psi_{II} = Ce^{ik_{II}x} + De^{-ik_{II}x}, \text{ where } K_{II} = \sqrt{\frac{2m(E - V_0)}{\hbar^2}}$$

As the incident particle comes from $x = +\infty$ (right), we may choose the coefficient $C = 0$. This is legitimate because, by the construction of the given problem, particle in $Region - II$ cannot move along the increasing x-direction. Both incident and reflected waves may contribute to *Region* $-$ *I*, but only the transmitted wave may contribute to the *Region* $-$ *II*. As $E < V_0$, classically, this particle cannot be found at $x < 0$ (in *Region* $-$ *II*); however, quantum mechanically, the particle may be found at $x < 0$ (in *Region* $-$ *II*). This is called **quantum mechanical tunneling/ penetration.**

Therefore, the general solution to the given problem should be written as follows:

$$Region - I : \psi_I = Ae^{ik_I x} + Be^{-ik_I x}, \text{ where } k_I = \sqrt{\frac{2mE}{\hbar^2}}$$

$$Region - II : \psi_{II} = De^{-ik_{II} x}, \text{ where } k_{II} = \sqrt{\frac{2m(E - V_0)}{\hbar^2}}$$

Here, as $E < V_0$, interestingly, k_{II} is a complex (imaginary) quantity, which can be re-expressed as follows:

$$k_{II} = \sqrt{\frac{2m(E - V_0)}{\hbar^2}} = \sqrt{-\frac{2m(V_0 - E)}{\hbar^2}} = i\sqrt{\frac{2m(V_0 - E)}{\hbar^2}} = ik'_{II},$$

$$\text{where } k'_{II} = \sqrt{\frac{2m(V_0 - E)}{\hbar^2}} \text{ is a real quantity} \tag{6.21}$$

This leads us to rewrite the wavefunction for *Region* $-$ *II* as

$$\Psi_{II} = De^{-i(ik'_{II}x)} = De^{k'_{II}x}$$

As x is negative in *Region* $-$ *II*, $De^{k'_{II}x}$ represents a **decreasing exponential.** k'_{II} determines how rapidly the wavefunction in the classically forbidden *Region* $-$ *II* decays to zero. $\frac{1}{k'_{II}}$, for which the amplitude of the wavefunction decreases to $\frac{1}{e}$ of its value at the boundary, represents a characteristic distance, which is called the **penetration depth**. From the relation $k'_{II} = \sqrt{\frac{2m(V_0-E)}{\hbar^2}}$, note that the penetration depth decreases with an increased mass of the particle and the height of the barrier above the energy of the incident particle $(V_0 - E)$.

Now, the continuity of the wavefunction, $\psi(x)$, at $x = 0$ demands that

$$\psi_I(0) = \psi_{II}(0)$$
$$\text{or } A + B = D \tag{6.22}$$

In addition, the continuity of the space derivative of the wavefunction, $\frac{d\psi(x)}{dx}$, at $x = 0$ demands that

$$\left[\frac{d\psi_I(x)}{dx}\right]_{x=0} = \left[\frac{d\psi_{II}(x)}{dx}\right]_{x=0}$$
$$\text{or } Aik_I - Bik_I = Dk'_{II} \tag{6.23}$$

Inserting equation (6.22) into equation (6.23), we get

$$\frac{A}{B} = \frac{(ik_I + k'_{II})}{(ik_I - k'_{II})} \tag{6.24}$$

The above ratio gives the probability of the particle arriving from $x = +\infty$ (right) to turn back at $x = 0$ (called **reflection**). The reflection coefficient is given by

$$R = \left|\frac{A}{B}\right|^2 = \frac{|ik_I + k'_{II}|^2}{|ik_I - k'_{II}|^2} = \frac{k_I^2 + k'^2_{II}}{k_I^2 + k'^2_{II}} = 1$$

This shows that the particle is totally reflected (just like a classical particle). However, a significant difference exists between classical and quantum scattering. Due to the existence of a decaying wave in *Region* $-$ *II* (called **evanescent wave**), the particle has a non-zero probability of its presence in *Region* $-$ *II* (classically forbidden region). This quantum mechanical effect results in a certain phase shift of the wavefunction upon reflection. This phase shift occurs due to the fact that the particle is delayed when it penetrates into *Region* $-$ *II*.

As any algebraic form of a complex quantity can be represented by its polar form, for mathematical convenience, we can express the relation $A = B\frac{(ik_I + k'_{II})}{(ik_I - k'_{II})}$ given in equation (6.24) in a complex exponential form: $A = B\,e^{i\eta}$. Here, although B is real, A is always complex. Thus, mathematically, a more convenient general solution for *Region* $-$ *I* should be given as follows:

$$\psi_I = Be^{-ik_I x} + Be^{i\eta}\, e^{ik_I x}$$

Here, the first term represents the incident wave, and the second term represents the reflected wave (which contains an additional phase factor $e^{i\eta}$, where η represents the **phase shift** due to scattering, as schematically depicted in Figure 6.17). One can easily find out the unknown phase shift η as follows.

From equation (6.24), we get $\frac{A}{B} = \frac{Be^{i\eta}}{B} = e^{i\eta} = \frac{(ik_I + k'_{II})}{(ik_I - k'_{II})}$. This can be further reduced to

$$\begin{aligned} e^{i\eta} &= \frac{(ik_I + k'_{II})}{(ik_I - k'_{II})} = \frac{(ik_I + k'_{II})(ik_I + k'_{II})}{(ik_I - k'_{II})(ik_I + k'_{II})} \\ &= \frac{-k_I^2 + 2ik_I k'_{II} + k'^2_{II}}{-(k_I^2 + k'^2_{II})} = \frac{(k_I^2 - k'^2_{II})}{(k_I^2 + k'^2_{\mathrm{II}})} - i\frac{2k_I k'_{II}}{(k_I^2 + k'^2_{II})} \end{aligned}$$

$$\text{or} \quad \cos\eta + i\sin\eta = \frac{(k_I^2 - k'^2_{II})}{(k_I^2 + k'^2_{II})} - i\frac{2k_I k'_{II}}{(k_I^2 + k'^2_{II})}$$

We can now equate the real parts of the above equation and obtain

$$\cos\eta = \frac{(k_I^2 - k'^2_{II})}{(k_I^2 + k'^2_{II})}$$

$$\text{or} \quad \frac{1}{1 + \cos\eta} = \frac{(k_I^2 + k'^2_{II})}{2k_I^2}$$

Using the trigonometric identity $1 + \cos 2\theta = 2\cos^2\theta$, we can write

$$\cos\left(\frac{\eta}{2}\right) = \frac{k_I}{\sqrt{k_I^2 + k'^2_{II}}} \qquad (6.25)$$

For further mathematical simplification, we may rewrite equation (6.21) as follows:

$$k'_{II} = \sqrt{\frac{2m\,(V_0 - E)}{\hbar^2}} = \sqrt{\frac{2mV_0}{\hbar^2} - \frac{2mE}{\hbar^2}} = \sqrt{\frac{2mV_0}{\hbar^2} - k_I^2} = \sqrt{K_0^2 - k_I^2} \qquad (6.26)$$

where we define a constant, $K_0^2 = \frac{2mV_0}{\hbar^2}$. We may now insert equation (6.26) into equation (6.25) to obtain

$$\cos\left(\frac{\eta}{2}\right) = \frac{k_I}{\sqrt{k_I^2 + K_0^2 - k_I^2}} = \frac{k_I}{K_0}$$

$$\text{or } \eta = 2\cos^{-1}\left(\frac{k_I}{K_0}\right) \tag{6.27}$$

So, finally, the phase factor is given by: $e^{i\eta} = e^{2i\cos^{-1}\left(\frac{k_I}{K_0}\right)}$.

When $\eta = 0$, i.e., when the particle experiences no phase shift, we can write

$$\cos(0) = 1 = \frac{k_I}{K_0}$$

$$\text{or } k_I = K_0$$

$$\text{or } \sqrt{\frac{2mE}{\hbar^2}} = \sqrt{\frac{2mV_0}{\hbar^2}}$$

$$\text{or } E = V_0$$

Therefore, when the particle's kinetic energy E is equal to V_0, it experiences no phase shift. For kinetic energy less than V_0, $\frac{k_I}{K_0} < 1$, the particle experiences a phase shift after reflection (see Figure 6.17).

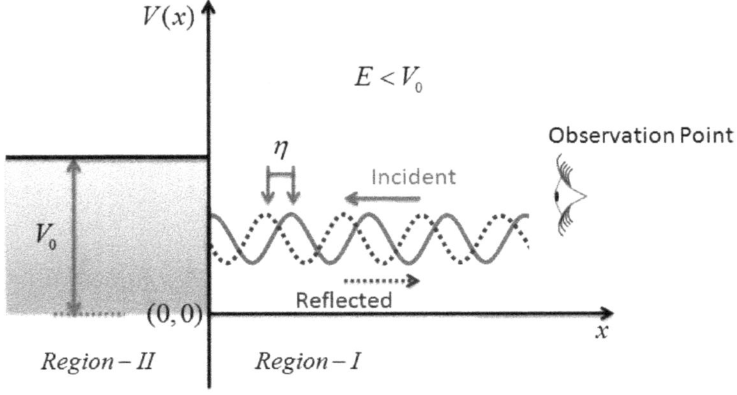

Figure 6.17. Phase shift due to scattering is schematically depicted.

Here carefully note that so far, the entire derivation of the phase shift due to elastic scattering has been developed based on the **plane wave solution** of a free particle experiencing scattering at the step potential.

6.4.3 *One-dimensional elastic scattering of a wavepacket*

Since the plane wave form of the wavefunction of a free particle is not square-integrable (normalizable), such a wave function doesn't represent a physically realizable state of the particle. Only a wavepacket which can be formed by linearly superimposing plane wave forms represents a physically realizable state of a particle. This is why we must reevaluate the above-mentioned one-dimensional elastic scattering problem using a wavepacket. This reevaluation will unravel an important concept, called the **time delay in scattering.**

Note that we are considering a one-dimensional scattering of a particle of mass m and energy E from a step potential as defined by

$$V(x) = 0 \text{ when } x \geq 0$$

$$V(x) = V_0 \text{ when } x < 0$$

Again, assume that the particle is incident from the right on the potential. The plane wave form of the **incoming** free particle is given by (a time-dependent form is adopted to create the wavepacket)

$$\psi_i = Ae^{-ikx}e^{-i\frac{Et}{\hbar}}$$

The particle has a linear momentum of $p = -\hbar k$, where k is the magnitude of the wave vector of the free-particle wave and $E = \frac{k^2\hbar^2}{2m}$. The incoming wave is reflected back toward the increasing x-direction. As a result, the scattered wave can be expressed as (in the time-dependent form)

$$\psi_s = Ae^{ikx}e^{i\eta}e^{-i\frac{Et}{\hbar}}$$

We have already understood that elastic scattering leads to the entire reflection of the incoming wave, which adopts a phase shift η, preserving the probability amplitude (A) and total kinetic energy (E). Thus, the total wavefunction associated with the present scattering event can be written as (for $Region - I$)

$$\begin{aligned}\psi(x,t) = \psi_i + \psi_s &= A\left[e^{-ikx}e^{-i\frac{Et}{\hbar}} + e^{ikx}e^{i\eta}e^{-i\frac{Et}{\hbar}}\right]\\ &= A\left[e^{-ikx}e^{-i\omega t} + e^{ikx}e^{i\eta}e^{-i\omega t}\right]\end{aligned}$$

where $\omega = \frac{E}{\hbar} = \frac{k^2\hbar}{2m}$.

Note that the above expression still describes the particle (both incident and scattered) using respective plane wave forms. To obtain a physically realizable state of the particle, we must construct a wavepacket choosing any value of k from 0 to K_0. Each k represents one plane wave form, and all the constituent plane waves forming the wave packet undergo reflection. Thus, the total wavepacket associated with the given scattering problem should be written as

$$\psi_{\text{w.p.}}(x,t) = \int_0^{K_0} A(k)e^{-ikx}e^{-i\frac{Et}{\hbar}}dk + \int_0^{K_0} A(k)e^{ikx}e^{\eta(k)}e^{-i\frac{Et}{\hbar}}dk$$

The above equation is too complicated. Therefore, we consider a much simpler form without losing important information.

6.4.4 *The time delay*

It is often sufficient to consider a simple wavepacket composed of only two plane waves to develop a physically transparent understanding of the behavior of a wavepacket. For the given scattering problem, we consider two plane wave forms (namely 1 and 2) with slightly different angular frequencies, $\omega_0 \pm \Delta\omega$, wavenumbers $k_0 \pm \Delta k$, and phase shifts $\eta_0 \pm \Delta\eta$. As we have two incident waves, we also get two scattered waves. Consequently, the total wavefunction (associated with this scattering event) can be written as

$$\begin{aligned}\psi_{\text{w.p.}}(x,t) = \psi_1 + \psi_2 = A&\left[e^{-ik_1x}e^{-i\omega_1 t} + e^{ik_1x}e^{i\eta_1}e^{-i\omega_1 t}\right]\\ &+A\left[e^{-ik_2x}e^{-i\omega_2 t} + e^{ik_2x}e^{i\eta_2}e^{-i\omega_2 t}\right]\end{aligned}$$

where

$$k_1 = k_0 + \Delta k,\ k_2 = k_0 - \Delta k$$
$$\omega_1 = \omega_0 + \Delta\omega,\ \omega_2 = \omega_0 - \Delta\omega$$
$$\eta_1 = \eta_0 + \Delta\eta,\ \eta_2 = \eta_0 - \Delta\eta$$

k_0, ω_0, η_0 are the respective average quantities which appear after interference of two particle waves, where $k_0 = \frac{k_1+k_2}{2}, \omega_0 = \frac{\omega_1+\omega_2}{2}$, and $\eta_0 = \frac{\eta_1+\eta_2}{2}$. Furthermore, $\Delta k = \frac{k_1-k_2}{2}$, $\Delta\omega = \frac{\omega_1-\omega_2}{2}$ and $\Delta\eta = \frac{\eta_1-\eta_2}{2}$.

Thus, inserting these relationships, we get

$$\begin{aligned}\psi_{\text{w.p.}}(x,t) &\\ = A\Big[&e^{-i(k_0+\Delta k)x}e^{-i(\omega_0+\Delta\omega)t} + e^{i(k_0+\Delta k)x}e^{i(\eta_0+\Delta\eta)}e^{-i(\omega_0+\Delta\omega)t}\\ &+ e^{-i(k_0-\Delta k)x}e^{-i(\omega_0-\Delta\omega)t} + e^{i(k_0-\Delta k)x}e^{i(\eta_0-\Delta\eta)}e^{-i(\omega_0-\Delta\omega)t}\Big]\end{aligned}$$

$$\text{or } \psi_{\text{w.p.}}(x,t) = 2A\Big[e^{-i(k_0x+\omega_0t)}\cos(\Delta kx + \Delta\omega t) + e^{i(k_0x-\omega_0t)}e^{i\eta_0}\cos(\Delta kx - \Delta\omega t + \Delta\eta)\Big] \tag{6.28}$$

The first term in the above equation represents the incoming wavepacket, and the second term represents the asymptotic form of the outgoing wavepacket (after scattering). As we are interested only in the form of the wavepacket after scattering, we focus on the following expression:

$$\psi^s_{\text{w.p.}}(x,t) \sim e^{i(k_0x-\omega_0t)}e^{i\eta_0}\cos(\Delta kx - \Delta\omega t + \Delta\eta)$$

This is our familiar wavepacket expression, in which two components can be easily distinguished: fast-varying $e^{i(k_0x-\omega_0t)}e^{i\eta_0}$, which is responsible for the phase velocity of the wavepacket, and slowly-varying $\cos(\Delta kx - \Delta\omega t + \Delta\eta)$, which is responsible for the group velocity of the wavepacket. The term $\cos(\Delta kx - \Delta\omega t + \Delta\eta)$ is also called the envelope of the wavepacket.

The total phase of the envelope is given by $\varphi = \Delta kx - \Delta\omega t + \Delta\eta$ or, for an infinitesimal difference,

$$\varphi = dk\, x - d\omega t + d\eta$$

The envelope function has a maximum when $\varphi = dk\, x - d\omega t + d\eta = 0$ (because $\cos(0) = 1$, maximum), or, in other words,

$$x_{\max}(\text{in presence of scattering potential}) = \frac{d\omega}{dk}t - \frac{d\eta}{dk} = v_g t - \frac{d\eta}{dk}$$

(Refer to Chapter 3 for the definition of group velocity, v_g)

On the other hand, if we assume that there was no scattering potential, i.e., $\eta = 0$ or $\frac{d\eta}{dk} = 0$, we may write

$$x_{\max}(\text{without scattering potential}) = \frac{d\omega}{dk}t = v_g t$$

Thus, we get

$$\begin{aligned} &x_{\max}(\text{with scattering potential}) \\ &\quad = v_g t - \frac{d\eta}{dk} = v_g\left(t - \frac{1}{v_g}\frac{d\eta}{dk}\right) = v_g(t - \Delta t) \end{aligned}$$

Physical dimensional analysis of the above expression simply confirms that the term $\frac{1}{v_g}\frac{d\eta}{dk}$ represents a dimension of time. In fact, the scattering potential delays the particle after scattering by

$$\Delta t = \frac{1}{v_g}\frac{d\eta}{dk} = \frac{1}{v_g}\frac{d\eta}{dE}\frac{dE}{dk}$$

As $E = \frac{k^2\hbar^2}{2m}$ and $v_g = \frac{\hbar k}{m}$, one can express the time delay as

$$\Delta t = \hbar\frac{d\eta}{dE} \tag{6.29}$$

Wigner identified this delay in the context of the scattering of a wavepacket by a scattering potential.[14] This is why this delay is often called the Wigner delay. The Wigner delay is essentially the difference between the time the particle spends within a distance x in a scattering event under a scattering potential and the time the particle would have spent there in the absence of the scattering potential. It is obvious now that the time delay Δt of a scattering event can be calculated if the phase shift η associated with the scattering event is known.

Guiding Question

For one-dimensional elastic scattering by a step potential (considering the plane wave solution), we have already realized that

$$\cos\left(\frac{\eta}{2}\right) = \frac{k_I}{K_0}$$

Find an expression of the Wigner delay Δt for this scattering event.

6.4.5 *Attosecond time delay in photoionization*

Einstein's photoelectric or photoemission effect is one of the fundamental processes of light–matter interaction. From an atomic or a molecular perspective, it is characterized by the emission of an electron from an atom or a molecule upon absorption of a single (ionizing) photon by the atom or the molecule. For a long time, this emission event was assumed to be an instantaneous process: A free electron was assumed to be created instantly in response to the interaction of the atom or molecule with the ionizing radiation. However, recent attosecond spectroscopic experiments, using either streaking[15] or RABBITT,[16] have unraveled that the photoemission event is not an instantaneous process. The photoemission process is a delayed event: a very small but measurable time delay exists for the liberated electron to be completely free from the atom or molecule following the ionization event. Usually, by relating the photoemission process to a half-scattering event, the time delay involved in a photoemission can be calculated using the expression of the Wigner delay, which is defined by *the energy derivative of the phase of the liberated electron wavepacket* (refer to equation (6.29)).[17]

Time-domain wavepacket dynamics perspective: In time-domain, the Wigner delay is nothing but the difference between the time a particle spends within a distance x in a scattering event under the influence of a scattering potential and the time that a particle would have spent there in the absence of the scattering potential. Therefore, from a time-domain wavepacket dynamics perspective,

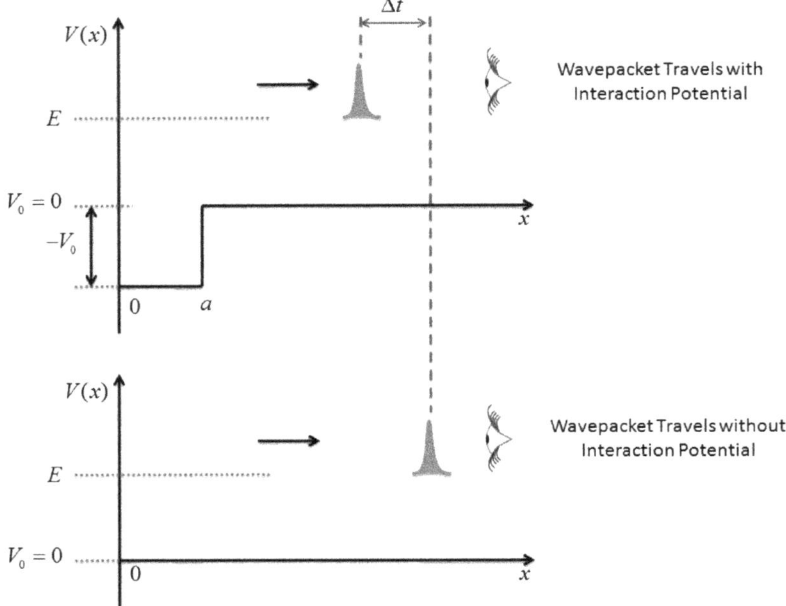

Figure 6.18. Wigner delay is explained from a time-domain wavepacket dynamics perspective.

it is quite clear that the time delay Δt in the photoemission process can also be calculated by comparing the wavepacket dynamics under a scattering potential and under a zero-interaction potential, as schematically depicted in Figure 6.18.

Guiding Question

From a scattering point of view, as stated already, the photoemission process can be considered as a half-scattering event. To obtain the simplest picture, let us consider one-dimensional (in the region $0 \leq x \leq \infty$) elastic scattering (i.e., the total translational kinetic energy of the particle remains unchanged after the scattering event) for a particle of mass m and energy $E(>0)$. The particle is scattered by a localized potential $V(x)$, defined by

$$V(x) = \begin{Bmatrix} 0 & (\text{when } x > a) \\ -V_0 & (\text{when } 0 \leq x \leq a) \\ \infty & (\text{when } x < 0) \end{Bmatrix}$$

(*Continued*)

(Continued)

The particle is assumed to be incident from the right on the potential (as depicted in the following).

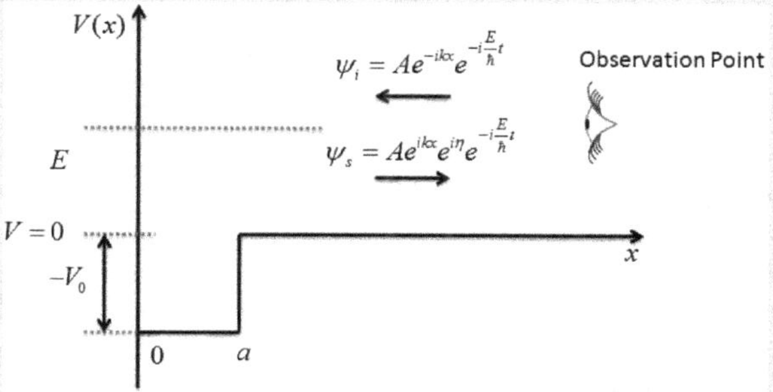

Find the Wigner delay experienced by an electron for a half-scattering event, considering the potential given above, using both (a) **energy derivative** and (b) **direct wavepacket dynamics** methods.

Notes, References, and Further Reading

1. The Born–Oppenheimer approximation has not been introduced in this textbook; however, one can go over this approximation in adequate detail in a lovely quantum chemistry textbook: L. Piela, Chapter 6: Separation of electronic and nuclear motions, *Ideas of Quantum Chemistry*, 2nd edn. Elsevier, Amsterdam (2007). Reference to the original work is M. Born and J. R. Oppenheimer, On the quantum theory of molecules. *Ann. Phys.* (Leipzig), 84, 457 (1927).
2. The breakdown of the Born–Oppenheimer approximation and its consequences on chemical reactivity are discussed in adequate detail in (a) G. A. Worth and L. S. Cederbaum, Beyond Born-Oppenheimer: Molecular dynamics through a conical intersection. *Annu. Rev. Phys. Chem.* 55, 127 (2004); (b) R. D. Levine, Chapter 7: Photoselective chemistry: Access to the transition state region and Chapter 9: State-changing collisions: Molecular energy transfer. *Molecular Reaction Dynamics*. Cambridge University Press, Cambridge, UK (2005). (c) W. Domcke, D. R. Yarkony and H. Koppel, *Conical Intersections: Electronic Structure, Dynamics and Spectroscopy*. World Scientific, New Jersey (2003). (d) I. Schapiro, F. Melaccio, E. N. Laricheva and M. Olivucci, *Photochem. Photobiol. Sci.* 10, 867 (2011).
3. A. Bhattacharya, Chapter 11: Ultrafast physical chemistry: Part I: Molecular photophysics and photochemistry. *Ultrafast Optics and Spectroscopy in Physical Chemistry*. World Scientific, Singapore (2018).

4. J. S. Baskin and A. H. Zewail, Freezing atoms in motion: Principles of femtochemistry and demonstration of laser stroboscopy. *J. Chem. Edu.* 78, 737–751 (2001).
5. G. Wald, Molecular basis of visual excitation. *Science*, 162, 230–239 (1968).
6. L. A. Peteanu, R. W. Schoenlein, Q. Wang, R. A. Mathies and C. V. Shank, The first step in vision occurs in femtoseconds: Complete blue and red spectral studies. *Proc. Natl. Acad. Sci. USA*, 90, 11762 (1993).
7. (a) H. Kandori, Y. Shichida and T. Yoshizawa, Absolute absorption spectra of batho- and photorhodopsins at room temperature. Picosecond laser photolysis of rhodopsin in polyacrylamide. *Biophys. J.* 56, 453–457 (1989); (b) H. Kandori, Y. Shichida and T. Yoshizawa, Photoisomerization in rhodopsin. *Biochemistry* (Moscow), 66, 1197–1209.
8. For a brief introduction to commercially available computer programs (including Gaussian or Molpro) which can calculate molecular properties with adequate accuracy, readers are referred to D. A. McQuarrie, Chapter 12: The Hartree-Fock-Roothaan method. *Quantum Chemistry*, 2nd edn. University Science Books (California) (2008).
9. Analytical fits are given in P. Schwendner, F. Seyl and R. Schinke, Photodissociation of Ar_2^+ in strong laser fields. *Chem. Phys.* 217, 233–247 (1997). Calculated potential energy curves are given in F. Y. Naumkin, P. J. Knowles and J. N. Murrell, Towards reliable modelling of large clusters: On the overall accuracy of the diatomics-in-molecule method for rare gas cluster ions. *Chem. Phys.* 193, 27–36 (1995).
10. (a) T. S. Rose, M. J. Rosker and A. H. Zewail, Femtosecond real-time probing of reactions. IV. The reactions of alkali halides. *J. Chern. Phys.* 91, 7415–7436 (1989); (b) V. Engel, H. Methu, R. Almeida, R. A. Marcus and A. H. Zewail, Molecular state evolution after excitation with an ultrafast laser pulse: A quantum analysis of NaI and NaBr dissociation. *Chem. Phys. Lett.* 152, 1–7 (1988); (c) V. Engel and H. Metiu, A quantum mechanical study of predissociation dynamics of NaI excited by a femtosecond laser pulse. *J. Chem. Phys.* 90, 6116–6128 (1989).
11. C. Cohen-Tannoudji, B. Diu and F. Laloe, Complements of Chapter IV: C. General study of two-level systems, Complements of Chapter I: J_I: Behaviour of a wavepacket at a potential step. *Quantum Mechanics*, Vol. 1 (English Translation). Hermann and John Wiley and Sons, New York (1977).
12. For further details, see (a) D. J. Tannor, Chapter 6: Correlation function and spectra. *Introduction to Quantum Mechanics: A Time-Dependent Approach.* University Science Books, Sausalito (CA) (2007); (b) E. J. Heller, The semiclassical way to molecular spectroscopy. *Acc. Chem. Res.* 14, 368 (1981).
13. For further details, see (a) D. J. Tannor, Chapter 7: One-dimensional barrier scattering. *Introduction to Quantum Mechanics: A Time-Dependent Approach.* University Science Books, Sausalito (CA) (2007); (b) G. C. Schatz and M. A. Ratner, Chapter 7: Quantum scattering theory. *Quantum Mechanics in Chemistry.* Dover Publications, Inc., New York (2015); (c) D. J. Griffiths, *Introduction to Quantum Mechanics*, 2nd edn. Pearson Education

Inc., New Jersey (2005); (d) P. Atkins and R. Friedman, *Molecular Quantum Mechanics*, 5th edn. Oxford University Press, New York (2011); (e) C. Cohen-Tannoudji, B. Diu and F. Laloe, Complements of Chapter I: J_I: Behavior of a wavepacket at a potential step. *Quantum Mechanics*, Vol. 1 (English Translation). Hermann and John Wiley and Sons, New York (1977).

14. (a) F. T. Smith, Lifetime matrix in collision theory. *Phys. Rev.* 118, 349 (1960); (b) E. P. Wigner, Lower limit for the energy derivative of the scattering phase shift. *Phys. Rev.* 98, 145 (1955).
15. M. Schultze *et al.* Delay in photoemission. *Science*, 328, 1658–1662 (2010).
16. K. Klünder *et al.* Probing single-photon ionization on the attosecond time scale. *Phys. Rev. Lett.* 106, 143002 (2011).
17. See recent combined experimental and theoretical works, for example, (a) M. Huppert, I. Jordan, D. Baykusheva, A. von Conta and A. H. J. Wörner, Attosecond delays in molecular photoionization. *Phys. Rev. Lett.* 117, 093001 (2016); (b) J. Vos *et al.* Orientation-dependent stereo Wigner time delay and electron localization in a small molecule. *Science*, 360, 1326–1330 (2018); (c) Biswas *et al.* Probing molecular environment through photoemission delays. *Nat. Phys.* 16, 778–783 (2020).

Appendix I

Multiple Choice Questions

1. Which of the following statements is true?
   a. Variable separation method can be used to solve the TDSE.
   b. Variable separation method can be used to solve the TDSE if the potential is time-dependent.
   c. Variable separation method can be used to solve the TDSE if the kinetic energy is time-independent.
   d. Variable separation method can be used to solve the TDSE if the potential energy is time-independent.

Ans: d

2. Let us assume that the energy difference between two stationary states is 11.84 eV. If we create a superposition state using these two stationary states, what would be the time period of oscillation of the probability density of the superposition state?
   a. 348 nanosecond
   b. 348 femtosecond
   c. 348 attosecond
   d. 348 microsecond

Ans: c

3. A quantum state is a non-stationary state only when
   a. the wavefunction of the state is time-dependent.
   b. the probability density of the state is time-dependent.

c. the wavefunction has a time-dependent complex phase factor.
d. the wavefunction of the state is time-dependent but the probability density of the state is time-independent.

Ans: b

4. An optical pulse travels with

a. group velocity.
b. phase velocity.
c. classical velocity.
d. both group and phase velocities.

Ans: d

5. Carefully go over the following Python program and pick the correct answer from the options that follow:

```
a=input("Enter Value of a")
b=a*2
Print(b)
```

a. The above program is executable.
b. The above program is wrong because "b" cannot be determined as "a" is a string.
c. The above program is wrong because a capital "P" is used for the print() functionality.
d. The above program is wrong because a capital "P" is used for the print() functionality and "b" cannot be determined as "a" is a string.

Ans: d

6. What is the nature of the ground state vibrational wavefunction (under a simple harmonic oscillator approximation) of a diatomic species?

a. Gaussian
b. exponential
c. sine
d. cosine

Ans. a

7. If the separation between two vibrational states is 100 meV, what is the typical oscillation timescale for the probability density of a superposition state created by those two states?

   a. 100 nanosecond
   b. 150 attoseconds
   c. 20 femtosecond
   d. 2 attosecond

Ans. c

8. Select the correct list which will be created by the arange(10,50,10) functionality of scipy:

   a. [10 20 30 40]
   b. [10 20 30 40 50]
   c. [20 30 40]
   d. [20 30 40 50]

Ans. a

9. What is the function of the plot() functionality of matplotlib.pyplot? Select the most appropriate one from the following options.

   a. It plots a graph and displays it.
   b. It plots a graph.
   c. It displays a graph.
   d. It plots but does not display a graph.

Ans: d

10. If for a certain one-dimensional particle of mass m, the wavefunction is given by $\psi(x,t) = Ae^{-ikx}e^{-\frac{iEt}{\hbar}}$, where $k = \sqrt{\frac{2mE}{\hbar^2}}$ and E is the total energy of the particle, find the potential energy for which the above wavefunction satisfies the TDSE.

    a. V = cos(x)
    b. V = 0
    d. V = −bx
    e. V = sin(x)

Ans. b

11. Which of the following statements is true?

    a. The wavefunction is an experimentally realizable quantity.
    b. The wavefunction is an experimentally realizable quantity, but with a certain degree of uncertainty, leading to the uncertainty principle.
    c. The wavefunction is not an experimentally realizable quantity, and only the probability density can be experimentally realized.
    d. Both the wavefunction and probability density are experimentally realizable quantities.

Ans. c

12. For a normalized wavefunction, which of the following options is true?

    a. $\frac{d}{dt}\left[\int_{-\infty}^{+\infty} |\psi(x,t)|^2 dx\right] = 0$
    b. $\int_{-\infty}^{+\infty} |\psi(x,t)|^2 dx = +\infty$
    c. $\int_{-\infty}^{+\infty} |\psi(x,t)|^2 dx = -\infty$
    d. $\int_{-\infty}^{+\infty} |\psi(x,t)|^2 dx = 0$

Ans. a

13. If an acceptable wavefunction in quantum mechanics is not normalized, which of the following options is true?

    a. $\frac{d}{dt}\left[\int_{-\infty}^{+\infty} |\psi(x,t)|^2 dx\right] = 0$
    b. $\int_{-\infty}^{+\infty} |\psi(x,t)|^2 dx = +\infty$
    c. $\int_{-\infty}^{+\infty} |\psi(x,t)|^2 dx = -\infty$
    d. $\int_{-\infty}^{+\infty} |\psi(x,t)|^2 dx = 0$

Ans. a

14. The expectation value of $3e^{-(x-2)^2}$ is

a. 0
b. 3
c. ∞
d. 2

Ans. d

15. For a superposition state which is expressed by the generic wavefunction $\psi(x,t) = A(x,t)e^{\frac{i}{\hbar}S(x,t)}$, the Bohmian velocity is given by

a. $\dfrac{\partial A(x,t)}{\partial x}$

b. $\dfrac{\partial S(x,t)}{\partial x}$

c. $\dfrac{\partial S(x,t)}{\partial t}$

d. $\dfrac{1}{m}\dfrac{\partial S(x,t)}{\partial x}$

Ans. d

16. A free particle expressed by the plane wave $\psi(x) = Ae^{ikx}$ exhibits a momentum of

a. $-\hbar k$

b. $\dfrac{\hbar k}{2m}$

c. $\hbar k$
d. k

Ans. c

17. If a wavepacket is created by the superposition of two plane waves, as expressed by $\psi(x,t) = Ae^{ik_1x}e^{-\frac{iE_1t}{\hbar}} + Ae^{ik_2x}e^{-\frac{iE_2t}{\hbar}}$, the phase velocity and group velocity of the wavepacket are given, respectively, by

    a. $\frac{\hbar k_0}{2m}$ and $\frac{\hbar k_0}{m}$

    b. $\frac{\hbar k_0}{m}$ and $\frac{\hbar k_0}{2m}$

    c. $\hbar k_0$ and $\frac{\hbar k_0}{2}$

    d. $\hbar k_0$ and $\frac{E_0}{2}$

Ans. a

18. The normalization constant of the $\psi(x,t) = e^{-ax^2}$ function is

    a. $\left(\frac{2a}{\pi}\right)^{1/4}$

    b. $\left(\frac{2a}{\pi}\right)^{1/2}$

    c. $\left(\frac{\pi}{2a}\right)^{1/4}$

    d. $\left(\frac{\pi}{2a}\right)^{1/2}$

Ans. a

19. Numerical integration in a one-dimensional space can be performed using Simpson's rule. Which functionality of scipy can be used to perform the integration?

    a. simps(X,Y) #X and Y are arrays.
    b. simps(Y,X) #X and Y are arrays.
    c. simps(X,Y) #X and Y are two numbers.
    d. simps(Y,X) #X and Y are two numbers.

Ans: b

20. Which of the following is returned by the abs(x) functionally if $x = (1 + 2i)$?

    a. 1
    b. 5
    c. 2.236
    d. 2

   Ans: c

21. Which of the following options is true for the two Gaussian functions e^{x^2} and $e^{(x-2)^2}$?

    a. They have different normalization constants.
    b. They have different widths.
    c. Their centers are different.
    d. They are the same Gaussians.

   Ans: c

22. Find out the direction of probability current for a particle expressed by the wavefunction $\psi(x,t) = Ae^{-i\left(\frac{Et}{\hbar}-kx\right)}$.

    a. Increasing x-direction.
    b. Decreasing x-direction.
    c. The probability current is zero.
    d. The probability current is constant, but its direction cannot be determined.

   Ans: a

23. Which of the following statements of a Gaussian probability density distribution is correct and complete?

    a. If we perform one experiment, the experimental result would correspond to the center of the Gaussian.
    b. If we perform one experiment, the experimental result would correspond to the center of the Gaussian, and the error in measurement would correspond to the FWHM of the Gaussian distribution.
    c. If we repeat the experiment several times (Avogadro number), the distribution of the experimental result would correspond to the Gaussian probability density distribution.

d. The Gaussian probability density distribution shows the shape of a quantum particle.

Ans. c

24. Which of the following statements is true for a free particle in one-dimensional space?

    a. A free particle can never come to rest in one-dimensional space; however, it can come to rest in three-dimensional space.
    b. A free particle never come to rest in one-dimensional space.
    d. A particle experiencing a linear potential is called a free particle.
    e. A particle experiencing a quadratic potential is called a free particle.

Ans. b

25. A free particle expressed by the plane wave $\psi(x) = Ae^{ikx}$ exhibits a momentum of

    a. $-\hbar k$
    b. $\dfrac{\hbar k}{2m}$
    c. $\hbar k$
    d. k

Ans. c

26. Which of the following forms represents a Gaussian wavepacket?

    a. $A_0 e^{-ax^2}$
    b. $A_0 e^{-ax^2} + B_0 e^{-bx^2}$
    c. $A_0 e^{-ikx^2}$
    d. $A_0 e^{-ikx} e^{-ax^2}$

Ans. d

27. If a wavepacket is created by superposition of two plane waves, as expressed by $\psi(x,t) = Ae^{ik_1x}e^{-\frac{iE_1t}{\hbar}} + Ae^{ik_2x}e^{-\frac{iE_2t}{\hbar}}$, the phase velocity and group velocity of the wavepacket are given, respectively, by

    a. $\dfrac{\hbar k_0}{2m}$ and $\dfrac{\hbar k_0}{m}$

    b. $\dfrac{\hbar k_0}{m}$ and $\dfrac{\hbar k_0}{2m}$

    c. $\hbar k_0$ and $\frac{\hbar k_0}{2}$

    d. $\hbar k_0$ and $\frac{E_0}{2}$

Ans. a

28. The normalization constant of the $\psi(x,t) = e^{-ax^2}$ function is

    a. $\left(\dfrac{2a}{\pi}\right)^{1/4}$

    b. $\left(\dfrac{2a}{\pi}\right)^{1/2}$

    c. $\left(\dfrac{\pi}{2a}\right)^{1/4}$

    d. $\left(\dfrac{\pi}{2a}\right)^{1/2}$

Ans. a

29. For a free particle, which of the following statements about its velocity is true?

    a. A free particle does not travel.
    b. A free particle travels with constant velocity.
    d. The velocity of a free particle increases linearly with time.
    e. The velocity of a free particle decreases linearly with time.

Ans. b

30. For a free particle, which of the following statements about its probability distribution width is true?
    a. The width of the probability density distribution does not change as a function of time.
    b. The width of the probability density distribution decreases as a function of time.
    c. The width of the probability density distribution increases linearly with time.
    d. The width of the probability density distribution increases as a function of time.

Ans. d

31. For a particle on linear potential, which of the following statements about its velocity is true?
    a. The velocity of the particle does not change as a function of time.
    b. The velocity of the particle increases or decreases as a function of time.
    c. The velocity of the particle increases as a function of time.
    d. The velocity of the particle decreases as a function of time.

Ans. b

32. A stationary Gaussian particle is defined by a probability density distribution for which
    a. the width does not change as a function of time.
    b. the velocity does not change as a function of time.
    c. both the width and velocity do not change with time.
    d. the expectation value of position does not change as a function of time, but its width does.

Ans. d

33. Which of the following statements about the energy of a free particle is true?
    a. The free particle has discrete energy levels.
    b. The free particle can have any kinetic energy.
    c. The free particle's energy can be negative.

d. The free particle's energy cannot be negative, but its energy levels are discrete.

Ans. b

34. For a normalized wavefunction, which of the following options is true?

a. $\frac{d}{dt}\left[\int_{-\infty}^{+\infty} |\psi(x,t)|^2 dx\right] = 0$

b. $\int_{-\infty}^{+\infty} |\psi(x,t)|^2 dx = +\infty$

c. $\int_{-\infty}^{+\infty} |\psi(x,t)|^2 dx = -\infty$

d. $\int_{-\infty}^{+\infty} |\psi(x,t)|^2 dx = 0$

Ans. a

35. If an acceptable wavefunction in quantum mechanics is not normalized, which of the following options is true?

a. $\frac{d}{dt}\left[\int_{-\infty}^{+\infty} |\psi(x,t)|^2 dx\right] = 0$

b. $\int_{-\infty}^{+\infty} |\psi(x,t)|^2 dx = +\infty$

c. $\int_{-\infty}^{+\infty} |\psi(x,t)|^2 dx = -\infty$

d. $\int_{-\infty}^{+\infty} |\psi(x,t)|^2 dx = 0$

Ans. a

36. Comment on whether $\psi(x) = x^{\frac{1}{2}}$ is an acceptable wavefunction in the interval $[0, 1]$.

a. This is not acceptable because it is not a normalized wavefunction.

b. This is not acceptable because it is not a normalizable wavefunction.

c. This is acceptable because it is a normalized wavefunction.
d. This is acceptable because it is a normalizable wavefunction.

Ans. d

37. Comment on whether $\psi(x) = x^{-\frac{1}{2}}$ is an acceptable wavefunction in the interval $[0, 1]$.

a. This is not acceptable because it is not a normalized wavefunction.
b. This is not acceptable because it is not a normalizable wavefunction.
c. This is acceptable because it is a normalized wavefunction.
d. This is acceptable because it is a normalizable wavefunction.

Ans. b

38. Find the adjoint of the operator $\frac{d}{dx}$.

a. $-\dfrac{d}{dx}$

b. $\dfrac{d}{dx}$

c. x

d. $i\dfrac{d}{dx}$

Ans. a

39. The eigenvalues of the matrix $\mathbf{A} = \begin{pmatrix} 2 & 1 \\ 1 & 2 \end{pmatrix}$ are

a. 2, 1
b. 4, 2
c. 3, 1
d. 3, 4

Ans. c

40. The eigenvalues of the matrix $\mathbf{A} = \begin{pmatrix} 2 & 0 \\ 0 & 5 \end{pmatrix}$ are

a. 2, 5
b. 0, 0

c. 4, 10
d. 3, 1

Ans. a

41. The kinetic energy operator under grid representation and finite difference method adopts the following matrix form:

    a. diagonal matrix
    b. off-diagonal matrix
    c. band-diagonal matrix
    d. square matrix

Ans. c

42. The potential energy operator under grid representation adopts the following matrix form:

    a. diagonal matrix
    b. off-diagonal matrix
    c. band-diagonal matrix
    d. square matrix

Ans. a

43. The exponential of a matrix is nothing but exponentiating each element of the matrix. For which matrix is this statement true?

    a. any matrix
    b. tridiagonal matrix
    c. identity matrix
    d. off-diagonal matrix

Ans. c

44. Which statement is true?

    a. The exponential of a Hamiltonian matrix is nothing but exponentiating each element of the matrix.
    b. The exponential of the kinetic energy part of a Hamiltonian matrix in the momentum domain is nothing but exponentiating each element of the matrix.
    c. The exponential of the potential energy part of a Hamiltonian matrix in the position domain cannot be obtained by exponentiating each element of the matrix.

d. The exponential of the kinetic energy part of a Hamiltonian matrix in the momentum domain cannot be obtained by exponentiating each element of the matrix.

Ans. b

45. Which of the following statements is true for the time-evolution operator?

a. The time-evolution operator is a unitary operator.
b. The time-evolution operator is irreversible in time.
c. The time-evolution operator does not preserve the norm of the wavefunction.
d. The time-evolution operator is not an exponential operator.

Ans. a

46. Which of the following statements is true for a split operator of the form $e^{\frac{-i\hat{H}\Delta t}{\hbar}} \approx e^{\frac{-i\hat{T}\Delta t}{\hbar}} e^{\frac{-i\hat{V}\Delta t}{\hbar}}$?

a. This expression does not have any error.
b. Error of this expression is proportional to Δt.
c. Error of this expression is proportional to Δt^2.
d. Error of this expression is proportional to Δt^3.

Ans. c

47. Which of the following statements is true for a split operator of the form $e^{\frac{-i(\hat{T}+\hat{V})\Delta t}{\hbar}} \approx e^{-i\frac{\hat{T}}{2}\frac{\Delta t}{\hbar}} e^{\frac{-i\hat{V}\Delta t}{\hbar}} e^{-i\frac{\hat{T}}{2}\frac{\Delta t}{\hbar}}$?

a. This expression does not have any error.
b. Error of this expression is proportional to Δt.
c. Error of this expression is proportional to Δt^2.
d. Error of this expression is proportional to Δt^3.

Ans. d

48. Find the value of the integration $\int_{-\infty}^{+\infty} f(x)\delta(x - x_0)dx$.

a. $\int_{-\infty}^{+\infty} f(x)\delta(x - x_0)dx = f(x_0)$

b. Cannot be determined

c. $\int_{-\infty}^{+\infty} f(x)\delta(x-x_0)dx = \delta(x-x_0)$

d. $\int_{-\infty}^{+\infty} f(x)\delta(x-x_0)dx = \pi f(x_0)$

Ans. a

49. Find the value of the integration $\int_0^t e^{\frac{i(E_n-E_m)\tau}{\hbar}} d\tau$.

a. $\int_0^t e^{\frac{i(E_n-E_m)\tau}{\hbar}} d\tau \approx \pi\hbar$

b. $\int_0^t e^{\frac{i(E_n-E_m)\tau}{\hbar}} d\tau \approx \pi$

c. $\int_0^t e^{\frac{i(E_n-E_m)\tau}{\hbar}} d\tau = \pi\hbar\, \delta(E_n - E_m)$

d. $\int_0^t e^{\frac{i(E_n-E_m)\tau}{\hbar}} d\tau \approx \delta(E_n - E_m)$

Ans. c

Index

www.ingramcontent.com/pod-product-compliance
Lightning Source LLC
Chambersburg PA
CBHW050538190326
41458CB00007B/1834

9789811277160